DRUGS

DRUGS
From Discovery to Approval

Second Edition

RICK NG, PhD, MBA
A-Bio Pharma Pte Ltd, Singapore

(W)WILEY-BLACKWELL

A John Wiley & Sons, Ltd., Publication

Wiley-Blackwell is an imprint of John Wiley & Sons, formed by the merger of Wiley's global Scientific, Technical, and Medical business with Blackwell Publishing.

Published by John Wiley & Sons, Inc., Hoboken, New Jersey.
Published simultaneously in Canada.

For general information on our other products and services or for technical support, please contact our Customer Care Department within the United States at (800) 762-2974, outside the United States at (317) 572-3993 or fax (317) 572-4002.

Wiley also publishes its books in a variety of electronic formats. Some content that appears in print may not be available in electronic format. For information about Wiley products, visit our web site at www.wiley.com.

Library of Congress Cataloging-in-Publication Data:

Ng, Rick.
   Drugs : from discovery to approval / Rick Ng. – 2nd ed.
      p.   ;   cm.
   Includes bibliographical references and index.
   ISBN 978-0-470-19510-9 (cloth)
   1. Drug development.   I. Title.
      [DNLM: 1. Technology, Pharmaceutical.   2. Chemistry, Pharmaceutical.
   3. Clinical Trials as Topic–methods.   4. Drug Approval–legislation & jurisprudence.
   5. Drug Approval–methods.   6. Drug Design.   QV 778 N576d 2009]
      RM301.25.N5 2009
      615'.19—dc22

                                                                    2008035472

Printed in the United States of America.

10 9 8 7 6 5 4

To
Cherry, Shaun and Ashleigh

CONTENTS

PREFACE   **xiii**

1   INTRODUCTION   **1**

1.1   Aim of This Book / 1
1.2   An Overview of the Drug Discovery and Development Process / 2
1.3   The Pharmaceutical Industry / 5
1.4   Economics of Drug Discovery and Development / 10
1.5   Trends in Drug Discovery and Development / 12
1.6   Case Study #1 / 14
1.7   Summary of Important Points / 16
1.8   Review Questions / 16
1.9   Brief Answers and Explanations / 16
1.10   Further Reading / 17

2   DRUG DISCOVERY: TARGETS AND RECEPTORS   **19**

2.1   Drug Discovery Processes / 20
2.2   Medical Needs / 21
2.3   Target Identification / 23
2.4   Target Validation / 28
2.5   Drug Interactions with Targets or Receptors / 30
2.6   Enzymes / 34

2.7   Receptors and Signal Transduction / 38
2.8   Assay Development / 45
2.9   Case Study #2 / 46
2.10   Summary of Important Points / 50
2.11   Review Questions / 50
2.12   Brief Answers and Explanations / 51
2.13   Further Reading / 51

3   DRUG DISCOVERY: SMALL MOLECULE DRUGS   53

3.1   Introduction / 54
3.2   Irrational Approach / 55
3.3   Rational Approach / 60
3.4   Antisense Approach / 79
3.5   RNA Interference Approach / 81
3.6   Chiral Drugs / 83
3.7   Closing Remarks / 84
3.8   Case Study #3 / 84
3.9   Summary of Important Points / 88
3.10   Review Questions / 89
3.11   Brief Answers and Explanations / 90
3.12   Further Reading / 91

4   DRUG DISCOVERY: LARGE MOLECULE DRUGS   93

4.1   Introduction / 94
4.2   Vaccines / 95
4.3   Antibodies / 106
4.4   Cytokines / 113
4.5   Hormones / 121
4.6   Gene Therapy / 124
4.7   Stem Cells and Cell Therapy / 126
4.8   Case Study #4 / 128
4.9   Summary of Important Points / 131
4.10   Review Questions / 132
4.11   Brief Answers and Explanations / 133
4.12   Further Reading / 134

5   DRUG DEVELOPMENT AND PRECLINICAL STUDIES   136

5.1   Introduction / 137
5.2   Pharmacodynamics / 139

5.3   Pharmacokinetics / 143
5.4   Toxicology / 155
5.5   Animal Tests, *In Vitro* Assays, and *In Silico* Methods / 158
5.6   Formulations and Delivery Systems / 161
5.7   Nanotechnology / 168
5.8   Case Study #5 / 169
5.9   Summary of Important Points / 171
5.10   Review Questions / 172
5.11   Brief Answers and Explanations / 173
5.12   Further Reading / 174

6 CLINICAL TRIALS   176

6.1   Definition of Clinical Trial / 177
6.2   Ethical Considerations / 177
6.3   Clinical Trials / 181
6.4   Regulatory Requirements for Clinical Trials / 186
6.5   Role of Regulatory Authorities / 199
6.6   Gene Therapy Clinical Trial / 199
6.7   Case Study #6 / 200
6.8   Summary of Important Points / 204
6.9   Review Questions / 205
6.10   Brief Answers and Explanations / 205
6.11   Further Reading / 206

7 REGULATORY AUTHORITIES   208

7.1   Role of Regulatory Authorities / 209
7.2   US Food and Drug Administration / 210
7.3   European Medicines Agency / 214
7.4   Japan's Ministry of Health, Labor and Welfare / 216
7.5   China's State Food and Drug Administration / 217
7.6   India's Central Drugs Standard Control Organization / 219
7.7   Australia's Therapeutics Goods Administration / 219
7.8   Canada's Health Canada / 220
7.9   Other Regulatory Authorities / 220
7.10   Authorities Other than Drug Regulatory Agencies / 221
7.11   International Conference on Harmonization / 222
7.12   World Health Organization / 222
7.13   Pharmaceutical Inspection Cooperation Scheme / 223

7.14   Case Study #7 / 225
7.15   Summary of Important Points / 227
7.16   Review Questions / 228
7.17   Brief Answers and Explanations / 228
7.18   Further Reading / 229

8   REGULATORY APPLICATIONS   **231**

8.1   Introduction / 232
8.2   Food and Drug Administration / 233
8.3   European Union / 250
8.4   Japan / 263
8.5   China / 264
8.6   India / 266
8.7   Australia / 269
8.8   Canada / 269
8.9   Case Study #8 / 269
8.10   Summary of Important Points / 273
8.11   Review Questions / 274
8.12   Brief Answers and Explanations / 274
8.13   Further Reading / 275

**9   GOOD MANUFACTURING PRACTICE:
     REGULATORY REQUIREMENT**   **278**

9.1   Introduction / 279
9.2   United States / 279
9.3   Europe / 283
9.4   International Conference on Harmonization / 283
9.5   Core Elements of GMP / 287
9.6   Selected GMP Systems / 297
9.7   The FDA's New cGMP Initiative / 310
9.8   Case Study #9 / 313
9.9   Summary of Important Points / 315
9.10   Review Questions / 316
9.11   Brief Answers and Explanations / 316
9.12   Further Reading / 317

**10 GOOD MANUFACTURING PRACTICE:
     DRUG MANUFACTURING**   **319**

10.1   Introduction / 320
10.2   GMP Manufacturing / 322

10.3 GMP Inspection / 325

10.4 Manufacture of Small Molecule APIs (Chemical Synthesis Methods) / 332

10.5 Manufacture of Large Molecule APIs (Recombirant DNA Methods) / 340

10.6 Finished Dosage Forms / 348

10.7 Case Study #10 / 352

10.8 Summary of Important Points / 355

10.9 Review Questions / 356

10.10 Brief Answers and Explanations / 356

10.11 Further Reading / 357

11 FUTURE PERSPECTIVES   **359**

11.1 Past Advances and Future Challenges / 360

11.2 Small Molecule Pharmaceutical Drugs / 360

11.3 Large Molecule Biopharmaceutical Drugs / 362

11.4 Traditional Medicine / 364

11.5 Individualized Medicine / 366

11.6 Gene Therapy / 366

11.7 Cloning and Stem Cells / 367

11.8 Old Age Diseases and Aging / 369

11.9 Lifestyle Drugs / 371

11.10 Performance-Enhancing Drugs / 373

11.11 Chemical and Biological Terrorism / 376

11.12 Transgenic Animals and Plants / 376

11.13 Antimicrobial Drug Resistance / 379

11.14 Regulatory Issues / 380

11.15 Intellectual Property Rights / 381

11.16 Bioethics / 382

11.17 Concluding Remarks / 384

11.18 Case Study #11 / 387

11.19 Further Reading / 389

APPENDIX 1 HISTORY OF DRUG DISCOVERY AND DEVELOPMENT   **391**

A1.1 Early History of Medicine / 391

A1.2 Drug Discovery and Development in the Middle Ages / 394

A1.3 Foundation of Current Drug Discovery and Development / 394

A1.4 Beginnings of Modern Pharmaceutical Industry / 395

A1.5 Evolution of Drug Products / 396
A1.6 Further Reading / 397

**APPENDIX 2 CELLS, NUCLEIC ACIDS, GENES,
AND PROTEINS**   398

A2.1 Cells / 398
A2.2 Nucleic Acids / 400
A2.3 Genes and Proteins / 404
A2.4 Further Reading / 410

**APPENDIX 3 SELECTED DRUGS AND THEIR MECHANISMS
OF ACTION**   411

APPENDIX 4 A DHFR PLASMID VECTOR   414

APPENDIX 5 VACCINE PRODUCTION METHODS   416

**APPENDIX 6 PHARMACOLOGY/TOXICOLOGY
REVIEW FORMAT**   418

APPENDIX 7 EXAMPLES OF GENERAL BIOMARKERS   424

APPENDIX 8 TOXICITY GRADING   428

APPENDIX 9 HEALTH SYSTEMS IN SELECTED COUNTRIES 434

ACRONYMS   436

GLOSSARY   441

INDEX   445

PREFACE

This second edition has been completely revised to include the latest advances in drug discovery, development, clinical trials, manufacturing, and regulatory processes. At the end of each chapter is a case study to provide more in-depth perspectives and current issues facing the pharmaceutical industry. A summary of important points and questions and answers are added to each chapter for reference.

In writing this edition, I am grateful to Dr. Loh Kean Chong, Dr. Ng Kok Chin, and Dr. Matthias Brand for their comments and suggestions. Ms. Lim Bee Ting and Mr. Joash Chong meticulously checked through the manuscript, and Ms. Tang Meiyuat helped to prepare all the graphics. My daughter, Ashleigh, managed the electronic files throughout the editing process.

RICK NG

CHAPTER 1

INTRODUCTION

1.1	Aim of This Book	1
1.2	An Overview of the Drug Discovery and Development Process	2
1.3	The Pharmaceutical Industry	5
1.4	Economics of Drug Discovery and Development	10
1.5	Trends in Drug Discovery and Development	12
1.6	Case Study #1	14
1.7	Summary of Important Points	16
1.8	Review Questions	16
1.9	Brief Answers and Explanations	16
1.10	Further Reading	17

1.1 AIM OF THIS BOOK

The pharmaceutical industry is perhaps one of the most regulated industries in the world. From discovering a new drug to registering it for marketing and commercialization, pharmaceutical organizations have to negotiate through very complex and lengthy processes.

The intention of this book is to provide an overview of how a drug is discovered, the amount and types of laboratory tests that are performed, and the conduct of clinical trials before a drug is ready to be registered for human use. Of importance is the role of regulatory authorities in these processes. Through

Drugs: From Discovery to Approval, Second Edition, By Rick Ng
Copyright © 2009 John Wiley & Sons, Inc.

Exhibit 1.1   FDA Definition of a Drug

"An active ingredient that is intended to furnish pharmacological activity or other direct effect in the diagnosis, cure, mitigation, treatment, or prevention of a disease, or to affect the structure of any function of the human body, but does not include intermediates used in the synthesis of such ingredient."

legislation the regulatory authorities oversee the safety and efficacy of drugs. This book aims to integrate, in a simplified manner, the relationships between all these complex processes and procedures.

To establish a frame of reference, it is appropriate to commence with a definition for the term "drug." Generally, a drug can be defined as a substance that induces a response within the human body, whether the response is beneficial or harmful. In this context, toxins and poisons can be classified as drugs. However, the term "drug" used in this book is strictly reserved for a medicinal substance, which provides favorable therapeutic or prophylactic pharmaceutical benefits to the human body. Readers are referred to Exhibit 1.1 for a definition of drug according to the Food and Drug Administration (FDA) of the United States.

It should be noted that the descriptions in this book on discovery and regulatory processes are mainly for ethical drugs, as opposed to over-the-counter (OTC) drugs. Ethical drugs require prescriptions by physicians, whereas OTC drugs can be purchased from pharmacies without prescriptions. The OTC drugs are mainly established drugs with long histories of use and are deemed to be safe enough to be taken without supervision by physicians.

There is a further differentiation of ethical drugs into new drugs (those covered by patents) and generics (copies of drugs that have expired patents— see Case Study #10). Most of the descriptions in this book apply to new drugs.

1.2   AN OVERVIEW OF THE DRUG DISCOVERY AND DEVELOPMENT PROCESS

Although human civilization has been experimenting and consuming drugs for many centuries, it is only in the past hundred years that the foundation was laid for the systematic research and development of drugs. Readers are referred to Appendix 1 for a brief description of the history of drug development since ancient times.

Today, personnel from a myriad of fields are involved in the process of drug discovery and development, from scientists, clinicians, and medical practitioners to statisticians. Even persons from seemingly disparate occupations, such

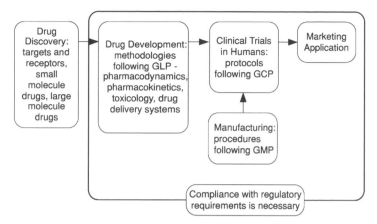

Figure 1.1   The stages from drug discovery to marketing approval.

as economists, lawyers, and regulatory staff, play a vital role as well. Previously, the main scientific personnel in the discovery process have been synthetic chemists. Now molecular biologists, biochemists, microbiologists, engineers, and even computer scientists play equally important roles in the drug discovery and development processes. The reason for this is that drug discovery and development has made a quantum leap forward in recent times with progress in genomics/proteomics and biotechnology. In addition, advances in laboratory equipment automation and high-speed computing have assisted in analyzing and processing large data sets. Personnel with different disciplines and expertise are needed to discover and develop drugs targeting diseases at the cellular and molecular levels.

It is estimated that, on average, a drug takes 10–12 years from initial research to reach the commercialization stage. The cost of this process is estimated to be more than US$1 billion. From discovery to marketing approval of a drug, the stages involved are shown in Fig. 1.1.

Drug Discovery: This process involves finding the target that causes or leads to the disease. Next, chemical or biological compounds are screened using specific assays and are tested against these targets to find leading drug candidates for further development. Many new scientific approaches are now used to determine targets (most targets are receptors or enzymes) and obtain the lead compounds, including the use of genomic and proteomic technology, synthetic chemistry, recombinant DNA (rDNA) technology, laboratory automation, and bioinformatics.

Drug Development: Tests are performed on the lead compounds in test tubes (laboratory, *in vitro*) and on animals (*in vivo*) to check how they affect the biological systems. The tests, often called preclinical research activities, include toxicology, pharmacodynamics, and pharmacokinetics, as well as optimization of drug delivery systems. Many iterations are carried out, and the leading compounds are modified and synthesized to improve their interactions with the targets, to reduce toxicity, or to improve pharmacokinetic performance. At the end of this process, an optimized compound is found and this becomes a potential drug ready for clinical trial in humans. The development work has to follow Good Laboratory Practice (GLP) to ensure that proper quality system and ethical considerations are established. Only compounds that satisfy certain performance and safety criteria will proceed to the next stage of clinical trial.

Clinical Trials: These are trials conducted on human subjects. The pertinent parameters for clinical trials are protocols (methods about how trials are to be conducted), safety and respect for human subjects, responsibilities of the investigator, institutional review board, informed consent, trial monitoring, and adverse event reporting. Clinical trials must follow regulations and guidelines from the FDA, the European Medicines Agency (EMEA) of the European Union (EU) or European Member States, Japan's Ministry of Health, Labor and Welfare (MHLW), or regulatory authorities in other prospective countries where the drug is intended to be registered and commercialized. Clinical trials are conducted in accordance with Good Clinical Practice (GCP).

Manufacturing: The drug designated for clinical trials and large-scale production has to be manufactured in compliance with current Good Manufacturing Practice (cGMP; the word "current" denotes that regulations change from time to time and the current regulations have to be applied) following US FDA requirements, EU Regulations or Directives, or International Conference on Harmonization (ICH) guidelines. Regulatory authorities have the right to conduct inspections on pharmaceutical manufacturing plants to ensure they follow cGMP guidelines so that the manufactured drug is safe and effective. A quality system has to be set up such that the drug is manufactured in accordance with approved procedures. There must also be an audit trail (i.e., traceability of materials and processes) as well as appropriate tests being conducted on the raw materials, intermediates, and finished products. The emphasis is that drugs should be safe, pure, effective, and of consistent quality to ensure that they are fit to be used for their intended functions.

Marketing Application: A drug is not permitted for sale until the marketing application for the new drug has been reviewed and approved by regulatory authorities such as the US FDA, the EU EMEA, or Japan's MHLW. Extensive dossiers and samples, if required, are provided to the authorities to demonstrate the safety, potency, efficacy, and purity of the drug. These are provided in the form of laboratory, clinical, and manufacturing data, which comply with GLP, GCP, and cGMP requirements. After the drug has been approved and

Exhibit 1.2   Did You Know?

Total drug development time grew from an average of 8.1 years in the 1960s to 11.6 years in the 1970s, to 14.2 in the 1980s, to 15.3 years for drugs approved from 1990 through 1995. Another report in 2003 has put the figure at 11.8 years. Pharmaceutical companies and regulatory authorities are working together to reduce this time span.

The average cost of developing a new drug is estimated to be about US$1–1.2 billion, including expenditures on failed projects. This amount is about four times the price of an Airbus A380 at US$270 million, or five times that of a Boeing B-787 Dreamliner at US$200 million.

Typically, tens of thousands of compounds are screened and tested, and only a handful make it onto the market as drug products. The statistics are such that, of the 5000–10,000 compounds that show initial promise, five will go into human clinical trials, and only one will become an approved drug.

Sources: (1) PhRMA (Pharmaceutical Research and Manufacturers of America) Press Release dated November 14, 2006. http://www.who.int/mediacentre/factsheets/fs310/en/index.html [accessed April 19, 2007]. (2) CNN.com. *Largest Passenger Jet Unveiled*, January 18, 2005. http://www.cnn.com/2005/BUSINESS/01/18/airbus.380/ [accessed April 19, 2007]. (3) Tufts Center for the Study of Drug Development. http://www.bizjournals.com/sanfrancisco/stories/2006/12/04/newscolumn3.html [accessed September 26, 2007].

marketed, there is continuous monitoring of the safety and performance of the drug to ensure that it is prescribed correctly and adverse events (side effects) are reported and investigated. The advertising of drugs is also scrutinized by regulatory authorities to ensure that there are no false representations or claims for the drugs.

Subsequent chapters will elaborate on each of these processes. An overview of the complexity, time, and cost of developing a new drug is shown in Exhibit 1.2.

1.3   THE PHARMACEUTICAL INDUSTRY

The pharmaceutical industry as we know it today started in the late 1800s. It started with the synthetic versions of natural compounds in Europe (refer to Appendix 1).

Drug discovery and development is mainly carried out by pharmaceutical companies, universities, and government research agencies, although there are increasing activities in the start-up and smaller companies that specialize in particular fields of research. A substantial number of the research findings and potential drugs from the start-ups, smaller companies, universities, and research organizations are, however, licensed to the multinational pharmaceutical companies for clinical trials, manufacturing, marketing, and distribution. Alternatively, alliances are formed with the multinational pharmaceutical companies to develop or market the drugs. A primary reason is the huge cost involved in drug development and commercialization.

In 2006, the combined worldwide pharmaceutical market was around US$643 billion. The distribution of the market (in US$ billion) is shown in Table 1.1. From this data, it is evident that the United States, Europe, and Japan account for almost 85% of the worldwide pharmaceutical market. The regulatory authorities in these countries are hence very important to the pharmaceutical companies to ensure their products are approved for commercialization.

Table 1.2 shows the top 10 drugs in 2006; with the exception of Enbrel (a biopharmaceutical, large molecule drug—see Chapter 4), all the others are

TABLE 1.1 Global Pharmaceutical Sales by Region, 2006

World	2006 Sales (US$ billion)	Global Sales (%)	Growth from Previous Year (%)
North America	290.1	45.1	+8.3
European Union (France, Germany, Italy, Spain, and the UK)	123.2	19.1	+4.4
Rest of Europe	66.1	10.3	+7.8
Japan	64.0	10.0	−0.4
Asia, Africa, and Australia	66.0	10.3	+10.5
Latin America	33.6	5.2	+12.7
Total	*643.0*	*100.0*	*+7.0*

Source: IMS. *IMS Health Reports Global Pharmaceutical Market Grew 7.0 Percent in 2006 to $643 Billion.* http://www.imshealth.com/ims/portal/front/articleC [accessed April 20, 2007].

TABLE 1.2 Top 10 Best-Selling Products, 2006

Product	Therapy	Company
Lipitor	Cholesterol reducer	Pfizer
Nexium	Antiulcerant	AstraZeneca
Advair/Seretide	Antiasthmatic	GSK
Plavix	Antiplatelet	BMS
Norvasc	Calcium antagonist	Pfizer
Enbrel	Antirheumatic	Amgen
Singulair	Antiasthmatic	Merck
Prevacid/Ogastro	Antiulcerant	Abbott
Zyprexa	Antipsychotic	Eli Lilly
Stilnox	Hypnotic	Sanofi-Aventis

Source: IMS Health. http://open.imshealth.com/dept.asp?dept%5Fid=4 [accessed April 20, 2007].

small molecule synthetic drugs (see Chapter 3). However, biopharmaceuticals have become increasingly important in the last two decades, when the first one was introduced. The biopharmaceutical market has grown substantially compared to the small molecule drugs. For comparison, Table 1.3 presents the top 10 biopharmaceuticals (biologics) in 2006. Out of the 101 "blockbuster" drugs (sales greater than US$1 billion) in 2006, 18 are biopharmaceuticals.

Most of us are familiar with the effects of cholesterol. Exhibit 1.3 provides an explanation of how cholesterol is formed and the mechanism of action for Lipitor and Zocor in lowering the sterol.

Acid reflux and heartburn are conditions that affect many of us from time to time. Exhibit 1.4 describes the proton pump inhibitor, Nexium, and its predecessor, Prilosec, and how they work as antiulcerants.

TABLE 1.3   Top 10 Best-Selling Biopharmaceuticals, 2006

Product	Therapy	Company
Enbrel	Arthritis	Amgen, Wyeth
Aranesp	Anemia	Amgen
Rituxan	Non-Hodgkin's lymphoma	Biogen Idec, Genentech, Roche
Remicade	Crohn disease, arthritis	Johnson & Johnson, Schering-Plough
Procrit/Eprex	Anemia	Johnson & Johnson
Herceptin	Breast cancer	Genentech, Roche
Epogen	Anemia	Amgen
Neulasta	Neutropenia	Amgen
Human insulin	Diabetes	Novo Nordisk
Avastin	Colon cancer	Genentech, Roche

Source: Biologic Drug Report. http://www.biologicdrugreport.com/leading.htm [accessed May 25, 2007].

Exhibit 1.3   Cholesterol and Cholesterol-Lowering Drugs

Cholesterol is a fat-like substance (a sterol) that is present in our blood and all the cells. It is synthesized within the body or derived from our diet. Cholesterol is an important constituent of the cell membrane and hormones.

Cholesterol is carried in the bloodstream by lipoproteins such as low density lipoprotein (LDL, or "bad cholesterol") and high density lipoprotein (HDL, "good cholesterol"). LDL carries cholesterol from the liver to other parts of the body. LDL attaches to receptors (see Chapter 2) on the cell surface and is taken into the cell interior. It is then degraded and the cholesterol is used as a component for the cell membrane. When there is excessive cholesterol inside the cell, it leads to a reduction in the synthesis of LDL receptors.

The number of active LDL receptors is also affected by a condition called familial hypercholesterolemia, in which there is a defective gene coding for the receptor. In either case, the reduction of active receptors means that the LDL carrying cholesterol is unable to enter the cell interior; instead, it is deposited in the arteries leading to the heart or brain. These deposits build up over time and may block blood supply to the heart muscle or brain, resulting in a heart attack or stroke. In contrast, HDL transports cholesterol from other parts of the body to the liver, where it is degraded to bile acids.

An enzyme (see Section 2.6) called HMG-CoA reductase is involved in the biosynthesis of cholesterol. Drugs such as atorvastatin (Lipitor) and simvastatin (Zocor) are competitive inhibitors of HMG-CoA reductase. They inhibit cholesterol synthesis by increasing the number of LDL receptors to take up the LDL.

Exhibit 1.4   Nexium

Nexium (esomeprazole magnesium, AstraZeneca) is a drug termed as a proton pump inhibitor. It turns off the secretions of acid into the stomach. When less acid is produced, there is a reduced amount of acid that can flow back up from the stomach into the esophagus to cause reflux symptoms.

Esomeprazole is the S-isomer of omeprazole (Prilosec), which is a mixture of the S- and R-isomers. Prilosec, for many years a best selling drug, is   5-methoxy-2-[[(4-methoxy-3,5-dimethyl-2-pyridinyl)methyl]sulfinyl]-1H-benzimidazole. Its empirical formula is $(C_{17}H_{18}N_3O_3S)_2Mg \cdot 3H_2O$ with a molecular weight of 767.2 as a trihydrate and 713.1 on an anhydrous basis.

It should be noted that there are normally three names associated with a drug: the trade or proprietary name (e.g., Nexium), generic or nonproprietary name (esomeprazole), and a specific chemical name for the active ingredient. In the case of esomeprazole, the active ingredient is the S-isomer of benzimidazole.

Sources: Food and Drug Administration. *Nexium*. http://www.fda.gov/cder/foi/label/2006/021153s022lbl.pdf [accessed July 27, 2007].

The top 10 pharmaceutical companies in July 2007 are shown in Table 1.4. These 10 companies account for almost half the global sales of drugs. Of significance is that, in the same period, the companies collectively spent in excess of US$40 billion in research and development (R&D). This amount is more than 10% of their sales revenues, showing the importance of R&D for these companies.

TABLE 1.4 Top 10 Pharmaceutical Companies, July 2007

Rank	Company	Market Capitalization (US$ billion)
1	Pfizer	173
2	GlaxoSmithKline	143
3	Novartis	127
4	Roche	123
5	Sanofi-Aventis	114
6	Johnson & Johnson (Diversified Healthcare)	178
7	Merck	112
8	AstraZeneca	83
9	Abbott	81
10	Wyeth	68

Source: Yahoo! Finance. http://biz.yahoo.com/p/510conameu.html [accessed July 27, 2007].

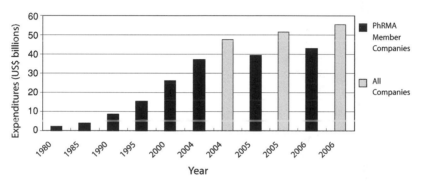

Figure 1.2 R&D investments by research-based US pharmaceutical companies. *Source*: The Pharmaceutical Research and Manufacturers of America (PhRMA). *Pharmaceutical Industry Profile 2006*. http://www.phrma.org/files/2006%20Indusry%20 Profile.pdf, and http://www.phrma.org/news_room/press_releases/r%26d_spending_ by_u.s._biopharmaceutical_companies_reaches_a_record_%2455.2_billion_in_2006/ [accessed July 7, 2007].

Further examples of R&D investments into drug research by research-based US pharmaceutical companies from 1980 to 2006 are shown in Fig. 1.2. The enormous spending on R&D has escalated in recent years. According to reports by The Pharmaceutical Research and Manufacturers of America (PhRMA), US pharmaceutical companies have almost doubled their R&D spending every five years since 1980. Out of every five dollars earned in sales, a dollar is put back into R&D. In 2006 the US pharmaceutical industry spent $55.2 billion to develop new drugs.

TABLE 1.5 Number of New Drugs in Development (Clinical or Later Development) in 2006

Disease Target	Number in Development	Disease Target	Number in Development
Cardiovascular disorders	303	Cancer	682
Neurologic disorders	531	Psychiatric disorders	190
HIV/AIDS	95	Diabetes	62
Arthritis	88	Infections	341
Alzheimer's disease and dementia	55	Asthma	60

Source: The Pharmaceutical Research and Manufacturers of America (PhRMA). *Pharmaceutical Industry Profile 2006*. http://www.phrma.org/files/2006%20Indusry%20Profile.pdf [accessed July 27, 2007].

TABLE 1.6 Top Five Biopharmaceutical Companies, July 2007

Rank	Company	Market Capitalization (US$ billion)
1	Genentech	79
2	Amgen	66
3	Gilead Science	24
4	Celgene	22
5	Biogen Idec	19

Source: Yahoo! Finance. http://biz.yahoo.com/p/515conameu.html [accessed July 27, 2007].

Pharmaceutical firms have to ensure that there is a pipeline of new and better drugs to return the substantial investments made. It is estimated that large pharmaceutical firms need 4–5 new drugs approved every year to maintain their premium positions. However, most firms are far short of this target, with only about 1–2 new drugs approved per year. Table 1.5 presents a snapshot of the number of new drugs being developed by US companies in 2006.

Biopharmaceutical products make up around 10% of the total pharmaceutical markets of US$643 billion. However, the growth rate for biopharmaceuticals is high, and it is expected that half the total pharmaceutical market will be biopharmaceuticals within the next 10–20 years. The top five biopharmaceutical companies are listed in Table 1.6.

1.4 ECONOMICS OF DRUG DISCOVERY AND DEVELOPMENT

The pharmaceutical market is very competitive. It is imperative that pharmaceutical companies (including biotechnology companies), large or small,

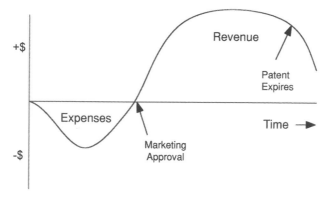

Figure 1.3 Expenses and revenues curve for a new drug.

discover and develop drugs efficiently and within the shortest time span to remain competitive.

Figure 1.3 shows the expenses versus revenues for a company's investment in developing a new drug. Up until the clinical stage, the investment is substantial in the discovery and development processes. The largest cash demand is in the clinical trial stages, where hundreds and thousands of human subjects have to be recruited to test the drug.

A positive return of revenue only occurs after the drug has been approved by regulatory authorities for marketing. The overall profitability of a drug is the difference between the positive returns and the negative expenses within the patent period of 20 years. After that period, if the patent is not extended, there is no further protection on the intellectual rights for the drug.

After patent expiry, generic drugs from other companies are unencumbered by patent rights infringement and can encroach into the profitability of the company that developed the original patented drug. It is thus crucial that drugs are marketed as quickly as possible to ensure there is a maximum patent coverage period and to be "first to market," to establish a premium position. When cimetidine (Tagamet, GlaxoSmithKline) came off patent in the United States, it lost almost 90% of sales within four years (from $2.085 billion in 1995, to $277 million in 1999).

Exhibit 1.5 provides a brief explanation of patents. Patents are the pillars that support the drug industry. In contrast, traditional medicines, which are mainly derived from natural products of plant or animal origins, are not patentable. This is because traditional medicines consist of a multitude of compounds and it is difficult to establish patent claims based on varying quantities of materials.

Exhibit 1.5   Patents

A patent is a right granted by a government for any device, substance, method, or process that is new, inventive, and useful. The patent discloses the know-how for the invention. In return for this disclosure, the owner of a patent is given a 20 year period of monopoly rights to commercial returns from exploiting the invention.

There are two ways to register patents: either through applying in individual countries, which means multiple applications for different countries, or through designating the desired countries in a single application using the Patent Cooperation Treaty (PCT) mechanism. There are more than 90 member countries belonging to the PCT, including major developed countries.

The PCT does not grant patents. Application under the PCT goes through two phases: an international phase and a national phase. The international phase is where the application is searched, published, and subjected to preliminary examination. Then the application enters into the national phase in each country. The application is subjected to examination and granting procedures in each country.

Another important item for a patent is the priority date. The priority date is established when a patent application is filed for the first time. If the invention is known before this date, then the patent is not granted. Most countries are first-to-file countries, which means that the patent is awarded to the person with the earliest filing date. In the United States, patents are awarded to the first person to invent. The inventor can attempt to show the invention was made before another person's filing date to claim priority.

Sources: World Intellectual Property Organization. *The Patent Cooperation Treaty*. http://www.wipo.int/pct/en/texts/articles/atoc.htm [accessed September 20, 2007].

1.5   TRENDS IN DRUG DISCOVERY AND DEVELOPMENT

The approach to drug discovery and development can generally be classified into the following areas:

- Irrational approach
- Rational approach
- Antisense approach
- RNA approach
- Biologics
- Gene therapy
- Stem cell therapy—both somatic cell and germ cell

Irrational Approach: This approach is the historical method of discovering and developing drugs. It involves empirical observations of the pharmacological effects from the screening of many chemical compounds, mainly those from natural products. The active component that gives rise to the observed effects is isolated. The chemical formula is determined, and modifications are made to improve its properties. This approach has yielded many drugs available today.

Rational Approach: This approach requires three-dimensional knowledge of the target structure involved in the disease. Drugs are designed to interact with this target structure to create a beneficial response. This is an emerging field in drug discovery started in the last 30 years.

Antisense Approach: This is a relatively new approach and it requires modifications to oligonucleotides that can bind to RNA and DNA (refer to Appendix 2 for a description of cell structure, genes, DNA, RNA, and proteins). The antisense drugs are used to stop transcriptional (from DNA) or translational (from RNA) pathways from proceeding, and so interfere with the process of disease.

RNAi Approach: This is the use of short interfering RNA (siRNA, sometimes called small interfering or silencing RNA) to interfere with the expression of a particular gene. The siRNAs are double-stranded RNAs of 20–25 nucleotides. It is envisaged that if the biological pathway of a disease is identified, siRNA could interfere to turn off the activity of the gene involved in the pathway and provide therapeutic effect.

Biologics: These are mainly protein-based drugs in the form of antibodies, vaccines, and cytokines. Their discovery generally starts from an understanding of the biological mechanistic pathways that cause specific diseases. Manufacturing of these drugs is based on recombinant DNA technologies using living organisms such as bacteria, yeast, and mammalian and insect cells.

Gene Therapy: The basis of this therapy is to remedy a diseased gene by inserting a missing gene or modified gene in the cells. This is a new topic that raises many unresolved ethical considerations. The cells with the diseased gene are removed from a patient, fixed outside the body (*ex vivo*), and then reinserted back into the body. In the case of a missing gene, a copy of the new gene is inserted into the patient's cells. The aim is for the inserted gene to influence the disease pathway or to initiate synthesis of missing proteins or enzymes.

Stem Cell Therapy: With stem cell therapy, the aim is to grow body parts to replace defective human organs and nerves. The stem cells are harvested from very early embryos or umbilical cord blood. Because of the very young age of these cells, they can be directed to grow into organ tissue to replace diseased tissue. The stem cell technology can provide an alternative to organ transplants with perhaps less rejection problems than the current practice of obtaining organs from a donor person. Stem cell therapy using germ cells involves cloning, and there are strict regulatory guidelines on how research is to be conducted.

Through the Human Genome Project, many novel disease targets have been discovered, which can be utilized to develop better and more effective drugs. Regardless of the approach used for discovering new drugs, pharmaceutical and biotechnology companies are now using a full suite of technologies to discover new drugs. These enabling technologies include the following:

- Microarray for disease target identification
- High throughput screening
- Combinatorial chemistry
- Structure–activity relationships: X-ray crystallography, nuclear magnetic resonance, computational chemistry
- Genomics and proteomics
- Metabolomics
- Systems biology
- Nanotechnology
- Bioinformatics: data mining
- Recombinant DNA technologies

Detailed discussions of these technologies are presented in Chapters 2–5.

1.6 CASE STUDY #1

Pfizer Inc.*

Pfizer is the largest pharmaceutical company in the world. Its corporate headquarters is located in New York and its R&D laboratories are spread around the globe.

In 2006 the company spent US$7.6 billion on R&D and its income for that year was US$48.4 billion. Table 1.7 clearly demonstrates that pharmaceutical companies are heavily research based; Pfizer's R&D expenditure is head and shoulders above other technology companies with much higher market capitalization.

---

* *Source*: Pfizer, http://www.pfizer.com/pfizer/are/mn_news.jsp [accessed July 23, 2007].

TABLE 1.7   R&D Expenditure in R&D-Based Companies

Company	Market Capitalization US$ billion (July 2007)	Revenue, US$ billion (2006)	R&D Expenditure, US$ billion (2006)
Pfizer	180	48.4	7.6
Microsoft	300	44.3	6.6
IBM	173	91.4	6.1
Intel	144	35.3	5.8
GE	420	163.4	3.4
Boeing	82	61.5	3.3

Source: Yahoo! Finance. http://biz.yahoo.com/p/515conameu.html [accessed July 27, 2007].

Pfizer employs more than 100,000 people, of which 12,000 are medical researchers. The company focuses its new drug development program in 11 therapeutic areas:

- Allergy and respiratory
- Cardiovascular, metabolic, and endocrine diseases
- Dermatology
- Gastrointestinal and hepatology
- Genitourinary
- Infectious diseases
- Inflammation
- Neuroscience
- Oncology
- Ophthalmology
- Pain

The top five drugs marketed by Pfizer are:

- Lipitor
- Norvasc
- Celebrex
- Lyrica
- Xalatan

To maintain its premier position, Pfizer has to remain vigilant to guard its market share. Recently, the company sued Ranbaxy of India, a generics manufacturer, on infringement of its atorvastatin patent. Atorvastatin is the active ingredient for Lipitor, the world's best-selling drug. The High Court in Dublin, Ireland ruled in favor of Pfizer, thus preventing Ranbaxy from launching a generic version of Lipitor until the expiry of the Irish patent in November 2011.

1.7 SUMMARY OF IMPORTANT POINTS

1. The process from drug discovery to approval is:

Discovery → Preclinical → Clinical → Regulatory Application → Approval

2. The overall process takes 10–12 years and costs more than US$1 billion.
3. Regulatory oversight is an integral part of the pharmaceutical industry, in order to ensure safety, efficacy, purity, and consistency of drugs for human use.
4. The global pharmaceutical market in 2006 was US$643 billion. Biopharmaceuticals account for more than 10% of the market, with higher growth rate compared to conventional pharmaceuticals.
5. The top-selling drug in 2007 was Lipitor with about US$12 billion in sales and the top biopharmaceutical was Enbrel, which brings in about US$2.5 billion revenue.
6. The pharmaceutical R&D expenditure, at more than 10% of revenue, is higher than many other technology-based industries.
7. Innovation is a hallmark of success for a pharmaceutical company.
8. Pharmaceutical companies rely on patents to protect their intellectual properties.

1.8 REVIEW QUESTIONS

1. Provide a definition for the term "drug" as adopted in this book.
2. Describe the process from drug discovery to approval.
3. Describe the role of regulatory bodies such as the FDA and EMEA. What are their main concerns about drugs?
4. Explain the terms GLP, GCP, and GMP. Why are these necessary?
5. Discuss how Lipitor and Nexium work in the body.
6. Explain the reason for the high R&D cost for drugs, and discuss how the cost can be reduced.
7. Explain why intellectual properties are important to pharmaceutical companies and how they can be protected. Give examples to illustrate.
8. List some of the approaches for drug discovery.

1.9 BRIEF ANSWERS AND EXPLANATIONS

1. Refer to Section 1.1 and Exhibit 1.1, but note that drugs should be used according to the indications and contraindications provided by the manufacturer.

2. This is explained in Section 1.2. The importance of regulatory control is an integral part of the process.

3. The FDA is the regulatory agency for drugs in the United States whereas the EMEA is the centralized agency for the EU countries. Refer to Section 1.7 for further explanations. Regulatory agencies are concerned with the safety, efficacy, purity, and consistency of drugs. Their role is to ensure that drugs are safe and fit for their purpose.

4. GLP stands for Good Laboratory Practice, GCP for Good Clinical Practice, and GMP for Good Manufacturing Practice. Together these practices ensure that there is planning, control, and monitoring of drug development all the way from preclinical to clinical and manufacturing stages—such that procedures are followed, records are kept, and processes are verified and tested.

5. Refer to Exhibits 1.3 and 1.4.

6. The high R&D cost stems from increasingly more stringent regulatory compliance requirements and failures of drugs at later clinical phases due mainly to adverse events. The introduction of risk-based approach, process analytical technology (refer to Section 9.7), and consolidation of regulatory documents (ICH, refer to Section 7.11) will reduce the regulatory burden. In addition, the development of more specific drugs and a better understanding of biochemical pathways, followed by focused evaluation using more representative assays and biomarkers, will lessen failures at late stage clinical trials (refer to Section 6.4).

7. Patent rights protect the intellectual properties and compensate the high R&D expenditure that pharmaceutical organizations spend on developing drugs. Without the protection of patents, it would be difficult for pharmaceutical companies to justify the expenditure and continue with innovations. The way forward for the pharmaceutical industry may include the need to review the patent law. This is particularly relevant with respect to the exclusivity period and the rules for revoking patent rights under compulsory licensing, whereby a government can force a patent holder to grant rights of the patent to the state or other parties without compensation in royalties.

8. Refer to Section 1.5.

1.10 FURTHER READING

Campbell JJ. *Understanding Pharma: A Primer on How Pharmaceutical Companies Really Work*, Pharmaceutical Institute, Raleigh, NC, 2005.

Center for Drug Evaluation and Research. *Drug Information: Electronic Orange Book*, FDA, Rockville, MD. http://www.fda.gov/cder/ob/default.htm [accessed July 2, 2007].

Center for Drug Evaluation and Research. *New Drug Development and Review Process*, FDA, Rockville, MD. http://www.fda.gov/cder/handbook/index.htm [accessed July 2, 2007].

Congressional Budget Office, The Congress of the United States. *Research and Development in the Pharmaceutical Industry*, October 2006.

Food and Drug Administration. *From Test Tube to Patient: New Drug Development in the US*, 2nd ed., FDA, Rockville, MD, 1995.

Food and Drug Administration. *The Drug Development Process: How the Agency Ensures that Drugs Are Safe and Effective*, FDA, Rockville, MD. http://www.fda.gov/opacom/factsheets/justthefacts/17drgdev.pdf [accessed July 10, 2007].

Food and Drug Administration, Center for Drug Evaluation and Research. *Drug Information: Electronic Orange Book*, FDA, Rockville, MD. http://www.fda.gov/cder/ob/default.htm [accessed July 2, 2007].

Food and Drug Administration, Center for Drug Evaluation and Research. *New Drug Development and Review Process*, FDA, Rockville, MD. http://www.fda.gov/cder/handbook/index.htm [accessed July 2, 2007].

Harvey AL, ed. *Advances in Drug Discovery Techniques*, Wiley, Hoboken, NJ, 1998.

Jurgen D. *In Quest of Tomorrow's Medicine*, Springer-Verlag, New York, 1999.

Lawrence S. Billion dollar babies—biotech drugs as blockbusters, *Nature Biotechnology* 25:380–382 (2007).

Pharmaceutical Research and Manufacturers of America. *Why Do Prescription Drug Cost So Much?* PhRMA, Washington, DC, 2000.

Smith CG, O'Donnell JT, eds. *The Process of New Drug Discovery and Development*, 2nd ed., Informa Healthcare, New York, 2006.

The Next Pharmaceutical Century: Ten Decades of Drug Discovery, 2007. http://pubs.acs.org/journals/pharmcent/Ch10.html [accessed September 29, 2007].

CHAPTER 2

DRUG DISCOVERY: TARGETS AND RECEPTORS

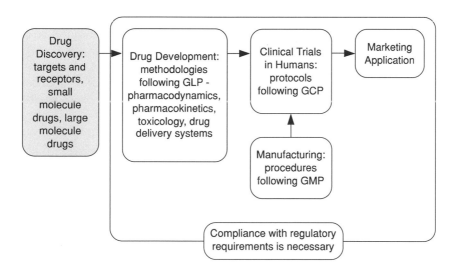

2.1	Drug Discovery Processes	19
2.2	Medical Needs	21
2.3	Target Identification	23
2.4	Target Validation	28

Drugs: From Discovery to Approval, Second Edition, By Rick Ng
Copyright © 2009 John Wiley & Sons, Inc.

2.5   Drug Interactions with Targets or Receptors   30
2.6   Enzymes   34
2.7   Receptors and Signal Transduction   38
2.8   Assay Development   45
2.9   Case Study #2   46
2.10 Summary of Important Points   50
2.11 Review Questions   50
2.12 Brief Answers and Explanations   51
2.13 Further Reading   51

2.1 DRUG DISCOVERY PROCESSES

For a drug to work, it has to interact with a disease target in our body and intervene in its wayward functions. An analogy is the lock and key comparison, with the lock being the disease target and the key representing the drug. The correct key has to be found to turn the lock and open the door to treat the disease.

The conventional method for drug discovery is the irrational approach. It involves scanning thousands of potential compounds from natural sources for a hit against specific assays that represent the target (more about this in Chapter 3). This procedure has been likened to finding a needle in a haystack. In our analogy, it is like trying out many keys to find a fit to a lock. As we can imagine, such a process is somewhat random and cumbersome. The chances of failure are high, although it should be kept in mind that many drugs on the market today were discovered in this manner.

Further advances in drug discovery led to the rational approach. This approach starts by finding out about the structure of the target and then designing a drug to fit the target and modify its functions. A comparison to the lock and key concept is to determine the construction of pin tumblers in the lock first and then design the key with the appropriate slots and grooves to open the lock. The latest progress in drug discovery is the contribution from genomics and proteomics research. Here the emphasis is to identify and validate targets *a priori* to drug discovery. This approach is to find out the target that causes the disease as the first step in drug discovery. After that, the rational approach would proceed. The analogy is to find out the exact diseased lock and then discover a drug to unlock the correct door.

With the foregoing in mind, the typical current drug discovery processes would proceed according to the flow chart in Fig. 2.1. This chapter focuses on the medical needs, identification, and validation of disease targets; followed by discussions on receptors, signal transduction, and assay development. Chapters 3 and 4 focus on lead compound generation and optimization, for small, synthetic drug molecules and large, protein-based macromolecules, respectively. In Chapter 5, we cover drug development and preclinical studies.

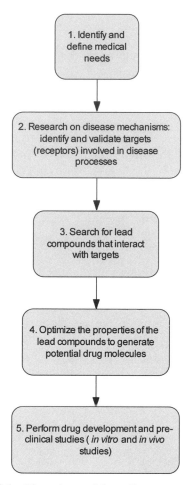

Figure 2.1   Flow chart of drug discovery processes.

2.2 MEDICAL NEEDS

A pharmaceutical organization has to determine which medical area has an unmet clinical need for an effective prophylactic or therapeutic intervention. Next, the organization has to evaluate its core competency, technological advantages, competitive barriers, and financial resources before committing to develop a drug to fulfill the unmet need. As discussed in Chapter 1, for a monetary outlay that averages around US$1–1.2 billion for each drug development, the organization has to weigh its options carefully. The important factors to consider are the following:

- Market potential
- Patent, intellectual property portfolio
- Competitive forces and regulatory status
- Core competencies

Overall, the organization needs to project the expected returns from such an investment and assess the competitive factors and barriers, including government regulations, before deciding which drug to develop.

Table 2.1 shows the major therapy classes in terms of sales and the number of prescriptions dispensed in the United States in 2003. The top three therapy classes are antidepressants, antihyperlipidemics, and antiulcerants. They account for 39% of sales in this list of 17 therapy classes. There are significant changes to the growth of some therapy classes within a single year. This is

TABLE 2.1 Leading Therapy Classes, 2003[a]

Therapeutic class[b] (Major Subclasses)	Number of Drugs in the Top 200	Total Sales (US$ billions)	Prescriptions (millions of units sold)
Antidepressants (SSRIs, SNRIs)	8	11.6	114.5
Antihyperlipidemics (statins)	6	11.1	108.4
Antiulcerants (proton-pump inhibitor)	5	10.4	70.0
Antihypertensives (ARBs, ACE inhibitors)	11	5.8	88.1
Antibiotics (broad and medium spectrum)	9	5.5	89.2
Diabetes therapies (oral, injectable)	6	4.9	63.5
Antiarthritics (COX-2 inhibitors)	4	4.8	48.4
Antipsychotics	3	4.2	20.2
Antihistamines (oral)	3	4.1	63.2
Neurological drugs (for seizures or pain)	5	4.0	36.2
Other vascular drugs (calcium- or beta-blockers)	7	3.7	68.7
Antiasthmatics	5	3.6	28.1
Analgesics (nonnarcotic)	3	2.8	20.1
Bone density regulators	4	2.3	32.0
Oral contraceptives	3	2.1	44.4
Antiallergy drugs (nasal steroids)	4	2.0	29.9
Analeptics (ADHD treatments)	3	1.3	16.9

[a]Therapeutic classes with three or more leading brand-name drugs in 2003.
[b]SSRI = selective serotonin reuptake inhibitor; SNRI = serotonin norepinephrine reuptake inhibitor; ARB = angiotensin receptor blocker; ACE = angiotensin converting enzyme; COX-2 = cyclooxygenase 2; ADHD = attention deficit hyperactivity disorder.
Source: Congress of the United States, Congressional Budget Office. *A CBO Study, Research and Development in the Pharmaceutical Industry*, October 2006.

especially true when new and more effective drugs are introduced; their sales can increase dramatically within a short time span and surpass the sales of more "established" drugs. A recent analysis shows that the drugs approved by the FDA target 394 human proteins in total. In the last 25 years, the average number of drugs approved was 19.5 per year, of which 6.3 are on novel targets.

Pharmaceutical companies have to be continuously vigilant and forecast the future directions of drug development and regulatory requirements. They have to use their core technical competencies to deliver a pipeline of products to remain competitive and profitable in the long term.

2.3   TARGET IDENTIFICATION

2.3.1   Genes

Most diseases, except in the case of trauma and infectious diseases, have a genetic connection. Genetic makeup and variations (see single nucleotide polymorphism in Section 11.5) determine a person's individuality and susceptibility to diseases, pathogens, and drug responses.

The current method of drug discovery commences with the study of how the body functions, in both normal and abnormal cases afflicted with diseases. The aim is to break down the disease process into cellular and molecular levels. An understanding of the status of genes and their associated proteins would help to pinpoint the cause of the disease. Drugs can be tailor-made to attack the "epicenter" of diseases. In this way, more specific (fewer side effects) and effective (high therapeutic index, see Section 5.2) drugs can be discovered and manufactured to intervene or restore the cellular or molecular dysfunction.

From the Human Genome Project, we know that there are approximately three billion base pairs that make up the DNA molecule (refer to Appendix 2). Only certain segments of the enormous DNA molecule encode for proteins. These segments are called genes. The estimate is that there are about 30,000 genes that encode proteins. Exhibit 2.1 provides some information about the number of genes and the complexity of life forms.

From these 30,000 genes, many thousands of proteins are produced. Drug targets are normally protein or glycoprotein molecules that make up the enzymes and receptors, with which drugs interact. To date, only about 500 proteins have been identified and targeted by the multitudes of drugs on the market. The opportunities that are opened up by genomics and proteomics research have paved the way for many more targets and new drugs to be discovered.

Exhibits 2.2, 2.3, and 2.4 provide examples of genetic causes of diseases, for example, cancer, sickle cell anemia, and cystic fibrosis. It should be noted that although some of these diseases are the result of mutations in a single gene (including Huntington's disease and Duchenne muscular dystrophy), most are due to the influence of multiple genes.

Exhibit 2.1   Genes and Molecular Complexity

The number of protein-coding genes in an organism provides a useful indication of its molecular complexity, although there is as yet no firm correlation between the number of genes and biological complexity.

Single-celled organisms typically have a few thousand genes. For example, *Escherichia coli* (a bacteria commonly found in the intestines of animals and humans) has 4300 genes, and *Saccharomyces cerevisiae* (a fungus commonly known as baker's or brewer's yeast) has 6000 genes. *Caenorhabditis elegans* (a small soil nematode about 1mm long) has 19,000 genes. *Drosophila melanogaster* (a 3mm fruit fly) has 13,600 genes. For human beings, the number of genes is estimated at around 30,000.

It was initially thought that the number of human genes was on the order of 100,000. The smaller number of 30,000–35,000 was surprising considering the complexity of human beings compared with smaller organisms. The latest view is that, although the number of genes indicates complexity, there are other factors involved in determining complexity. Each gene may code for more than one protein, to account for human complexity.

Source: Ewing B, Green P. Analysis of expressed sequence tags indicates 35,000 human genes, *Nature Genetics* 25:232–234 (2000).

Exhibit 2.2   The p53 Protein in Cancer

The *p53* gene is a tumor suppressor gene, which means that its activity stops the formation of tumors via the production of p53 protein. As shown in the picture below, the p53 protein has four identical chains, which are joined together by a central tetramerization domain. The p53 protein molecule wraps around and binds DNA. This wrapping action then turns on another gene, which codes for a 21-kDa protein that regulates DNA synthesis.

Normally, a cell grows by cell division and then dies through a process called apoptosis—programmed cell death. The p53 protein triggers apoptosis, which is a "stop signal" for cell division, to arrest cancer growth.

In the case of cancer growth, the gene that codes for p53 is mutated. The mechanism for programmed cell death becomes inactivated and no longer functions. Cancer cells then continue growing and dividing at the expense of surrounding cells, thus leading to tumor formation. It is also found that the oncogene, murine double minute (mdm2), overexpresses the mdm2 protein, which binds to the transactivation domain of p53 and blocks p53's transcription process, switching off the cell death program.

Sources: (1) Campbell MK, Farrell SO. *Biochemistry*, 5th ed., Thomson Brooks/Cole, Belmont, CA, 2006. (2) Vousden KH, Lane DP. p53 in health and disease, *Nature Reviews Molecular Cell Biology* 8:275–283 (2007). (3) Kirkpatrick P. Unleashing p53, *Nature Reviews Drug Discovery* 3:111 (2004).

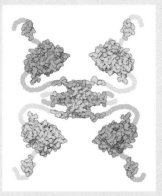

The p53 molecule. (*Source*: Goodsell DS. The Scripps Institute, *Featured Molecule: p53 Tumor Suppressor, Bio.Com*. http://www.bio.com/ [accessed September 7, 2002].)

Exhibit 2.3   Sickle Cell Anemia

Hemoglobin is a tetramer with four polypeptide chains: two identical α chains (141 residues) and two identical β chains (146 residues).

In people with sickle cell anemia, there is just one mutation in each of the β chains. The glutamic acid in position 6 is substituted by valine. This substitution, two residues out of a total of 474, is sufficient to cause the red blood cell to deform and constrict blood flow by blocking the capillaries.

Source: Campbell MK, Farrell SO. *Biochemistry*, 5th ed., Thomson Brooks/Cole, Belmont, CA, 2006.

2.3.2   Targets

There are a number of techniques used for target identification. Radioligand binding was a common technique until recently. Now DNA microarrays, expressed sequence tags, and *in silico* methods are used.

Radioligand Binding: The classic method to discover drug targets or receptors (Exhibit 2.5) is to bind the potential receptors with radioligands (Exhibit 2.6) so that targets can be picked out from a pool of other receptors. Bound

Exhibit 2.4 Cystic Fibrosis

Cystic fibrosis (CF) is a hereditary disease of abnormal fluid secretion. It affects cells of the exocrine glands, such as intestine, sweat glands, pancreas, reproductive tract, and especially the respiratory tract. The disease affects about 1 in 2500 infants of the Caucasian population to varying degrees of seriousness. Patients produce thickened mucus that is difficult to get out of the airway. This leads to chronic lung infection, which progressively destroys pulmonary function.

CF is caused by the absence of a protein called cystic fibrosis transmembrane conductance regulator (CFTR). This protein is required for the transport of chloride ions across cell membranes. On the molecular level, there is a mutation in the gene that encodes for CFTR. As a result, CFTR cannot be processed properly by the cell and is unable to reach the exocrine glands to assume its transport function.

Source: Karp G. *Cell and Molecular Biology, Concepts and Experiments*, Wiley, Hoboken, NJ, 1996.

Exhibit 2.5 Receptors

According to the International Union of Pharmacology Committee, a receptor is a cellular molecule, or an assembly of macromolecules, that is concerned directly and specifically in chemical signaling between and within cells. Combination of a hormone, neurotransmitter, drug, or intracellular messenger with its receptor(s) initiates a change in cell function.

Source: The American Society for Pharmacology and Experimental Therapeutics. *Pharmacological Reviews* 55:597–606 (2003).

receptors are then separated from the radioligands and sequenced and their nucleotide sequence is decoded. Potential drug molecules are then studied with these receptors or their nucleotide sequences to determine their interactions in terms of biochemical and functional properties.

DNA Microarray: DNA microarray, also known as DNA or gene chips, is a technology to investigate how genes interact with one another and how they control biological mechanisms in the body. The gene expression profile is dynamic and responds to external stimuli rapidly. By measuring the expression profile, scientists can assess the clues for the regulatory mechanisms, biochemical pathways, and cellular functions. In this way, microarrays enable scientists to find out the target genes that cause disease.

Exhibit 2.6   Radioligands

Ligands are molecules that bind to a target. They may be endogenous (i.e., produced by the body), such as hormones and neurotransmitters, or exogenous, such as drug molecules. Ligands (exogenous or endogenous) with high specificity for particular targets are labeled with radioisotopes. The tissue cells known to contain the target are mixed with a known quantity of the radioligands. Those targets bound with radioligands are separated by rapid filtration or centrifugation, followed by washing with cold buffers to remove unbound ligands. Scintillation counting techniques are used to reveal the amount of bound radioligands.

The target bound with radioligands can be isolated and its amino acid sequence determined. The sequence information enables classifications of the target based on previously known targets. Targets that do not appear to show homology to known ligands and have no known endogenous ligand are called "orphan" targets. Active research is ongoing to find molecules of compounds to interact with these orphan targets as possible sites for therapy.

Sequence information can be used to clone the target by using recombinant technology. In this way, biochemical pathways of the target can be studied in detail, rendering the development of a drug molecule with higher chances of success.

Figure 2.2   Microarray slides (Photo courtesy of Thermo Fisher Scientific).

The heart of the technology is a glass slide or membrane that consists of a regular array of genes (Fig. 2.2). Thousands of genes can be spotted on the array, using a photolithography method. DNA samples extracted from healthy and diseased cells are mixed with the genes on the array. In this way, many genes can be studied and their expression levels in healthy and diseased states can be determined within a short time. The gene that is responsible for a particular disease can be identified. Exhibit 2.7 presents a more detailed explanation of microarrays.

Expressed Sequence Tags and In Silico Methods: Expressed sequence tags (ESTs) are short nucleotide sequences of complementary DNA with about 200–500 base pairs. They are parts of the DNA that code for the expression of particular proteins. EST sequencing provides a rapid method to scan for all the protein coding genes and to provide a tag for each gene on the genome.

The scanning of nucleotide sequences is achieved through *in silico* (computer) methods. The premise is that all proteins, even those with sequences that appear considerably different, can be members of families sharing essentially similar structures and functions.

Scientists carry out searches on databases. Each EST of interest can be compared with sequences in proteins, and the degree of match can be determined. A technique called threading is used. This involves using data on three-dimensional (3D) protein structure, coupled with knowledge of the physicochemical properties of amino acids, to determine if the amino acid sequence is likely to fold in the same way as a sequence for which the structure is known. In this way, more information about the putative target protein can be assessed.

2.4 TARGET VALIDATION

Once a potential disease-causing target has been identified, a process of validation is carried out to confirm the functions and effects of the target. The ultimate target validation is a series of human clinical trials in which the effects of a drug on the target are evaluated. However, this kind of validation is at the other end of the drug discovery spectrum, when much time and commitment have already been expended on the drug. What is required at this early stage is validation of the identified target, to lay the path for developing appropriate drugs aimed at this target. This will ensure that time, resources, and investments can be optimized.

Some questions that target validation has to answer are the following:

- What is the function of the target?
- Which disease pathway does the target regulate?
- How important is this disease pathway?
- What is the expected therapeutic index if a drug is to interact with the target?

Exhibit 2.7   Microarrays

To use the microarray, a known sequence of short DNA is printed onto a solid support of membrane or glass slide. From healthy and diseased cells, mRNAs are isolated. The mRNAs are used to generate complementary DNAs (cDNAs). Fluorescent tags are attached to the cDNAs, and the cDNAs are then mixed and incubated with the microarray supports (slides).

Through a process called hybridization probing, the genes from the samples pair up with their complementary counterparts on the solid supports. When the hybridization step is completed, a scanner (laser beam and camera) is used to capture the fluorescence image of the array.

From comparison of the intensities and ratios of red and green fluorescence, the expression levels of genes from the diseased and healthy cells can be deciphered. For example, if the disease causes some types of genes to be expressed more, more of these genes will hybridize with the DNA on the solid support, providing a greater intensity of red fluorescence than of green. In this way, disease targets can be identified.

The flow chart shows a schematic representation of the use of a microarray for identification of disease genes.

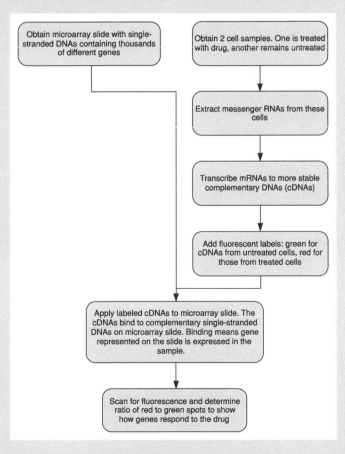

Source: Friend SH, Stoughton RB. The magic of microarrays, *Scientific American* February:44–53 (2002).

Exhibit 2.8   Knockout Mice

Genetic research, such as by microarray, reveals the possible genes that may cause the disease under study. Transgenic mice are bred with the putative gene modified or inactivated, giving rise to the term "knockout" mice models. The effects of gene knockouts are studied in relation to the progress of disease. It is also possible to study drug interactions by treating these mice with potential drug candidates.

Source: Harris S. Transgenic knockouts as part of high-throughput, evidence-based target selection and validation strategies, *Drug Discovery Today* 16:628–636 (2001).

Validations can be divided into two groups: *in vitro* laboratory tests and *in vivo* disease models using animals.

Typically, *in vitro* tests are cell- or tissue-based experiments. The aim is to study the biochemical functions of the target as a result of binding to potential drug ligands. Parameters such as ionic concentrations, enzyme activities, and protein expression profiles are studied.

For *in vivo* studies, animal models are set up and how the target is involved in the disease is analyzed. One such model is the use of knockout or transgenic mice (Exhibit 2.8). It should be borne in mind, however, that there are differences between humans and animals in terms of gene expression, functional characteristics, and biochemical reactions. Nevertheless, animal models are important for the evaluation of drug–target interactions in a living system.

More recently, *in silico* target validation has been used. This is similar to the method discussed for ESTs. The DNA sequence of the putative target is compared with those of known liganded receptors. If homologies (similarities) of sequences and structures are determined, they can provide clues to ligands that are likely to interact with the target.

2.5   DRUG INTERACTIONS WITH TARGETS OR RECEPTORS

It should be clarified that targets identified using microarrays are mainly the genes that regulate or contribute to diseases. These gene targets give us the clues to the proteins that are affected. In most situations, it is the proteins or receptors that drug molecules are developed to interact with to provide the therapy. The exceptions are in cases such as antisense and RNA interference drugs and gene therapy, where the nucleotides, RNA, and genes are targeted, respectively.

When presented to the target, drug molecules can elicit reactions to switch on or switch off certain biochemical reactions. The main drug targets in the human body can be classified into three categories:

- *Enzymes:* There are many different types of enzymes in the human body. They are required for a variety of functions. Drugs can interact with enzymes to modulate their enzymatic activities.
- *Intracellular Receptors:* These receptors are in the cytoplasm or nucleus. Drugs or endogenous ligand molecules have to pass through the cell membrane (a lipid bilayer) to interact with these receptors. The molecules must be hydrophobic or coupled to a hydrophobic carrier to cross the cell membrane.
- *Cell Surface Receptors:* These receptors are on the cell surface and have an affinity for hydrophilic binding molecules. Signals are transduced from external stimuli to the cytoplasm and affect cellular pathways via these surface receptors. There are three main superfamilies (groups) of cell surface receptors: G-protein coupled receptors, ion channel receptors, and catalytic receptors using enzymatic activities.

Hydrophilic or water-soluble drugs do not cross membranes. They stay in the bloodstream for durations that are normally short, lasting on the order of seconds, and mediate responses of short duration. In contrast, hydrophobic drugs require carrier molecules for transport through the bloodstream. Hydrophobic drugs remain in the bloodstream and can persist for hours and days, providing much longer effects.

When the action of the drug is to activate or switch on a reaction, the drug is called an "agonist." On the other hand, if the drug switches off the reaction, or inhibits or blocks the binding of other agonist components onto the receptor, it is called an "antagonist." When the interaction is with an enzyme, the terms "inducer" and "inhibitor" are used to denote drugs that activate or deactivate the enzyme.

Appendix 3 lists some common drugs and their mechanisms of action, showing their roles as agonists or antagonists, and inducers or inhibitors. Figures 2.3 and 2.4 provide schematic lock and key representations of the agonist and antagonist actions. It should be noted that the drug molecule (agonist or antagonist), receptor, and cell membrane are in fact complicated 3D structures. The analogy is that only certain keys can be inserted into the lock and activate or deactivate the lock. Some facts about interactions between drug molecules and targets to bear in mind are the following:

- Binding is specific.
- Binding occurs at particular sites in the target molecule.
- Binding is reversible.

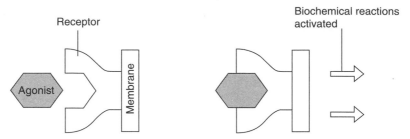

Figure 2.3   Agonist binding to receptor initiates biochemical reactions.

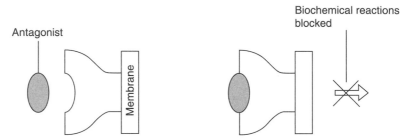

Figure 2.4   Antagonist binding to receptor blocks biochemical reactions.

Allosteric binding occurs when two molecules bind to different sites on the target. When the two molecules are identical, it is termed homotropic interaction. If the molecules differ from each other, it is termed heterotropic interaction. Binding is competitive when two different ligand molecules compete for the same site. We discuss ligand binding further in Chapter 3. The specificity of ligand–receptor interaction is illustrated in Exhibit 2.9.

2.5.1   Types of Interactions

Binding between drug molecule and receptor or enzyme is critically dependent on the shapes and sizes of the molecules. To deliver therapeutic actions, drug molecules with the right shapes and sizes have to be designed to fit into the binding sites (pockets) of the receptor or enzyme. Another important factor is the nature of the coupling. Before a drug can fit into the binding site, it has to overcome thermal and vibrational motions at the cellular level. The attractive forces must be strong enough for the drug to dock with the binding site. When molecules couple together, the type of bonding can be divided into covalent bonding and electrostatic interactions due to hydrogen bonding or van der Waals forces. The stronger the coupling between the drug and binding site, the more sustained is the interaction.

Covalent bonds are strong bonds. Actual bonds are formed between the interacting molecules via the sharing of electrons. Hence, this type of interac-

Exhibit 2.9   Aspirin (Acetylsalicylic Acid)

The enzyme prostaglandin H_2 synthase-1 (PGHS-1) manufactures prostaglandin H_2, which is converted to prostaglandin E_2 and causes fever and inflammation. PGHS-1 contains two protein subunits with long channels.

The chemical arachidonic acid enters these channels and is converted to prostaglandin H_2. Aspirin, with the correct shape and size, enters these channels and blocks entry of arachidonic acid. As a result, the agent for causing fever and inflammation cannot be manufactured. Unfortunately, an undesirable effect of aspirin is that it blocks other types of PGHS, including the types that protect the stomach lining, giving rise to potential for stomach bleeding.

Recent advances with other anti-inflammatory drugs, ibuprofen and naproxen, which only work by physically blocking the channel to arachidonic acid, mean that the adverse effect of stomach bleeding can be avoided.

See Case Study #2 for more discussion of anti-inflammatory drugs.

Source: Garavito M. Aspirin, *Scientific American* May: 108 (1999).

tion is expected to provide long-lasting effects, although not many drug–receptor bonds are of this nature.

Electrostatic forces are due to the ionic charges residing on the molecules, which attract or repel each other. The macromolecular structures of the receptors and enzymes mean that there are a number of ionic charges to attract the oppositely charged drug molecules. The forces of electrostatic interactions are weaker than covalent bonding. Electrostatic interactions are more common in drug–receptor interactions. There are two types of electrostatic interactions:

• Hydrogen bonding
• Van der Waals forces

Hydrogen bonds are due to the attractive forces between the distorted electron cloud of a hydrogen atom and other more electronegative atoms such as oxygen and nitrogen. The attractive forces are weaker than covalent bonds, but many hydrogen bonds can be formed in macromolecular protein molecules. Van der Waals forces are weaker attractive forces, due to the attraction between neutral atoms.

A third type of interaction is due to hydrophobic effects. These are the result of nonelectrostatic domains interacting. This type of interaction occurs mainly with the highly lipid-soluble drugs in the lipid part within the cytoplasm of the cell.

2.6 ENZYMES

Enzymes are biological molecules that catalyze biochemical reactions. The thermodynamics of biochemical enzymatic reactions are described in Exhibit 2.10.

Almost all enzymes are proteins. They provide templates whereby reactants (substrates) can bind and are favorably oriented to react and generate the products. The locations where the substrates bind are known as "active sites." Because of the specific 3D structures of the active sites, the functions of enzymes are specific; that is, each particular type of enzyme catalyzes specific biochemical reactions. Enzymes speed up reactions, but they are not consumed and do not become part of the products. Enzymes are grouped into six functional classes by the International Union of Biochemists (Table 2.2).

Exhibit 2.10   Thermodynamics of Enzymatic Reactions

In general, there are two types of biochemical reactions: exothermic and endothermic. Exothermic reactions are those where the energy states (free energy, labeled as G) of the reactants are higher than those of the products—they are energetically favorable. Endothermic reactions are those for which the products have higher energy states than the reactants—they are energetically unfavorable.

Regardless of whether the reaction is favorable, the reactants have to come together in close proximity to react. They have to overcome a potential energy barrier that may involve displacing solvating molecules around the reactants and reorienting the reactants. The energy needed to overcome this potential energy barrier is called the activation energy (see figure).

Enzymes bind to the reactants and provide an alternative mechanism of lower activation energy for the reaction to proceed. Hence, enzymes speed up biochemical reactions that are otherwise too sluggish to advance.

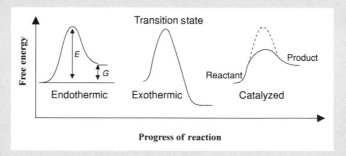

Note: E is the potential energy barrier, known as the activation energy. G is the change in free energy. The catalyzed reaction provides an alternative reaction pathway with lower activation energy.

TABLE 2.2   Classification of Enzymes

Number	Classification	Biochemical Properties
1	Oxidoreductases	Remove or add hydrogen atoms in oxidation or reduction reactions.
2	Transferases	Transfer functional groups from one molecule to another. Kinases are specialized transferases that transfer phosphate from ATP to other molecules.
3	Hydrolases	Hydrolyze various functional groups.
4	Lyases	Add water, ammonia, or carbon dioxide across double bonds, or remove these elements to produce double bonds.
5	Isomerases	Convert between different isomers.
6	Ligases	Form a bond between molecules.

In some cases, enzymes require the assistance of coenzymes (cofactors) to ensure the reactions proceed. Coenzymes include vitamins, metal ions, acids, and bases. They can act as transporters or electron acceptors or be involved in oxidation–reduction reactions. At the completion of the reaction, coenzymes are released, and they do not form part of the products. For some reactions that are energetically unfavorable, an energy source provided by the compound adenosine triphosphate (ATP) is needed to ensure the reactions proceed, as shown in the following reactions:

$$ATP + H_2O \rightarrow ADP + Ph + H^+ \qquad \Delta G = -30.54\,kJ/mol$$

$$ATP + H_2O \rightarrow AMP + PhPh + H^+ \qquad \Delta G = -45.60\,kJ/mol$$

where ADP is adenosine diphosphate, Ph is a phosphate group, PhPh are two phosphate groups, and ΔG is the energy given off.

Enzymatic reactions can be impeded by the addition of exogenous molecules. This is how drugs are used to control biochemical reactions, and most drugs are used for inhibitory functions. Drugs may function as competitive inhibitors or as noncompetitive inhibitors. Competitive inhibitors compete with the substrates for binding to the active sites, whereas noncompetitive inhibitors bind to another location (allosteric site) but affect the active site and its consequential interactions with the substrates. Some drugs used as enzyme inhibitors are the following:

- Esomeprazole (Nexium, AstraZeneca): proton pump inhibitor for the prevention of relapse in reflux esophagitis
- Captopril (Capoten, Bristol-Myers Squibb): angiotensin converting enzyme (ACE) inhibitor for the treatment of hypertension
- Imatinib mesylate (Gleevec, Novartis): tyrosine kinase inhibitor for the treatment of chronic myeloid leukemia (refer to Exhibit 7.3)

- Sertraline (Zoloft, Pfizer): selective serotonin (5-hydroxytryptamine; 5HT) uptake inhibitor for treating major depression and obsessive compulsive disorder
- Atorvastatin (Lipitor, Pfizer) and simvastatin (Zocor, Merck): HMG-coenzyme A inhibitors for the reduction of cholesterol level in blood (refer to Exhibit 1.3).

Exhibit 2.11 shows two selected drugs, celecoxib (a COX-2 inhibitor) and orlistat (a lipase inhibitor), and their actions on disease targets.

Drugs are also used to inhibit the enzymatic reactions of foreign pathogens that enter the human body. An example is the use of reverse transcriptase inhibitor and protease inhibitor for combating the human immunodeficiency virus (HIV), as shown in Exhibit 2.12. Some new inhibitors are used to block HIV from attaching to the human cell, CD4, thus stopping replication and infection of other cells, as presented in Exhibit 2.13.

Exhibit 2.11   Two Selected Drugs

COX-2 Inhibitor: Celecoxib (Celebrex, Pfizer) inhibits the enzyme COX-2, which is involved in pain and inflammation, but it has no effect on the COX-1 enzyme, which helps to maintain stomach lining. It is prescribed for the relief of pain and symptoms of osteoarthritis and rheumatoid arthritis. Previously, nonsteroidal anti-inflammatory drugs (NSAIDs) were used. NSAIDs inhibit both COX-1 and COX-2 enzymes and cause stomach bleeding (see Case Study #2).

Lipase Inhibitor: Orlistat (Xenical, Roche) is prescribed for the treatment of obesity. It inhibits the gastrointestinal lipase enzymes by binding to the lipase through the serine site and inactivates the enzyme. Fat in the form of triglycerides cannot be hydrolyzed by the lipase and converted to free fatty acids and monoglycerides. Thus, there is no uptake of fat molecules into the cell tissue.

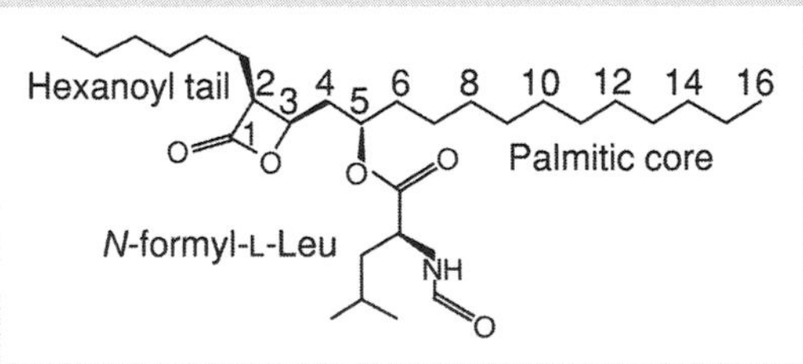

Source: Pemble CP IV, Johnson LC, Kridel SJ, Lowther WT. Crystal structure of the thioesterase domain of human fatty acid synthase inhibited by Orlistat, *Nature Structural & Molecular Biology* 14:704–709 (2007).

Exhibit 2.12   Drugs Against HIV

The diagram below shows the various stages of HIV infection of the CD4 cell (see also Exhibit 4.7).

Reverse transcriptase is an enzyme that makes a DNA copy of the virus RNA (Step 3). Once made, the DNA enters the cell nucleus and replicates many times. Another enzyme, the protease, is required to cut the HIV proteins into proper sizes and assemble into the viral particles (Step 7).

The drug zidovudine (AZT; Retrovir, GlaxoSmithKline) is a reverse transcriptase inhibitor. It is structurally similar to a nucleotide called thymidine except for an azido ($-N_3$) group in place of the –OH group at the 3' position of the thymidine nucleotide sugar. Thymidine is a building block for the viral DNA, and when AZT is incorporated into the DNA chain it blocks further chain linkages, as there is no –OH group available.

At Step 6, large molecules of the viral proteins are made. Protease cleaves the large protein molecules into smaller pieces. Another drug, a protease inhibitor (darunavir, Prezista), is used to inhibit the protease, thus stopping the formation of new viruses.

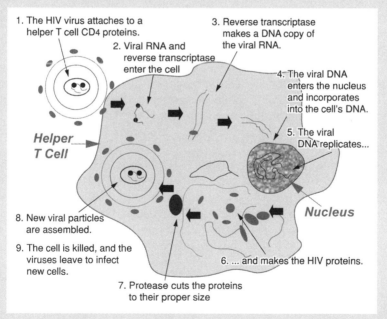

Source: Casidy R, Frey R. *Drug Strategies to Target HIV: Enzyme Kinetics and Enzyme Inhibitors*, Washington University. http://wunmr.wustl.edu/EduDev/LabTutorials/HIV/DrugStrategies.html [accessed May 21, 2003].

Exhibit 2.13   Latest Entry Inhibitor HIV Drugs

Existing treatment for HIV focuses mainly on the use of protease inhibitors and reverse transcriptase inhibitors. This strategy, however, is to inhibit the HIV from replicating after it has invaded the CD4 cell.

A more recent strategy is the use of entry inhibitors, which seek to stop the HIV from entering the CD4 cell. There are two such drugs: enfuvirtide (Fuzeon) by Roche and maraviroc (Selzentry) by Pfizer, approved by the FDA in March 2003 and August 2007, respectively.

Fuzeon works by attaching to the glycoprotein gp41 of the HIV and preventing the virus from using the gp41 to attach to the surface proteins of the CD4 cell. Selzentry, on the other hand, targets the CCR5 receptor on the CD4 surface. Once bound, the HIV cannot attach to CD4 as the CCR5 is not available to interact with the HIV.

There is ongoing research on another type of inhibitor, which targets the glycoprotein gp120 on the HIV.

Sources: (1) Food and Drug Administration, *Fuzeon*. http://www.fda.gov/medwatch/ SAFETY/2004/oct_PI/Fuzeon_PI.pdf [accessed September 28, 2007]. (2) Food and Drug Administration, *Selzentry*. http://www.fda.gov/cder/drug/mg/maravirocMG.pdf [accessed September 28, 2007].

2.7   RECEPTORS AND SIGNAL TRANSDUCTION

Cells communicate to coordinate the biochemical functions within the human body. If the communication system is interrupted or messages are not conveyed fully, our bodily functions can go haywire. An example of this is discussed in Exhibit 2.2: if the p53 protein is mutated, cell growth is unchecked and cancer can form.

There are hundreds of receptors on the cell surface. They act as "antennas" to receive signals from the extracellular environment. These signals may be from endogenous sources, such as neurotransmitters, cytokines, and hormones, or exogenous sources, such as viruses and drugs. On receiving the signals, receptors transduce these signals to the cell interior. Within the cell, the signal may cause a cascade of reactions to proceed. Fig. 2.5 illustrates this signal transduction process.

Signals may be relatively straightforward, as in the case of ion channels for opening and closing of channel gates to control migration of ions. There are also signals that are more complex, involving the binding of ligand to the receptor. A consequence of the binding is a conformational (shape) change in the receptor, which leads to further amplifying processes.

We discuss later a number of receptor classes and analyze how signals are transduced. These receptors are G-protein coupled receptors (GPCRs), ion channel receptors, tyrosine kinases, and intracellular receptors. A list of selected

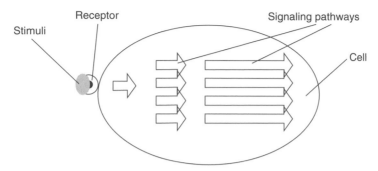

Figure 2.5   Signal transduction showing cascades of reaction occurring inside the cell.

drugs and target receptors is shown in Table 2.3, and Table 2.4 presents the mechanisms of action for several drugs.

2.7.1   G-Protein Coupled Receptors

G-protein coupled receptors (GPCRs) represent possibly the most important class of target proteins for drug discovery. They are always involved in signaling from outside to inside the cell. The number of diseases that are caused by a GPCR malfunction is enormous, and therefore it is not surprising that most commonly prescribed medicines act on a GPCR. It is estimated that more than 30% of drugs target this receptor superfamily.

The common feature of this superfamily (there are many different families and subtypes of receptors in this group) of receptors is that there are seven domains that cross the cell membrane (Fig. 2.6). These seven transmembrane receptors are often referred to as serpentine receptors.

The serpentine receptors are coupled to the G-proteins (guanine nucleotide regulatory proteins) inside the cell. There are three subunits that make up the G-proteins: α, β, and γ. When a ligand, for example, a drug or neurotransmitter, binds to the receptor on the cell surface, the shape of the receptor changes. This induces an activated change in the trimeric clusters of α, β, and γ subunits within the cell. A phosphorylation (the addition of a phosphate group, such as PO_3H_2, to a compound) reaction occurs, in which guanosine diphosphate (GDP) changes to guanosine triphosphate (GTP):

$$GDP + Phosphate \rightarrow GTP$$

This reaction then switches on the effector molecule, and the signal is relayed along the pathway (see Fig. 2.7). When the enzyme GTPase hydrolyzes GTP to GDP and removes the phosphate group, the trimeric subunits change back to the inactivated state. This receptor is once again ready to receive and transmit further signals.

TABLE 2.3 Selected Drugs and Target Receptors

Drug	Therapeutic Category	Drug Target
Amlodipine (Norvasc, Pfizer)	Cardiovascular	Ion channel
Atorvastatin (Lipitor, Pfizer)	Cardiovascular	Enzyme inhibitor
Augmentin/amoxicillin plus clavulanic acid (GlaxoSmithKline)	Anti-infective	Enzyme inhibitor
Bevacizumab (Avastin, Genentech)	Cancer	Vascular endothelial growth factor inhibitor
Celecoxib (Celebrex, Pharmacia)	Musculoskeletal	Enzyme inhibitor
Clopidogrel (Plavix, BMS/Sanofi-Aventis)	Hematology	Platelet receptor inhibitor
Erythropoietin (Epogen, Amgen)	Hematology	Transmembrane agonist
Erythropoietin (Procrit, Ortho Biotech)	Hematology	Transmembrane agonist
Esomeprazole (Nexium, AstraZeneca)	Gastrointestinal/ metabolism	Ion channel
Etanercept (Enbrel, Amgen)	Rheumatoid arthritis	TNF-α
Fluoxetine (Prozac, Eli Lilly)	Central nervous system	GPCR[a]
Lansoprazole (Takepron, Takeda)	Gastrointestinal/ metabolism	Ion channel
Loratadine (Claritin, Schering)	Respiratory	GPCR
Olanzapine (Zyprexa, Eli Lilly)	Central nervous system	GPCR
Omeprazole (Losec, AstraZeneca)	Gastrointestinal/ metabolism	Ion channel
Paroxetine (Seroxat, GlaxoSmithKline)	Central nervous system	GPCR
Sertraline (Zoloft, Pfizer)	Central nervous system	GPCR
Simvastatin (Zocor, Pfizer)	Cardiovascular	Enzyme inhibitor
Trastuzumab (Herceptin, Genentech)	Cancer	Overexpressed HER2 protein
Venlafaxine (Effexor, Wyeth)	Central nervous system	Serotonin-norepinephrine reuptake inhibitor

[a]GPCR = G-protein coupled receptor.

Source: Adapted from Renfrey S, Featherstone J. From the analyst's couch: structural proteomics, *Nature Reviews Drug Discovery* 1:175–176 (2002). A comprehensive list of receptors is given by Imming P, Sinning C, Meyer A. Drugs, their targets and the nature and number of drug targets, *Nature Reviews Drug Discovery* 5:821–834 (2006).

TABLE 2.4   Mechanisms of Action for Selected Drugs

Drug	Mechanism
Antihypertensive	
Hydrochlorothiazide	Increases sodium and water excretion, decreases blood volume, thereby reduces cardiac output
Prazosin	Alpha-adrenergic receptor antagonist, inhibits symphathetic stimulation of arteriolar contraction
Atenolol	Beta-adrenergic receptor anatagonist, reduces cardiac output by decreasing heart rate and contraction
Captopril	Angiotensin converting enzyme inhibitor, decreases arterial and venous pressure, reduces cardiac load
Verapamil	Calcium channel blocker, relaxes vascular smooth muscle
Hyperlipidemia	
Atorvastatin	HMG-CoA reductase inhibitor, inhibits conversion of HMG-CoA to mevalonic acid for the synthesis of cholesterol
Anticoagulant	
Warfarin	Inhibits synthesis of clotting factors II (prothrombin), VII, IX, and X
Heparin	Activates antithrombin III, inhibitor of thrombin and factor X
Clopidogrel	Antiplatelet, inhibits expression of glycoprotein receptors to reduce fibrinogen binding and platelet aggregation
Central Nervous System	
Benzodiazepines	Acts as sedative-hypnotic, opens ion channel, chloride ion influx, leading to neuronal membrane hyperpolarization
Bupivacaine	Anesthetic, binds to sodium channel, decreases sodium permeability, stops action potential from propagating and thus sensory input not available
Phenothiazine	Functions as antipsychotic by blocking dopamine and 5-HT receptors
Fluoxetine	Reduces neurotransmitter uptake by acting as selective serotonin reuptake inhibitor
Albuterol	A bronchodilator that blocks selective β_2-adrenergic receptor, increases cyclic adenosine monophosphate concentration in smooth muscle, causes muscle to relax
Antiulcerant	
Cimetidine	Histamine H_2 receptor antagonist, reduces volume of gastric acid produced
Antineoplastic (Cancer)	
Methotrexate	Inhibits dihydrofolate reductase, enzyme that converts folate to tetrahydrofolate for thymidine and purine synthesis
Fluorouracil	Generates two active metabolites: one prevents synthesis of thymidine, the other interferes with RNA function
Doxorubicin	Binds to DNA and uncoils DNA

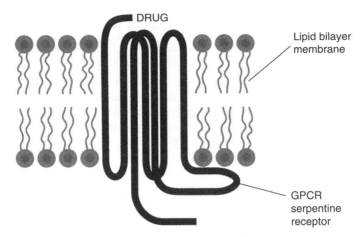

Figure 2.6 A G-protein coupled receptor.

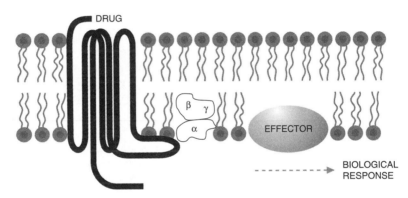

Figure 2.7 Signal cascade in GPCR.

GPCRs are involved in a wide range of diseases, including asthma, hypertension, inflammation, cardiovascular disease, cancer, and gastrointestinal and central nervous system diseases. From the Human Genome Project, it is estimated that there are about 1000 GPCRs. Current therapeutic drugs are only targeting about 50 of these GPCRs. Thus, there are many possibilities of developing new drugs to target this family of receptors.

2.7.2 Ion Channel Receptors

There are two main types of ion channel receptors: ligand-gated and voltage-gated. In addition, some ion channels are regulated through GPCRs or via activation by amino acids.

The ligand-gated family consists of receptors of the so-called cys-loop superfamily (nicotinic receptor, gamma-aminobutyric acid (GABA$_A$ and GABA$_C$) receptors, glycine receptors, 5-HT$_3$ receptors, and some glutamate activated anionic channels). The common feature is that they are made up of five subunits (designated as two α, one β, one γ, and one δ subunits—Fig. 2.8). Natural ligands for this family of ion channels include acetylcholine, GABA, glycine, and aspartic acid. They are, in general, synaptic transmitters.

Normally, in the resting state, the channel is impermeable to ions. When a ligand binds to the receptor, it becomes activated and opens a channel to a diameter of about 6.5 Å (6.5×10^{-10} m). This action allows the migration of, for example, extracellular sodium ions to the interior of the cell. A cascade of further changes then proceed within the cell to amplify the signal.

Voltage-gated ion channels depend on changes of transmembrane voltage to regulate the opening and closing of channel gates. A common feature of this type of receptor is the presence of four domains, where each domain consists of six membrane-spanning regions. Some of these channels are the sodium, calcium, and potassium channels, and they regulate the influx of these ions into the cell interior to propagate the signal.

Diseases mediated through ion channel receptors include cardiovascular disease, hypertension, and central nervous system dysfunctions. A voltage-gated ion channel is a key to the treatment of cystic fibrosis (see Exhibit 2.4).

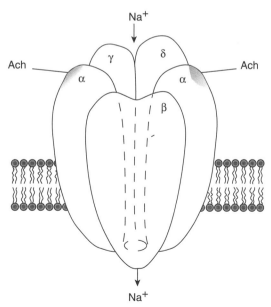

Figure 2.8 A ligand-gated ion channel receptor. ACh = acetylcholine. The binding of ACh to the α subunits opens the ion channel, allowing Na$^+$ ions to flow through the channel into the cell.

> **Exhibit 2.14   Neurotransmitters Binding to GPCR and Ion Channel**
>
> Acetylcholine: Binds to cholinergic receptor, which are of two types—muscarinic and nicotinic.
>
> Gama-aminobutyric acid (GABA): Binds to both GPCR and ion-channel receptors.
>
> Dopamine: Binds to dopamine receptors D1 and D5, activating adenylyl cyclase; binds to receptors D2, D3, and D4, inhibiting adenylyl cyclase.
>
> Norepinephrine: Binds to α- and β-adrenergic receptors, causing vasoconstriction and increasing blood pressure for the treatment of hypotension and shock.
>
> Serotonin: Binds to 5-hydroxytryptamine (5-HT) receptor to act as excitatory and inhibitory neurotransmitter; in its inhibitory function, it can treat anxiety and depression; in its excitatory function, it is an antipsychotic.

Some examples of the effects of neurotransmitters of the central nervous system binding to the GPCR and ion channel are given in Exhibit 2.14.

2.7.3   Tyrosine Kinases

This class of receptors transmits signals carried by hormones and growth factors. The structure consists of an extracellular domain for binding ligands and a cytoplasmic enzyme domain. The function of kinases is to enable phosphorylation. Phosphorylation regulates most aspects of cell life.

When a ligand binds to the receptors, the receptors dimerize and join together. This action activates the enzyme within the cell. As a result, protein molecules are phosphorylated (Fig. 2.9).

Other kinase receptors are serine/threonine kinases, protein kinases, and mitogen-activated protein (MAP) kinases. Insulin, transforming growth factor-beta (TGF-β), and platelet-derived growth factor (PDGF) are the natural ligands that interact with kinase receptors.

2.7.4   Intracellular Receptors

Intracellular (nuclear) receptors are located inside cells, in the cytoplasm or nucleus. Endogenous ligands such as hormones and drugs bind to these receptors to either activate or inhibit transcription messages from genes. There is a large superfamily of these intracellular receptors. The common feature is that they all have a single polypeptide chain consisting of three distinct domains:

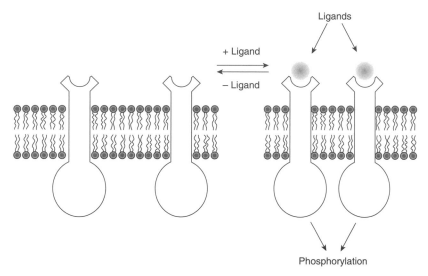

Figure 2.9 A tyrosine kinase receptor.

- Amino terminus: this region in most instances is involved in activating or stimulating transcription.
- DNA binding domain: amino acids in this region are responsible for binding of the receptor to specific sequences of DNA.
- Carboxy terminus, or ligand-binding domain: this is the region that binds ligands.

2.8 ASSAY DEVELOPMENT

To study drug–receptor/enzyme interaction, it is not always convenient or appropriate to use a living system of the target receptor. Instead, biochemical assays can be devised to mimic the target. Very often, the assays use multicolor luminescence or fluorescence-based reagents. In this way, the reaction path can be followed in space and time to enable quantitative evaluation of the reaction.

Many parameters can be monitored, for example, free-ion concentrations, membrane potentials, activities of specific enzymes, rate of proton generation, transport of signaling molecules, and gene expression.

Primary assays are devised to incorporate physiological or enzymatic targets for screening biological activity of potential drug compounds. The biological assays are then reconfirmed in specific biochemical and whole cell assays to characterize the target–compound interaction.

Exhibit 2.15 shows some current assays used in ligand–receptor studies.

Exhibit 2.15   Reporter Assays and Bioluminescence

Assays can be prepared with a reporter system containing, for example, the firefly luciferase gene. The reporter cells are coupled to receptor genes. When a ligand binds to the receptor, a luminescent glow can be observed. In this way, the effects of the signaling events are evaluated.

There are other reporter gene systems, such as β-galactosidase (a bacterial enzyme), chloramphenicol acetyltransferase (a bacterial enzyme), and aequorin (a jellyfish protein).

Using blue-light photoreceptors from *Bacillus subtilis* and *Pseudomonas putida* that contain light-oxygen-voltage sensing domains, flavin mononucleotide-based fluorescent proteins were produced that can be used as fluorescent reporters in both aerobic and anaerobic biological systems.

Source: (1) Naylor LH. Reporter gene technology: the future looks bright, *Biochemical Pharmacology* 58:749–757 (1999). (2) Drepper, T, et al. Reporter proteins for *in vivo* fluorescence without oxygen, *Nature Biotechnology* 25:443–445 (2007).

2.9   CASE STUDY #2

Anti-inflammatory Therapy*

Since 1898 aspirin **(1)** has been used to treat pain and inflammation. Although it is effective, it also causes adverse events in the form of gastrointestinal bleeding and ulceration. More potent nonsteroidal anti-inflammatory drugs (NSAIDs) such as ibuprofen **(2)** and naproxen **(3)** were developed in the 1960s and 1970s. Unfortunately, they too suffer from the same dilemma in causing varying degrees of bleeding and ulceration in prolonged use.

Studies in the 1970s revealed that an enzyme, cyclooxygenase (COX), converts arachidonic acid **(4)** to an intermediate, prostaglandin H_2 **(5)**, as shown below:

Arachidonic acid $\rightarrow$ Hydroperoxy endoperoxide prostaglandin G_2 (PGG_2) $\rightarrow$ Prostaglandin H_2 (PGH_2)

---

*Sources: (1) Michaux C, et al. A new potential cyclooxygenase-2 inhibitor, pyridinic analogue of nimesulide, *European Journal of Medicinal Chemistry* 40:1316–1324 (2005). (2) Stix G. Better ways to target pain, *Scientific American* January: 84–88 (2007).

(1)

(2)

(3)

(4)

(5)

Another enzyme converts prostaglandin H_2 to prostaglandin E_2 (PGE_2):

$$\text{Prostaglandin } H_2 \rightarrow \text{Prostaglandin } E_2$$

It is PGE_2 that is responsible for mediating pain and inflammation. NSAIDs such as aspirin, ibuprofen, and naproxen block the active site of the COX enzyme, preventing the arachidonic acid from docking to this site, and become converted to PGH_2 (see Exhibit 2.9).

In the early 1990s two isoforms of the COX enzyme were found: COX-1 and COX-2, with about 60% homology between them (there is actually a COX-3, but it is a variant of COX-1; there are differences in the two active sites between COX-1 and COX-2, with the latter possessing an additional hydrophilic pocket).

COX-1 is found in healthy individuals and is important in maintaining a balanced physiological role in kidneys and stomach. COX-2, on the other hand, is induced in the case of inflammation where it mediates the inflammation process. Aspirin, ibuprofen, and naproxen inhibit both COX-1 and COX-2 indiscriminately. While this reduces the production of PGE_2 through the inhibition of COX-2, it upsets the hemostasis function of COX-1, which has a protective function for the mucosal lining, and leads to bleeding and ulcer formation.

Specific COX-2 inhibitors were developed in the 1990s, culminating in the approval of three drugs in late 1990s and early 2000s: celecoxib (Celebrex), rofecoxib (Vioxx), and valdecoxib (Bextra) (see **(6)**, **(7)**, and **(8)** and Exhibit 2.11). Indeed, these drugs work well as anti-inflammatories and have no appreciable side effects with respect to bleeding and ulcer. But through prolonged use and under high dosage, both Vioxx and Bextra were found to have the potential to cause heart attacks and strokes. These two drugs were withdrawn from the market in 2004/5 although Celebrex is still on the market with a change in labeling to warn of potential cardiovascular problems for chronic usage.

(6)

(7)

(8)

Recent work in the early 2000s showed a more complicated role for the COX-2 enzyme; together with additional enzymes, COX-2 generates a number of other prostaglandin compounds besides PGE_2, which have other regulatory functions:

PGH_2 + Enzyme → PGD_2 (involved in sleep regulation and allergic reactions

PGH_2 + Enzyme → PGF_2 (controls contraction of the uterus during birth and menstruation)

PGH_2 + Enzyme → Thromboxane (TXA_2 stimulates contraction of blood vessels and induces platelet aggregation [clotting])

PGH_2 + Enzyme → PGI_2 (dilates blood vessels and inhibits platelet aggregation; may protect against arteriosclerosis and damage to stomach lining)

It is postulated that selective inhibition of COX-2 by Vioxx, for example, halts the production of PGE_2 as shown below:

PGH_2 + Enzyme (microsomal prostaglandin E synthase, mPGES) → PGE_2

The selective inhibition of COX-2 has no effect on COX-1, which continues to manufacture PGE_2 using additional cytosolic prostaglandin E synthase enzyme (cPGES), and hence the integrity of the stomach lining is maintained. But at the same time COX-2 inhibition reduces the production of PGI_2, leading to the possibility of cardiovascular problems.

These new findings undoubtedly would pave the way for other more selective drugs to be developed and we can anticipate better anti-inflammatory drugs to be available to treat pain and inflammation.

2.10 SUMMARY OF IMPORTANT POINTS

1. Pharmaceutical companies evaluate the future direction and R&D activities based on medical needs, market size, patent protection, and key competencies.
2. Most diseases, apart from trauma and infections, have their origins in the genes or the proteins associated with them.
3. The current approach to drug discovery starts with identification of a target or targets that cause or lead to disease.
4. Microarray is a technology used to study gene interactions and control of biochemical pathways.
5. Target validation is necessary to confirm the validity of a target as representative of a disease model before too much investment and time are expended on it.
6. The main drug targets are enzymes and receptors that are found on the cell surface or reside within the intracellular matrix.
7. Drugs interact with enzymes and receptors mainly through van der Waals forces and hydrogen bonding; they need the correct shapes and sizes to fit into the active sites of the targets.
8. Drugs work in two ways: as agonists and antagonists. Agonists activate the receptors whereas antagonists deactivate, block, or inhibit the receptors.
9. After binding of drug and receptor, a cascade of signal transductions occur within the cell and are manifested as a variety of effects on the diseased biochemical pathways.
10. The major receptors are GPCRs, ion channels, and tyrosine kinases.
11. Assays are devised to test biological systems in the laboratories as they are readily available and provide a means to evaluate the effects of potential drug candidates.

2.11 REVIEW QUESTIONS

1. Why is target validation an important process?
2. Explain how microarray works.
3. What are the main targets for drugs?
4. Explain how enzymes work and describe the different types of enzymes.
5. What is meant by signal transduction?
6. Describe the different types of receptors and explain how they function.
7. Why do scientists use assays in drug discovery and development?

2.12   BRIEF ANSWERS AND EXPLANATIONS

1. The importance of target validation is to confirm the role of the target and its effect on the biochemical process in altering the disease before a substantial investment is committed to the R&D.

2. Refer to Section 2.3 and Exhibit 2.7.

3. The main drug targets are enzymes, intracellular receptors, and extracellular (cell surface) receptors. Drugs are normally designed to interact with these entities either as agonists or antagonists to achieve control over the disease pathway.

4. Refer to Section 2.6. Explanation should include the blocking action of the drug on the enzyme and provision of examples of drugs in achieving this function.

5. Refer to Section 2.7 and Fig. 2.5. It should be noted that the signal transduction process is very dynamic and there are many cascading pathways. This explains the need to have drugs with specific interactions to reduce other reactions that can give rise to adverse events (side effects). Section 2.9 shows the effects of drug specificity.

6. The important receptors are GPCRs, ion channels, tyrosine kinases, and intracellular receptors. Refer also to Appendix 3 for specific functions of the drugs in interacting with receptors.

7. Assays provide a means to test the potential drug candidate quickly and in a cost-effective manner. Until such time as the efficacy and safety assays (including preclinicals in animals) are completed and show that the candidate has the potential to become a drug, it should never be tested on humans.

2.13   FURTHER READING

Brenner GM, Stevens CW. *Pharmacology*, 2nd ed., Saunders Elsevier, Philadelphia, 2006.

Cambridge Healthtech Institute. *Streamlining Drug Discovery with Breakthrough Technologies for Genomic Target Identification and Validation.* http://www.chireports. com/content/articles/targetart.asp [accessed May 7, 2002].

Campbell MK, Farrell SO. *Biochemistry*, 5th ed., Thomson Brooks/Cole, Belmont, CA, 2006.

Campbell NA, Reece JB, Mitchell LG. *Biology*, 5th ed., Benjamin/Cummings, Menlo Park, CA, 1999.

Deller MC, Jones EY. Cell surface receptors, *Current Opinion in Structural Biology* 10:213–219 (2000).

Dowell SJ. Understanding GPCRs—from orphan receptors to novel drugs, *Drug Discovery Today* 6:884–886 (2002).

Ezzell C. Beyond the human genome, *Scientific American* July:64–69 (2000).

Foreman JC, Johansen T, eds. *Textbook of Receptor Pharmacology*, CRC Press, Boca Raton, FL, 2002.

Friend SH, Stoughton RB. The magic of microarrays, *Scientific American* February:44–53 (2002).

Harris S. Transgenic knockouts as part of high-throughput, evidence-based target selection and validation strategies, *Drug Discovery Today* 6:628–636 (2001).

Hernandez MA, Rathinavelu A. *Basic Pharmacology: Understanding Drug Actions and Reactions*, CRC Press, Boca Raton, FL, 2006.

Imming P, Sinning C, Meyer A. Opinion: drugs, their targets and the nature and number of drug targets, *Nature Reviews Drug Discovery* 5:821–834 (2006).

Katzung BG, ed. *Basic and Clinical Pharmacology*, 10th ed., McGraw-Hill, New York, 2007.

Kirkpatrick P. G-protein-coupled receptors: putting the brake on inflammation, *Nature Reviews Drug Discovery* 1:99 (2002).

Leff P, ed. *Receptor-Based Drug Design, Drugs and the Pharmaceutical Sciences*, Volume 89, Marcel Dekker, New York, 1998.

Lindsay MA. Target discovery, *Nature Reviews Drug Discovery* 2:831–838 (2003).

Persidis A. Signal transduction as a drug-discovery platform, *Nature Biotechnology* 16:1082–1083 (1998).

Pharmocodynamics, "Drug–receptor interactions," in *The Merck Manual of Diagnosis and Therapy*, Section 22, Clinical Pharmacology. http://www.mercksource.com/pp/us/cns/cns_merckmanual_frameset.jsp [accessed September 29, 2007].

Schacter BZ. *The New Medicines: How Drugs Are Created, Approved, Marketed and Sold*, Praeger, Westport, CT, 2006.

Scott JD, Pawson T. Cell communication: the inside story, *Scientific American* June:72–79 (2000).

Smith CG, O'Donnell JT, eds. *The Process of New Drug Discovery and Development*, 2nd ed., Informa Healthcare, New York, 2006.

Wise A, Gearing K, Rees S. Target validation of G-protein coupled receptors, *Drug Discovery Today* 7:235–246 (2002).

Yildirim MA, et al. Drug–target network, *Nature Biotechnology* 25:1119–1126 (2007).

CHAPTER 3

DRUG DISCOVERY: SMALL MOLECULE DRUGS

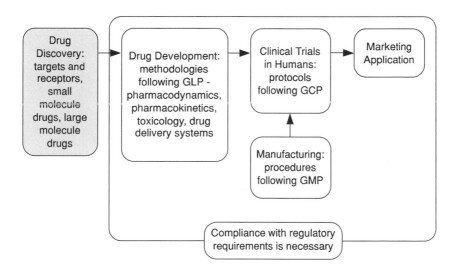

3.1   Introduction   54
3.2   Irrational Approach   55
3.3   Rational Approach   60
3.4   Antisense Approach   79
3.5   RNA Interference Approach   81

Drugs: From Discovery to Approval, Second Edition, By Rick Ng
Copyright © 2009 John Wiley & Sons, Inc.

53

3.6   Chiral Drugs   83
3.7   Closing Remarks   84
3.8   Case Study #3   84
3.9   Summary of Important Points   88
3.10 Review Questions   89
3.11 Brief Answers and Explanations   90
3.12 Further Reading   91

3.1   INTRODUCTION

The World Health Organization (WHO) in its May 2002 report estimated that currently up to 80% of the African people and a significant percentage of the worldwide population still practice some form of traditional medical treatment. Typically, these treatments are in the forms of decoctions, tinctures, syrups, or ointments with plant or animal products (see Exhibit 3.1).

However, for most readers of this book, the drugs that we are familiar with, such as analgesics (paracetamol), antibiotics (penicillin), hormones (insulin), and vaccines (influenza) are not part of the traditional medical armory. These drugs are either chemically synthesized (small molecule drugs with typical molecular weights of <500 Da) or produced using rDNA technology (protein-based large molecule drugs with molecular weights in excess of thousands of daltons). We will describe the discovery methodologies for the small molecule drugs in this chapter and that for the protein-based large molecule drugs in Chapter 4.

There are two main approaches to discovering small molecule drugs: the irrational approach or the more recent structured rational approach. Antisense, RNA interference, and chiral drugs are other drug discovery methodologies.

Exhibit 3.1   Forms of Traditional Medicine

Decoctions: These are liquid extracts of active components and volatile oils from natural products.

Tinctures: These are made by steeping fresh or dried herbs in alcohol or vinegar.

Syrups: These are made by combining tinctures or medicinal liquors with honey or glycerin.

Ointments: These are typically prepared by mixing floral or plant ingredients with essential oil and wax, such as beeswax.

3.2   IRRATIONAL APPROACH

The basic steps of the irrational or random scanning approach are shown in Fig. 3.1.

3.2.1   Natural Product Collection

This approach to drug discovery commences with the collection of natural materials from their habitats. Such collections typically gather 1–5 kg of materials, which may consist of leaves, shoots, bark, and roots of plants. Marine life forms are collected as well. The locations are recorded, to facilitate further collection of materials should that be required.

An important aspect of natural product collection is to provide biodiversity, that is, products with different and diverse chemical compositions so that many potential variations of chemical compositions and structures can be extracted for testing.

Collection of natural products, or bioprospecting, used to be relatively straightforward, with little formality or encumbrances. Now, however, there are new rules (Exhibit 3.2) to protect the natural habitats.

3.2.2   Extraction of Potential Drug Compounds

Compounds are extracted from natural materials using organic solvents such as alcohol. Tannins and chlorophylls from plant materials are normally removed using chromatographic columns, as they can interfere with the screening. In some cases, animal products, such as venoms from snakes, are gathered for screening. In other cases, collections of microorganisms are examined. Two examples are provided below:

- One of the first angiotensin converting enzyme (ACE) inhibitors was teprotide. It is an antihypertensive drug for use after heart attacks. The active ingredient was isolated from the venom of a South American viper snake. Other well-known ACE inhibitors such as captopril and analopril were developed based on modifications to the venom chemical structures.

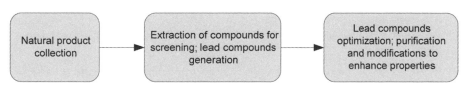

Figure 3.1   The basic steps of the irrational drug discovery process.

Exhibit 3.2   Regulations for Biodiversity Prospecting

Until recently, obtaining samples of plants, microorganisms, animals, and marine life forms was straightforward. Normally, a researcher would arrive at the collection site with permission from the local authority and collect samples without much restriction.

There are now new rules for biodiversity prospecting regarding the collection of natural products. The 1993 Convention on Biological Diversity (CBD) established sovereign national rights over biological resources and committed member countries to conserve them, develop them sustainably, and share the benefits resulting from their use.

The CBD requires that permission must be obtained before biological samples can be taken. To comply with the CBD, an Access and Benefit-Sharing Agreement (ABA) has to be agreed between the researcher and the source country providing the natural products. The ABA sets out the clauses with respect to the observation, development, and benefit sharing accruing from the use of natural products for medical applications.

With the ABA, the source country must know in advance how the natural products are to be exploited and the benefits that can be shared. If the CBD is not observed, the natural products can be treated as being poached and the patent based on these products may be invalidated.

Source: Gollin MA. New rules for natural products research, *Nature Biotechnology* 17:921–922 (1999).

- Tetracyclines are a group of antibiotics derived from bacteria. Chlortetracycline was isolated from *Streptomyces aureofaciens* and oxytetracycline from *Streptomyces rimosus*. Tetracyclines act by binding to receptors on the bacterial ribosome and inhibit bacterial protein synthesis.

Exhibit 3.3 shows the chemical formulas of some of the well-known drugs.

3.2.3   Screening Compounds to Find "Hits": Lead Compound Generation

The next step is the screening of thousands of these compounds to find lead compounds or potential drug molecules that bind with receptors and modulate disease pathways. When an interaction happens, it is referred to as a "hit."

In some cases, compounds are purchased from laboratories or other suppliers. The collections of various compounds are called "libraries," and libraries of large pharmaceutical organizations have from hundreds of thousands to millions of compounds. For example, the Developmental Therapeutics Program

Exhibit 3.3   Chemical Formulas of Selected Drugs

Lyrica	Prozac

Tamiflu	Advair

Lipitor

of the United States National Cancer Institute has a collection of more than 600,000 synthetic and natural compounds.

Lead compounds are those that have shown some desired biological activities when tested against the assays. However, these activities are not optimized. Modifications to the lead compounds are necessary to improve the physicochemical, pharmacokinetic, and toxicological properties for clinical applications.

3.2.4 Purification and Modifications to Optimize Lead Compounds

Following "hits," the lead compounds are purified using chromatographic techniques and their chemical compositions are identified via spectroscopic and chemical means. Structures may be elucidated using X-ray or nuclear magnetic resonance (NMR) methods.

Further tests are carried out to evaluate the potency and specificity of the isolated lead compounds. This is usually followed by modifications of the compounds to improve properties through synthesis of variations to the compounds via chemical processes in the laboratory and frequently with modifications to the functional groups. The optimized lead compounds go through many iterative processes to keep improving and optimizing the drug interaction properties to achieve improved potency and efficacy.

Only after all these exhaustive tests are a few candidates selected for preclinical *in vivo* studies using animal disease models. The current approach is to perform as many tests as possible based on tissue cultures or cell-based assays, as they are less costly and provide results more readily. At the end of this long process is the availability of selected drug candidates with sufficient efficacy and safety required for human clinical trials.

Drug discovery and development is a tortuous path—factors that constrain its development may sometimes arise from unexpected quarters, such as environmental groups (Exhibit 3.4).

Exhibit 3.4   Paclitaxel (Taxol)

Paclitaxel (Taxol, Bristol-Myers Squibb) is a chemotherapy drug for ovarian cancer, breast cancer, and certain lung cancers. It was discovered by the US National Cancer Institute in the 1960s. Originally, it was extracted from the bark of the North American yew tree (*Taxus brevifolia*). Clinical tests had necessitated the harvesting of the bark, and this method damaged the trees irreversibly.

Environmental groups objected to this practice, and many demonstrations were staged. A solution was eventually found when the needles of the European yew tree (*Taxus baccata*) provided a source for the paclitaxel precursor without destroying the bark.

Paclitaxel was introduced by Bristol-Myers Squibb in 1993. Today, paclitaxel or a precursor can be obtained from cell cultures of *Taxus* media formed by hybridizing *Taxus baccata* and *Taxus cuspidate*.

Source: Taxolog Inc. *The Taxol Story*, 2000–2001. http://www.taxolog.com/taxol.html [accessed August 15, 2002].

3.2.5   High Throughput Screening

As we can imagine, the screening of thousands of natural products using a wet laboratory chemistry process is extremely time consuming. The latest technology in screening is based on laboratory automation and robotics systems. This is termed high throughput screening (HTS) or ultra-HTS (UHTS). These two systems can screen thousands and hundreds of thousands of samples per day, respectively.

The heart of the HTS system is a plate, or tray, which consists of tiny wells where assay reagents and samples are deposited, and their reactions monitored. The configuration of the plate has changed from 96 wells (in a matrix of 8 rows by 12 columns) to 384, and now to a high-density 1536-well format, which enables large-scale screening to be undertaken. Assay reagents may be coated onto the plates or deposited in liquid form together with test samples into the wells. Both samples and assay reagents may be incubated, and those that interact show signals, which can be detected. A variety of detector systems are used, ranging from radioactive readouts to fluorescence and luminescence. These signals indicate "hits," and the strength of a signal shows the quality of the hit. Lead compounds with good quality hits warrant further evaluation as potential candidates for optimization or modification to become drug candidates. Exhibit 3.5 shows a schematic representation of HTS.

The aim of HTS and UHTS is cost effectiveness and speed of compound scanning. Hence, not only does the robotics system have to deliver fast and

Exhibit 3.5   High Throughput Screening

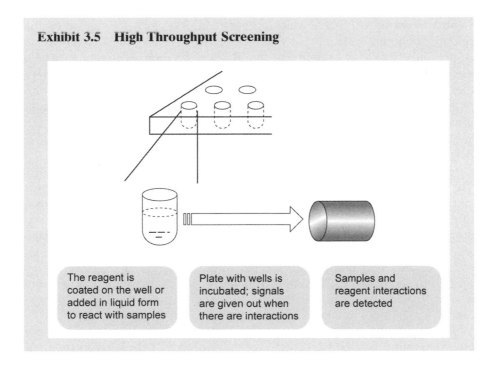

| The reagent is coated on the well or added in liquid form to react with samples | Plate with wells is incubated; signals are given out when there are interactions | Samples and reagent interactions are detected |

accurate liquid samples into the wells, but it has to be miniaturized to conserve the required volumes of the valuable samples and expensive assay reagents. For 1536 wells, the liquids being dispensed are in the nanoliter (10^{-9} L) to picoliter (10^{-12} L) range. Ink jet technology provided the technological basis for liquid dispensation in HTS.

The design of assay systems is another particularly important factor for testing the sample compounds. Assays have to be specific and sensitive. The assays used for HTS come in many forms. There are binding assays and enzyme-based or cell-based assays. Cell-based assays have become an important test compared with other *in vitro* assays, as they can provide information about bioavailability, cytotoxicity, and effects on biochemical pathway. Invariably, the enzyme-based and cell-based assay systems consist of receptors or mimetics of receptors (components that mimic active parts of receptors). Normally the assays are linked to an indicator that shows the ligand–receptor interaction as some form of signal. Radioligand binding assays were used previously. However, because of the lengthy processing and limited data provided, radioligand binding assays have been superseded by other assays. Scintillation proximity assays (SPAs) and reporter systems such as luciferase (an enzyme in firefly that gives off light) are common forms of signal generation for assay systems (see Section 2.8). The advantage of cell-based assays over biochemical assays is that cell-based assays enable the analysis of sample compound activity in an environment that is similar to the one in which a drug would act. It also provides a platform for toxicity studies.

The NIH has set up a consortium called the Molecular Libraries Screening Center Network (MLSCN), which performs HTS on assays provided by the research community. It currently has more than 100,000 chemically diverse compounds. This is an initiative of the Molecular Libraries Roadmap, which also has another two components: Cheminformatics and Technology Development. The aim is to generate a comprehensive database of chemical compounds and their bioactivities to enhance the capability for the development of new drug entities.

3.3   RATIONAL APPROACH

The premise for the rational approach is that drug discovery based on knowledge of the three-dimensional (3D) structure as well as the amino acid sequence of the chosen receptor molecule would reveal potential binding sites for drug molecules. Structural information and modeling simulations will help to design a drug that will fit precisely within the binding site, similar to the lock and key concepts discussed in Chapter 2. Such information about the receptor or protein structure will significantly improve the probability of obtaining a successful drug and eliminate unlikely drug candidates at a very early stage of drug discovery. The steps in rational drug discovery are summarized in Fig. 3.2.

The standard techniques used for 3D structural determinations are X-ray crystallography and NMR spectroscopy. Modeling of drug–receptor inter-

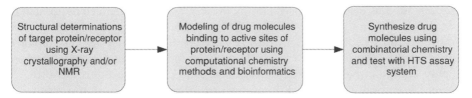

| Structural determinations of target protein/receptor using X-ray crystallography and/or NMR | Modeling of drug molecules binding to active sites of protein/receptor using computational chemistry methods and bioinformatics | Synthesize drug molecules using combinatorial chemistry and test with HTS assay system |

Figure 3.2 The basic steps of the rational drug discovery process.

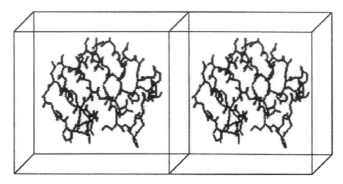

Figure 3.3 Crystalline molecules in three dimensions. Unit cells are imaginary blocks used to represent the regular arrangement of molecules in 3D space.

actions is studied using computational chemistry (*in silico*) methods and mining data using bioinformatics. The modeled drug is then synthesized using combinatorial chemistry and screened against assays in an HTS system. These enabling technologies are described next.

3.3.1 X-Ray Crystallography

To determine the structures of drug compounds or protein molecules using X-ray crystallography, it is necessary to have these compounds or molecules available in crystalline form. For example, when crystals of protein are formed, the protein molecules are arranged in orderly fashions like tiny imaginary "cubes" stacked on top of each other. Each of these building blocks contains a molecule of protein and is termed a unit cell (Fig. 3.3).

For examining atomic structures with bond lengths of 1–2 Å, the interrogating beams ideally should have wavelengths of the same dimensions, to resolve atomic details. X-rays fulfill this criterion because their wavelength, for example, CuK_α (X-ray using copper target) is 1.5418 Å (1.5418×10^{-10} m), which is similar to atomic dimensions.

When X-ray beams are focused on different orientations of these unit cells, the regular lattice arrangement scatters the X-ray and a 3D diffraction pattern consisting of thousands of diffraction spots is created. The intensity of each diffraction spot is the summation of constructive interference by the atoms (represented as electron density) in the molecule at a certain orientation, and the dimensions and pattern arrangement are due to the geometric positions of these atoms within the protein molecule. From the intensities and diffraction pattern, together with solving the phase angles of the diffracted beams, highly complex mathematical functions are used to determine the protein structure. Figure 3.4 shows a diagrammatic representation of the structure determination process. A more detailed description of X-ray crystallography is presented in Exhibit 3.6.

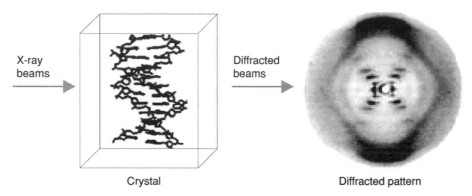

X-ray beams		Diffracted beams	
Crystal			Diffracted pattern

Figure 3.4 Steps in X-ray structure determination. X-ray scattering by the crystal gives rise to a diffraction pattern. From the diffraction pattern, the molecular structure can be determined using Fourier transformation mathematical calculations. (*Source*: For diffraction photograph, Nicholls H. Double helix photo not taken by Franklin, *BioMedNet News and Comments*, 2003. http://news.bmn.com/news/story?day=030425 &story=1#caption_name [accessed April 28, 2003].)

Exhibit 3.6   X-Ray Crystallography

Electrons are the components in atoms that scatter X-rays. The magnitude of scattering is measured in terms of the atomic scattering factor, f, which is proportional to the number of electrons in the atom. Hence, heavier atoms have higher atomic scattering factors. For a molecule composed of many atoms, the combined scattering of X-rays by a group of atoms is known as the structure factor, which is the summation of all the atomic scattering factors in space from the unit cell. This is given by the equation

$$\mathbf{F}(hkl) = \sum f_j \cos 2\pi (hx_j + ky_j + lz_j) + \sum f_j \sin 2\pi (hx_j + ky_j + lz_j)$$

where $\mathbf{F}$ is the structure factor, f is the atomic scattering factor, h, k, l are the indices for imaginary diffracting planes, and x, y, z gives the position of the scattering atom.

It should be noted that **F** is a vector quantity with magnitude and direction (phase angle).

The diffraction pattern provides us with intensity and geometric information. The equation for intensity of each diffracting plane, I_{hkl}, is given by

$$I(hkl) = |\mathbf{F}(hkl)|^2 \cdot LP \cdot A$$

where LP is a combined geometry and polarization factor and A is the absorption correction factor.

To determine the structure, we have to locate the atoms, which are given by the electron density equation:

$$\rho(xyz) = \frac{1}{V} \sum_h \sum_k \sum_l F(hkl) \exp[-2\pi i(hx + ky + lz)]$$

where V is the volume of the unit cell.

Note that we need **F**, a complex quantity, to find the electron density. However, from the intensity we can only derive the amplitude of **F**. The lack of phase information requires specific methods for solution, beyond the scope of this book.

From the intensity, the value for the observed **F** is obtained. This is substituted into the electron density equation for locating the atoms that determine the structure of the protein. Through an iterative process, the observed and calculated **F** values are compared to determine the "goodness of fit" and hence the quality of the structure.

The quality function is given by the crystallographic reliability factor, R_f:

$$R_f = \sum |F_o - F_c| / F_o$$

where F_o is the observed structure amplitude from the real molecule, and F_c is the structure factor calculated from the derived structure.

A perfect match has $R_f = 0$. Most protein structures have R_f values of 0.05–0.10, and a completely random structure gives an R_f of 0.59.

A major drawback with X-ray crystallography is the requirement to obtain crystals of proteins, which is a difficult process. However, techniques are being improved and many structures of proteins have been solved—more than 15,000 at the time of this writing. The very nature of crystallization also means that the protein molecules are "frozen" in space, rather than in the natural liquid state as found in the human body.

When a ligand is cocrystallized with protein, the active binding site is easily discernible from the determined structure. In situations where it is not possible to insert the ligand, the active site on the protein, which is normally in the form of a clef or pocket on the surface, can be inferred from comparison with other known structures. Once the structure of the active site is known, potential drug molecules can be designed using computational chemistry methods (Exhibit 3.7). Examples of some of the well-known drugs discovered using X-ray crystallography are the human immunodeficiency virus (HIV) drugs, such as amprenavir (Agenerase) and nelfinavir (Viracept). They were designed by studying the interactions of potential drug compounds using the crystal structure of HIV protease. The flu drugs zanamivir (Relenza) and oseltamivir (Tamiflu) were developed with extensive modeling of the crystal structure of neuraminidase, as described in Exhibit 3.7.

Exhibit 3.7   Development of Zanamivir (Relenza) and Oseltamivir (Tamiflu)

Flu virus has two types of spikes on the surface: neuraminidase and hemagglutinin. The neuraminidase and hemagglutinin undergo mutations, and these mutations account for the different types of flu viruses (Exhibit 4.2).

The virus uses its hemagglutinin to bind to a human cells by interacting with the sialic acid on the human cell surface. The cell then takes up the virus. The virus eventually enters the nucleus, where it replicates to produce many new virus genes. The virus genes combine to become multiple copies of viruses, which are released from the cell. When viruses are released, there is a coating of sialic acid on the hemagglutinin, rendering it unable to bind to a new cell surface. But the neuraminidase of the virus is able to cleave the sialic acid, thereby letting the hemagglutinin loose to attach and infect other cells.

There is a conserved part on neuraminidase, and this does not mutate or bind to sialic acid. X-ray crystallography revealed that this conserved part is a cleft with four parts. Drug molecules were designed to fit into this cleft and jam the neuraminidase, so that it is not available to cleave the sialic acid. When the sialic acid remains intact on the hemagglutinin, the virus is unable to attach to new cells and propagate the infection.

Two drugs were designed: zanamivir (Relenza) by Glaxo Wellcome and oseltamivir (Tamiflu) by Roche. Zanamivir is a powder that has to be inhaled, and oseltamivir is an oral drug.

Source: Laver WG, Bischofberger N, Webster RG. Disarming flu viruses, *Scientific American* January:78–87 (1999).

3.3.2 Nuclear Magnetic Resonance Spectroscopy

NMR is another powerful tool for determining the 3D structures of compounds. In contrast to X-ray crystallography, NMR requires that the compound be in solution, rather than crystalline. It provides information about the number and types of atoms in the molecule and the electronic environment around these atoms.

The principle behind NMR is that, when a molecule is placed in a strong external magnetic field, certain nuclei of atoms within the molecule, such as ^1H, ^{13}C, ^{15}N, ^{19}F, and ^{31}P, will resonate as they absorb energy at specific frequencies that are characteristics of their electronic environment. Because most drug and protein molecules are composed of hydrogen, carbon, nitrogen, fluorine, and phosphorus, NMR is ideally suited to unravel structural information of drugs and proteins.

An NMR spectrum shows the types of environment around the nuclei (atoms) and the ratios of these nuclei. Compared with X-ray crystallography, NMR has the advantage of being carried out in concentrated solutions rather than requiring crystal samples. The solution states are more representative of the native environment of receptor proteins. NMR can be used to study ligand–receptor interactions. A receptor protein is labeled with isotopes such as ^{13}C or ^{15}N, and changes in their spectra when bound with ligands can be monitored.

However, NMR is limited to molecules with molecular weights of less than 35 kDa. Both techniques, X-ray crystallography and NMR, when combined can provide invaluable information for drug design. With the precise binding site topologies derived from X-ray crystallography and dynamic properties obtained from NMR, tailor-made drug molecules can be designed to fit in the binding sites.

Exhibit 3.8 presents a more detailed description of NMR, and Exhibit 3.9 illustrates the use of NMR in drug discovery.

3.3.3 Bioinformatics

Bioinformatics is the use of information technology for the collation and analysis of biological data. This field of study started in the 1980s; it was initially set up by the US Department of Energy for the storage and retrieval of short sequences of DNA. The database is called GenBank. Now GenBank has been transferred to the National Institutes of Health's National Center for Biotechnology Information (NCBI). Many more databases, both public and private, have been set up to enable scientists to deposit, revise, retrieve, and analyze biological information. The Human Genome Project is a prime example of bioinformatics. Terabytes (2^{40} bytes) of capacity are used to store the sequence information of billions of DNA base pairs.

As bioinformatics evolves and matures, more and more information beyond sequences of DNAs and amino acids is added to the database. The amount

Exhibit 3.8   Principles of NMR

It is a fundamental property of atomic particles, such as electrons, protons, and neutrons, to have spins. Spins can be classified as $+\frac{1}{2}$ or $-\frac{1}{2}$ spin. For example, a deuterium atom, ^{2}H, has one unpaired electron, one unpaired proton, and one unpaired neutron. The total nuclear spin = $\frac{1}{2}$ (from the proton) + $\frac{1}{2}$ (from the neutron) = 1. Hence, the nuclear spins are paired and result in no net spin for the nucleus. For atoms such as ^{1}H, ^{13}C, ^{15}N, ^{19}F, and ^{31}P, the nuclei consist of protons with unpaired spins.

In the presence of an external magnetic field, the spin of the nucleus can align in two energy states: with or against the field. When energy is applied at the right frequency, resonance occurs and the spin flips from one energy state to another according to the formula

$$v = \gamma B$$

where v is the frequency, γ is the gyromagnetic ratio, and B is the external field strength.

Nuclei are affected by the microenvironment around them. Electrons around the nuclei shield the magnetic field experienced by the nuclei. If the electrons are withdrawn, the nuclei will experience a stronger magnetic field and require more energy (higher resonance frequency) to flip the spins, and vice versa. For ^{1}H NMR, the hydrogen nuclei of a compound can resonate downfield (higher frequency) or upfield (lower frequency) relative to a standard called tetramethylsilane (TMS).

The NMR spectrum also provides information about the number of nuclei within each distinct environment. This is given by the area under each resonant peak representing the relative number of nuclei of each type. Furthermore, the surrounding nuclei also cause a splitting pattern. For example, one H surrounded by n other H neighbors will have its resonance peak split into $n + 1$ peaks.

Source: Hornak JP. *The basics of NMR.* http://www.cis.rit.edu/htbooks/nmr/ [accessed October 31, 2002].

Exhibit 3.9   NMR in Drug Discovery

In addition to being used for structural determinations of protein targets, NMR is increasingly being used to examine the dynamic interactions of ligand–receptor binding. Two NMR properties are particularly important: chemical shift and nuclear spin relaxation.

When a ligand binds to a protein receptor, it perturbs the microchemical environment of the protein nuclei through bond formation, hydrogen bonding, and/or van der Waals forces. The shifting of the resonance frequency reflects the strength of the interaction.

The nucleus absorbs the magnetic energy and flips to another spin state. After a finite time, the spin state reverts to the original state through an equilibrium process. Generally, small molecules with fast rotational motions have slow relaxation rates. When a ligand binds to a target, the interaction slows the rotational motions. The relaxation time thus changes. This is observed via the nuclear Overhauser effects (NOEs), which are a measure of Brownian motions (rotational motions of molecules). A negative NOE for a small molecule is indicative of binding to the protein target.

An example of the use of NMR to design inhibitors of the protein kinase p38 is shown below. The first NMR spectrum shows the resonance peaks of nicotinic acid (a) and 2-phenoxy benzoic acid (b) in the absence of a target enzyme. When a target enzyme is added, in this case the p38 MAP kinase, binding of the ligand and the enzyme causes line broadening and attenuation of the resonance peaks. This is shown by the second NMR spectrum, in which the affected peaks are those of the 2-phenoxy benzoic acid (from 7.2 ppm to 6.6 ppm), indicating the interactions between p38 MAP kinase and 2-phenoxy benzoic acid.

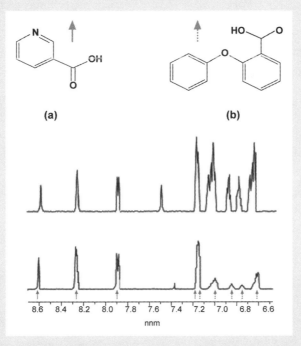

Example of the use of NMR to design inhibitors. (*Source*: Pellecchia M, Sem DS, Wuthrich K. NMR in drug discovery, *Nature Reviews Drug Discovery* 1:211–219 (2002). Used with permission.)

of data that can be generated is phenomenal. It is reported that the growth in bioinformatics data exceeded even the well-known Moore's law for electronics, which states that the number of transistors on a chip doubles every 18 months. The Internet has played a central role in the growth of bioinformatics. It provides a comprehensive and easily accessible means for information storage, retrieval, and analysis.

A process called data mining is used to extract the ever-expanding valuable information from the databases. Data mining consists of complex computer algorithms and mathematical functions with testable hypotheses for a range of analyses. The analyses include homology comparison, gene identification, RNA transcription, and protein translation. Some of these are discussed in Chapter 2 under microarrays and expressed sequence tags for target identification. Newly sequenced DNA can be compared with previously sequenced DNA segments of model organisms. Sequences that match or closely resemble model systems enable scientists to predict the likely proteins being produced. Bioinformatics also assists scientists in assessing the probable 3D structures of proteins. Bioinformatics, in conjunction with structural data from X-ray crystallography and NMR, helps scientists focus on the likely target and the binding sites for drugs to be designed.

In addition, bioinformatics databases have been expanded to integrate data on absorption, distribution, metabolism, excretion, and toxicity of drugs. Through these comprehensive sets of data, scientists have at their disposal powerful means to relate disease targets and their cellular functions to physiological and pathological processes.

Another application of bioinformatics is the use of pharmacogenomics. There are some diseases, such as sickle cell anemia (Exhibit 2.3), in which the difference of one amino acid group can have drastic consequences. These differences in nucleotides are termed single nucleotide polymorphisms (SNPs). SNPs, whether due to genetic origins or environmental factors, translate to individual differences. By understanding these SNPs using bioinformatics, more individualized medicines with better efficacy and less adverse effects can be prescribed.

In essence, bioinformatics is applied as follows:

- Scan the DNA sequences to determine locations of genes.
- Analyze transcription and translations of genetic codes to proteins (see Appendix 2, Section A2.3).
- Determine possible functions and structures for proteins.
- Predict binding sites for drug interactions and modulations of causative effects of disease pathway.
- Provide information for drugs to be designed to fit the binding sites.
- Analyze SNPs for tailored prescriptions to individuals.

A simplified bioinformatics process for provision of information is shown in Fig. 3.5.

Figure 3.5 Bioinformatics flow of information.

3.3.4 Computational Chemistry

Computational chemistry is an *in silico* method (computational approach) that is used to determine the structure–activity relationships (SARs) of ligand–protein receptor binding. It encompasses a number of techniques, such as computer-assisted drug design, computer-aided molecular design, and computer-assisted molecular modeling. There are many software algorithms written for computational chemistry, with different emphasis on modeling and SAR functions.

When a ligand (drug molecule) interacts with a protein, the protein binds with the drug and undergoes varying degrees of conformational change to accommodate the drug. As a result, the biological activity regulated by the protein is modified, as shown by the equation below, relating SAR as a function of the interaction:

$$SAR = f \text{ (ligand–protein bonds and conformational change)}$$

The aim of computational chemistry is to perform virtual screening using computer-generated ligands via a *de novo* drug design method—the design of drug compounds by incremental construction of a ligand model within a model of the receptor or enzyme active site. Libraries of virtual ligands are generated by computer based on certain building blocks or frameworks (scaffolds) of chemical compounds. The method uses a genetic algorithm, which simulates the genetic evolutionary process to produce "generations" of virtual compounds. The new structures may have improved ability to bind to the receptor protein, similar to the concept of "survival of the fittest" in the biological process.

To maximize the chances of success, it is necessary to build in "drug-like" properties. One of the drug-like criteria adopted is the "Lipinski Rule of 5"— so named because of its emphasis on the number 5 and multiples of 5. The rule states that potential drug candidates are likely to have poor absorption and permeability if they have:

- More than 5 hydrogen bond donors (the sum of –OH and NH_2 groups)
- Molecular weight >500
- Log P (the octanol/water partition coefficient, which indicates lipophilicity) >5, or
- More than 10 hydrogen bond acceptors (the sum of nitrogen and oxygen atoms).

Using information about the 3D shape of a protein receptor active site, which is derived from X-ray crystallography or NMR, ligands from the virtual library can be selected and fitted into the site. This is a modeling process known as docking simulation (Fig. 3.6).

Ligands are selected based on their drug-like properties, shapes, and orientations and distributions of chemical functional groups complementary to those of the protein. For example, a hydrogen bond donor of the ligand matches with a hydrogen bond acceptor of the protein, where positive electrostatic charge aligns with negative electrostatic charge and so on. Constituent side chains or functional groups of the ligands are varied to provide many different configurations for the docking analysis, with a view to optimize the best ligands that can be used as potential drug candidates.

Docking simulation is distinct from wet laboratory chemistry, where chemical reactions are performed using real rather than virtual compounds. The docking approach is more cost effective and efficient than the conventional chemical synthesis route. It allows a large database of virtual compounds to be screened and matched up with the binding site of the targeted protein.

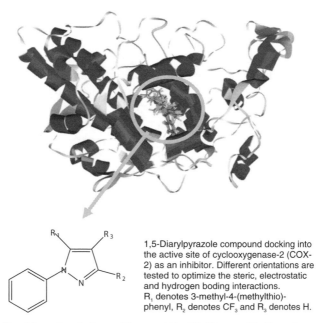

1,5-Diarylpyrazole compound docking into the active site of cyclooxygenase-2 (COX-2) as an inhibitor. Different orientations are tested to optimize the steric, electrostatic and hydrogen boding interactions. R_1 denotes 3-methyl-4-(methylthio)-phenyl, R_2 denotes CF_3 and R_3 denotes H.

Figure 3.6 Docking simulations. (*Source*: Liu H, Huang X, Shen J, et al. Inhibitory mode of 1,5-diarylpyrazole derivatives against cyclooxygenase-2 and cyclooxygenase-1: molecular docking and 3D QSAR analyses, *Journal of Medicinal Chemistry* 45:4816–4827 (2002). Used with permission.)

Scoring systems are set up to quantitatively calculate how well the ligand docks with the active site in terms of alignment, hydrogen bonding, van der Waals forces, and electrostatic and hydrophobic interactions. In addition, flexibilities of both the ligands and protein in the binding process as they accommodate each other have to be considered.

The affinity of interactions can be calculated in a number of ways. One example is the force field method to calculate the free energy of binding for the ligand–protein system before and after the docking, as given by the equation

$$\Delta G = T\Delta S_{rt} + n_r E_r + \sum n_x E_x$$

where ΔG is the free energy of binding, $T\Delta S_{rt}$ is the loss of overall rotational and translational entropy upon binding, n_r is the number of internal degrees of conformational freedom lost on binding, E_r is the energy equivalent of the entropy loss, n_x is the number of functional groups in the ligand, and E_x is the binding energy associated with each ligand functional group (Andrews, 2000).

The energy calculations include the rotational and translational changes and torsional angular effects of the ligands and protein, as well as solvation and desolvation energies because ligands have to displace water molecules normally residing in the active site. An analogy is fitting a hand into a rubber glove. The fingers have to be extended and the glove stretched to accommodate the fit similar to the rotational, translational, and torsional changes required for a good fit. Entrapped air inside the glove has to be expelled, much like the ligand replacing the water molecules at the active site. A schematic view is shown in Fig. 3.7.

There are other scoring functions for rating the docking of ligands to the protein binding site. These functions include *ab initio* (from first principle) quantum mechanical calculations, which take into account the electronic populations of the entire ligand–protein system and the bonding scheme, and molecular Monte Carlo iterative processes, which consider thermodynamic properties, minimum energy structures, and kinetic coefficients.

The result of computational chemistry is some potential drug candidates. These can be synthesized using combinatorial or wet laboratory techniques, and then tested with assays. Screening an array of ligands virtually is cost effective and compresses the discovery timeline. Exhibit 3.10 shows a typical workflow process for virtual screening.

3.3.5 Combinatorial Chemistry

Combinatorial chemistry is a laboratory chemistry technique to synthesize a diverse range of compounds through methodical combinations of building block components. These building blocks (reagents) are added to reaction vessels, and the reactions proceed simultaneously to generate an almost infi-

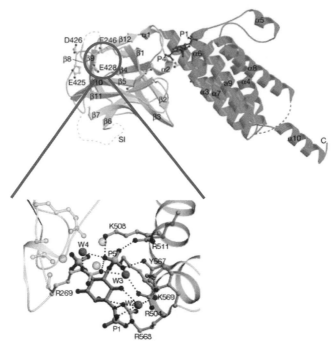

Figure 3.7 A ligand fitting into a binding site: binding of inositol 1,4,5-trisphosphate (InsP$_3$) with its receptor. The InsP$_3$ receptor plays a key role in cellular and physiological processes. (*Source*: Bosanac I, Alattia JR, Mal TK, et al. Structure of the inositol 1,4,5-trisphosphate receptor binding core in complex with its ligand, *Nature* 420:696–700 (2002). Used with permission.)

Exhibit 3.10   Virtual Screening Process

Presented below is a pictorial description of the workflow of a virtual screening run against a specific target. The typical workflow consists of a preparation of the virtual library database and the target. Docking simulations are next taken, and various scoring functions are used to rate the "goodness of fit" for the potential candidate to the target.

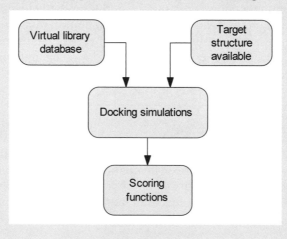

nite array of compounds, limited only by the imaginations of the scientists. This technique is in contrast to traditional methods, in which compounds are synthesized sequentially by mixing one reagent with another and with further reagents to build up the compound. By using combinatorial chemistry techniques, large libraries of many thousands of compounds can be prepared very quickly, unlike the laborious task needed to collect natural compounds from the field. However, there is a debate as to which method, combinatorial chemistry or natural collection, would provide more diverse range of compounds and biodiversity to be tested.

The selection of building blocks is based on information derived from, for example, computational chemistry, where potential virtual ligand molecules are modeled to fit the receptor–protein binding site. Combinatorial chemistry commences with a scaffold or framework to which additional groups are added to improve the binding affinity. Compounds are prepared and later screened using HTS. In this way, many compounds are tested within a short time frame to speed up drug discovery.

There are two basic combinatorial chemistry techniques: (1) parallel synthesis and (2) split and mix methods. They are illustrated next.

Parallel Synthesis: We start the reaction by using two sets of building blocks, amines (A) and carboxylic acids (B). The amines are first attached to solid supports, normally polystyrene beads coated with linking groups, in separate reaction vessels for each amine. After the amines have been attached, excess unreacted amines are washed off. Next, the carboxylic acids are added to the amines to form the desired amides. We illustrate these steps in Fig. 3.8. Assuming there are 8 amines to react with 12 carboxylic acids in a 96-well plate with 8 rows and 12 columns of tiny wells, the amines, A1 to A8, are added across the rows to each well containing the polystyrene beads. Different types of carboxylic acids, B1 to B12, are added to the wells in each column.

A1 B1	A1 B2	A1 B3	A1 B4	A1 B5	A1 B6	A1 B7	A1 B8	A1 B9	A1 B10	A1 B11	A1 B12
A2 B1	A2 B2	A2 B3	A2 B4	A2 B5	A2 B6	A2 B7	A2 B8	A2 B9	A2 B10	A2 B11	A2 B12
A3 B1	A3 B2	A3 B3	A3 B4	A3 B5	A3 B6	A3 B7	A3 B8	A3 B9	A3 B10	A3 B11	A3 B12
A4 B1	A4 B2	A4 B3	A4 B4	A4 B5	A4 B6	A4 B7	A4 B8	A4 B9	A4 B10	A4 B11	A4 B12
A5 B1	A5 B2	A5 B3	A5 B4	A5 B5	A5 B6	A5 B7	A5 B8	A5 B9	A5 B10	A5 B11	A5 B12
A6 B1	A6 B2	A6 B3	A6 B4	A6 B5	A6 B6	A6 B7	A6 B8	A6 B9	A6 B10	A6 B11	A6 B12
A7 B1	A7 B2	A7 B3	A7 B4	A7 B5	A7 B6	A7 B7	A7 B8	A7 B9	A7 B10	A7 B11	A7 B12
A8 B1	A8 B2	A8 B3	A8 B4	A8 B5	A8 B6	A8 B7	A8 B8	A8 B9	A8 B10	A8 B11	A8 B12

Figure 3.8 Additions of amines (A) and carboxylic acids (B) in a 96-well plate.

After the reactions, the compounds are separated from the beads, for instance, by using UV light, which severs the linking groups. Purification steps are applied to separate the enantiomeric compounds (see Section 3.6). From a mere 20 reagents, 8 amines plus 12 carboxylic acids, we end up with 96 different compounds. By using different types of reagents, for example, X, Y, and Z, we generate $X \times Y \times Z$ compounds. Hence, very large libraries are obtained through such combinations.

Split and Mix: Here we use eight amines and eight carboxylic acids as our example. The amines are added to eight different reaction vessels and attached to polystyrene beads. Next, all the amines bound to polystyrene are taken and mixed in one reaction vessel. The mixed amines are then split into eight vessels of equal portions. Each of these vessels contains amines of A1 to A8 bound to the beads. Carboxylic acids are separately added to each vessel: B1 to vessel 1, B2 to vessel 2, and so on. The compounds prepared would be as follow:

Vessel 1: A1B1, A2B1, A3B1, ..., A8B1
Vessel 2: A1B2, A2B2, A3B2, ..., A8B2
$\vdots$
Vessel 8: A1B8, A2B8, A3B8, ..., A8B8

Compounds can be tagged via "coding" groups on the polystyrene beads. The coding can be performed for each reaction step. At the completion of the reactions, each compound can be uniquely identified through a decoding process. All the compounds are screened and tested against target assays, and the potent ones ("hits") are selected for analysis, which may include further synthesis to refine the hits and optimization to yield lead compounds.

Exhibit 3.11 gives a synopsis of the development of Gleevec using the rational approach.

3.3.6 Genomics and Proteomics

Genomics is the use of genetics and molecular biology to study an organism's entire genome. From the sequence of the derived genome, patterns of gene expression in cells under various conditions, healthy or diseased, can be discerned.

DNA sequencing includes the following steps:

Identify region of genome of interest → Generate clones of the region →
Purify DNA from clones → Sequence purified DNA

The Human Genome Project (see Appendix 2, Section A2.3.3) was completed in 2006. One method it used to improve the speed and quality of the sequencing was capillary array electrophoresis (Exhibit 3.12).

Exhibit 3.11   A Rational Approach to the Development of Imatinib Mesylate (Gleevec)

Imatinib mesylate (Gleevec, Novartis; Glivec in countries other than the United States) is a drug for the treatment of chronic myeloid leukemia (CML). CML is a result of a chromosomal problem and gives rise to high levels of white blood cells. An enzyme called *BCR-ABL* is involved. The *BCR-ABL* gene encodes a protein with elevated tyrosine kinase activity (see Exhibit 7.3).

The lead compound for Gleevec was identified in screening of a combinatorial library. This compound is a phenylaminopyrimidine derivative that inhibits protein kinase C (PKC). It is a signal transduction inhibitor. Using docking studies and X-ray crystallography, different groups were introduced into the basic phenylaminopyrimidine template. Stronger PKC inhibition was obtained with a 3′-pyridyl group, compound (a). An amide group provided an inhibitory effect on *BCR-ABL* tyrosine kinase, compound (b). Compound (c) lost PKC activity but improved tyrosine kinase inhibition. Solubility and bioavailability were studied, and finally a methylpiperazine compound (d), code name ST1571, was selected for clinical trial.

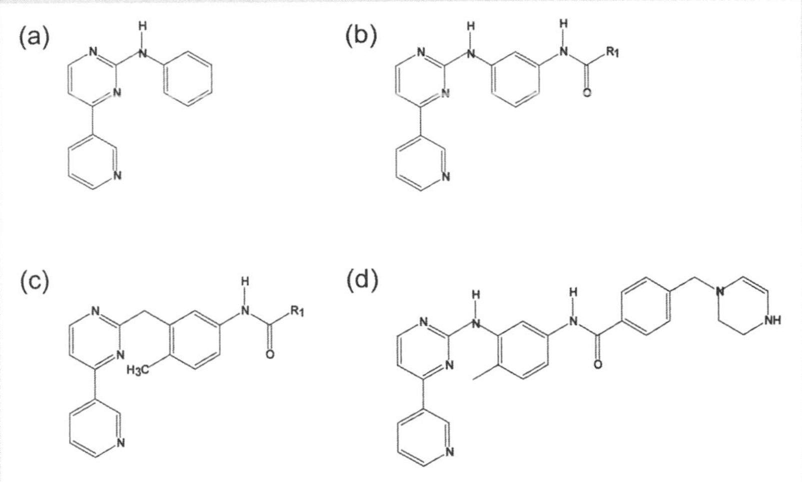

Development of Gleevec. (*Source*: Capdeville R, Buchdunger E, Zimmermann J, Matter A. Glivec (ST1571, Imatinib), a rationally developed, targeted anticancer drug, *Nature Reviews Drug Discovery* 1:493–502 (2002). Used with permission.)

Exhibit 3.12 Capillary Array Electrophoresis

DNA samples are introduced into the 96-capillary array. When the samples are separated through the capillaries, the fragments are irradiated with laser light. A charge coupled device measures the fluorescence and acts as a multichannel detector. The bases are identified in order in accordance to the time required for them to reach the laser-detector region.

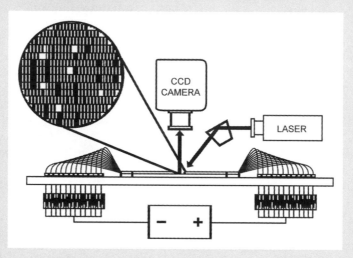

Source: Oak Ridge National Laboratory. *Facts About Genome Sequencing*. http://www.ornl. gov/sci/techresources/Human_Genome/faq/seqfacts.shtml [accessed August 12, 2007].

Proteomics, on the other hand, is the application of molecular biology, bio-chemistry, and genetics to study the structures and functions of proteins expressed by cells. Unlike the genome, which is reasonably static, the proteome changes constantly in response to intra- and extracellular signals. As proteins are vital to our cells and the biological pathways, an in-depth understanding of proteomics would help to elucidate the processes of disease and to devise means to counteract errant cells and processes.

It is estimated that there are more than two million different proteins in our body. Proteins are involved in a whole host of functions vital to our well-being:

- As enzymes responsible for catalytic reactions
- As messengers for signaling and transmission
- As defense systems against microorganisms
- As components for oxygen transport and blood clotting
- As controls to regulate growth
- As ingredients of tissues and muscles

When a disease occurs, it is often the defective proteins that are involved. Most drugs also target receptors and enzymes, which are themselves proteins. Through an understanding of the proteins and their functions, better and more specific drugs can be developed (refer to Appendix 2, Section A2.3 for more information about proteins and Exhibit 3.13 for protein extraction and studies).

3.3.7 Metabolomics

The NIH refers to the term metabolomics as a means to "identify, measure and interpret the complex, time-related concentration, activity and flux of endogenous metabolites in cells, tissues and other biosamples such as blood, urine and saliva."

Together with genomics and proteomics, metabolomics—by tracking the changes to the metabolites—helps to study the multivariate ways in the interactions between cells, tissues, and organs via many complex biochemical pathways. The quantitative studies of the substrates, intermediates, and products from biochemical reactions can yield useful data about healthy and diseased states and allow for the effects of potential drug candidates to be assessed.

3.3.8 Systems Biology

The advances of drug discovery technologies have been spectacular in the last two decades; yet they failed to improve the discovery of more disease-ameliorating molecules. The productivity of new drugs did not match the concomitant increased investment and technological efforts.

Exhibit 3.13   Protein Extraction and Studies

To study protein from a particular cell type, the cells are grown in nutrients. After a few days, millions of cells are collected and detergents are added to rupture the cell membranes, thus enabling proteins to be released into solution. The proteins are separated from the cell debris by centrifugation, where the proteins remain in solution and cell debris settles to the bottom.

Proteins can be separated using the 2D electrophoresis method. The first dimension is separation according to the pH of the proteins. The proteins are placed on a gel strip in a buffer solution. An electrical current is applied and the proteins separate and migrate to their isoelectric points (pI).

Next, the proteins are separated according to size. A detergent solution is added to the proteins gel strip to confer a negative charge to the proteins. Then the gel strip is placed on a precast gel where a voltage is applied and the proteins separate in accordance with their sizes, the larger ones moving through the gel at slower rates than the smaller ones.

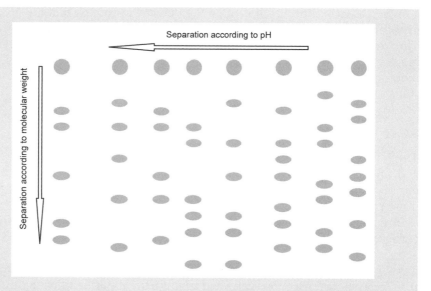

At this stage each spot on the gel may contain several proteins. The spot of interest is removed by cutting the gel and dissolving with an enzyme. Then the cleaved proteins are studied by using chromatography and mass spectroscopy techniques to determine the amino acids in the proteins and their sequences. The results are compared with database and the identities of proteins are revealed.

Critical to the success of drug discovery is the understanding of the complex biological and disease systems. Systems biology starts from the premise to relate the complex biological systems down to the level of organs, tissues, cells, and their molecular pathways and regulatory networks. It is perceived that biological systems are robust against various perturbations such as mutations, toxins, and environmental changes but are ill-equipped to deal with perturbations against which they are not optimized.

Systems biology uses computational analysis to integrate genomics, proteomics, and metabolomics data with disease physiology information. A knowledge of the type of genes, protein expression levels, and metabolite production can identify the specific molecular pathways switched on during certain disease states. This provides the *in silico* framework for constructing testable biological systems.

Cell, tissue, and organ-type models consisting of networks of signaling pathways, nodes, and regulatory points can be probed with putative drug molecules. Through understanding the interactions between robustness of biological systems, diseases, and drug effects, it is postulated that systems biology

Exhibit 3.14   Systems Biology

Several approaches are utilized to study systems biology. The bottom–up approach starts from the molecular level, the "omics," to identify and evaluate the genomic and proteomic basis of diseases. The top–down approach attempts to integrate human physiology and diseases to provide models to understand disease pathways at organ levels.

Another way is the intermediate method to bridge the two approaches above. This method determines biologically multiplexed activity profile data. It integrates biological complexity at multiple levels: pathways, signal transductions, and environmental factors.

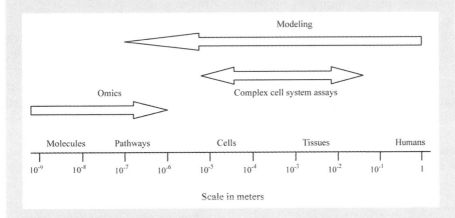

Source: Butcher EC, Berg EL, Kunkel EJ. Systems biology in drug discovery, *Nature Biotechnology* 22:1253–1259 (2004).

could provide a better discovery approach for drugs with the desired efficacy and less adverse effects.

Exhibit 3.14 provides a diagrammatic visualization of the approaches to tackle this new field of study.

3.4   ANTISENSE APPROACH

Genetic information is transcribed from the genes in the DNA to mRNA. The information is then translated from the mRNA to synthesize the protein (refer to Appendix 2). This process is depicted as follows:

DNA $\xrightarrow[\text{transcription}]{}$ mRNA $\xrightarrow[\text{translation}]{}$ Protein

The aim of antisense therapy is to identify the genes that are involved in disease pathogenesis. Short lengths of oligonucleotides of complementary sense (hence "antisense") are bound to DNA or mRNA (Fig. 3.9). These antisense drugs are therefore used to block expression activity of the gene. Information (the sense) from either the gene (DNA) or the mRNA is blocked from being processed (transcribed or translated), and the manufacture of protein is thus terminated. This technology differs from conventional drugs whereby the drugs interfere with the disease-causing protein, rather than stopping its production. Antisense drugs have high specificity since they can match their targets by countering their genetic codes.

A strategy for antisense therapy is based on the binding of oligodeoxyribonucleotides to the double helix DNA. This stops gene expression either by restricting the unwinding of the DNA or by preventing the binding of transcription factor complexes to the gene promoter. Another strategy centers on the mRNA. Oligoribonucleotides form a hybrid with the mRNA. Such a duplex formation ties up the mRNA, preventing the encoded translation message from being processed to form the protein.

Although these seem like elegant ways to stop the disease at the source, at the DNA or mRNA level, there are practical problems. First, the antisense drug has to be delivered to the cell interior, and the polar groups of oligonucleotides have problems crossing the cell membrane to enter the cytoplasm and nucleus; second, the oligonucleotides have to bind to the intended gene sequence through hydrogen bonding; and, third, the drug should not exert

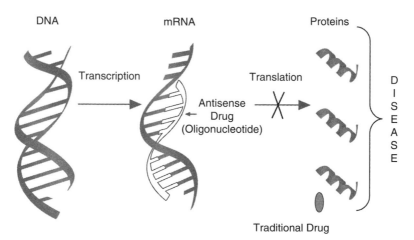

Figure 3.9   Mode of action for antisense drugs. An example is Fomivirsen (Vitravene, Isis Pharmaceuticals), which is a 21-nucleotide phosphorothioate that binds to the complementary mRNA of cytomegalovirus and blocks the translation process. Cytomegalovirus is a virus that belongs to the herpes group. (*Source*: Chang YT. *Keyword of the Post Genomic Era—Library*, New York University, New York, 2002. http://www.nyu.edu/classes/ytchang/book/e003.html [accessed September 10, 2002].)

Exhibit 3.15   Antisense Drugs

Bcl-2: B cell lymphoma protein 2 (Bcl-2) is a family of proteins that regulate apoptosis (programmed cell death). Apoptosis is a necessary process whereby aged or damaged cells are replaced by new cells. Dysfunction of the apoptosis process results in disease: inhibition of apoptosis results in cancer, autoimmune disorder, and viral infection, whereas increased apoptosis gives rise to neurodegenerative disorders, myelodysplastic syndromes, ischemic injury, and toxin-induced liver disease.

Some lymphomas, for example, are related to overexpression of Bcl-2. Antisense oligonucleotides are specially designed to target the overexpression of Bcl-2. Oblimersen (Genasense) is an antisense drug by Genta to block Bcl-2 production and enhance the efficacy of other standard chemotherapy drugs such as paclitaxel, fludarabine, irinotecan, and cyclophosphamide.

ICAM-1: Intracellular adhesion molecule 1 (ICAM-1), an immunoglobulin, plays an important role in the transport and activation of leukocytes. In Crohn's disease (inflammation of the alimentary tract), there is an overexpression of ICAM-1, which causes inflammation. Laboratory studies show that antisense oligonucleotides can reduce the expression of ICAM-1 and hence inflammation.

Source: Opalinska JB, Gewirtz AM. Nucleic-acid therapeutics: basic principles and recent applications. *Nature Reviews Drug Discovery* 1:503–514 (2002).

toxicities or side effects as a result of the interaction. For these reasons, there have been difficulties in bringing antisense drugs to market.

Currently, there is only one antisense drug on the market—Vitravene (active ingredient: fomivirsen) for the treatment of cytomegalovirus (CMV)-induced retinitis (inflammation of the retina) in AIDS patients. Fomivirsen has 21 nucleotides complementary to a CMV mRNA sequence, which is necessary for the production of infectious virus. Two examples of experimental antisense drugs are provided in Exhibit 3.15, while Table 3.1 lists other antisense drugs in clinical phase.

Another antisense drug, Genasense, was refused filing by the FDA for insufficient evidence to demonstrate its efficacy, although an appeal is ongoing from early 2007.

3.5   RNA INTERFERENCE APPROACH

RNA interference (RNAi) is a cellular defense mechanism through which double stranded RNAs (dsRNAs) are processed into short lengths of small interfering RNAs (siRNAs) of 20–25 nucleotides by an enzyme called Dicer (Fig. 3.10).

TABLE 3.1   Selected Antisense Therapies In Clinical Development

Company	Drug	Target	Disease or Indication	Clinical Status
Isis Pharmaceuticals	ISIS 301012	ApoB	High cholesterol	Phase II
	ISIS 113715	PTP-1B	Diabetes	Phase II
AVI BioPharma	Resten-NG	c-Myc	Re-stenosis	Phase II
	AVI-4065	Hepatitis C virus	Hepatitis C	Phase Ib
	AVI-4557	Cytochrome P-450	Drug metabolism	Phase I
	AVI-4020	West Nile virus	West Nile virus	Phase I
Genta	Genasense	BCl-2	Solid tumors, blood cancers	Phase III
	G4460	c-Myb		Phase I
Lorus Therapeutics	GTI-2040	R2 subunit of RNR	Renal cell carcinoma, acute myeloid leukemia	Phase II
	GTI-2501	R1 subunit of RNR	Prostate and kidney tumors	Phase I/II
Topigen	TPI-ASM8	Chemokine receptor-3/ IL-3,-5	Asthma	Phase II
	TPI-1100	Phosphodiesterases PDE4 and PDE7	COPD	Pre-IND
Oncogenex	OGX-011	Clusterin	Prostate and breast cancer, non-small-cell lung cancer	Phase II
VIRxSYS	VRX496	HIV	Chronic HIV infection	Phase I

Source: Potera C. Antisense—down but not out, *Nature Biotechnology* 25:497–499 (2007).

The siRNAs assemble into a complex called RNA-induced silencing complex (RISC) and unwind in the process. The single stranded siRNAs then attach to complementary RNA molecules, thus targeting these RNAs for destruction—a process that is called gene knockdown.

Using these principles, it is postulated that when genes causing disease pathways are identified, therapeutic siRNAs in the form of small drug molecules can be introduced into cells. Through these the siRNAs have high specificity and only target those errant genes and knockdown the disease pathways.

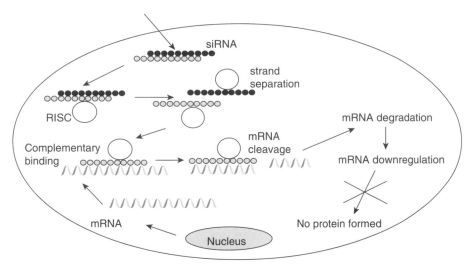

Figure 3.10   siRNA cellular mechanism.

While this technique has its appeal, there are several challenges. Delivery of the siRNA to the cells has to be devised. One method is the use of plasmid as a delivery vehicle. Another challenge is to overcome the possibility of invoking the innate immune response. To alleviate this problem, the use of microRNA rather than siRNA has been proposed.

Several clinical trials are in progress and it will be of interest to see how this new technology develops.

3.6   CHIRAL DRUGS

Most drugs and biological molecules are chiral. Chirality means "handedness," that is, left-hand and right-hand mirror images. This is because of the existence within the molecules of chiral centers. For example, a carbon atom attached to four different groups can be oriented in such a way that two different molecules that are mirror images are obtained (Fig. 3.11).

The two forms of mirror images are called enantiomers, or stereoisomers. All amino acids in proteins are "left-handed," and all sugars in DNA and RNA are "right-handed." Drug molecules with chiral centers when synthesized without special separation steps in the reaction process result in 50/50 mixtures of both the left- and right-handed forms. The mixture is often referred to as a racemic mixture.

As we can imagine, putting the right hand into a left-handed glove is not going to give a good fit. Similarly, the presentation of a racemic mixture of drug to a chiral binding site in a protein will not result in effective therapeutic treatment. One drug isomer is the actual effective component, while the other

Figure 3.11   Chiral molecules. Black triangular bond projects out of the page, gray triangular bond projects into the page.

isomer may have varying degree of activity. The other isomer may have little or no net effect, or it may nullify the activity of the active isomer. Worse still, it may cause an adverse reaction.

Before the 1980s, most drugs were manufactured in racemic mixtures. These drugs are being rediscovered so that only the active isomers are synthesized. The reason is that the active isomer is more effective in the absence of its mirror image, or it can be prescribed in higher dose without the adverse reaction due to the inactive isomer. Another reason is that pharmaceutical companies are rediscovering the active isomer to extend the life cycle of blockbuster drugs. This is illustrated in Exhibit 3.16.

3.7   CLOSING REMARKS

Drug discovery is extremely challenging and demanding. The attrition rate for failures is very high. Although the described approaches for drug discovery afford a higher probability of success, the astute observations and inventiveness of the scientists are critical ingredients for success. Exhibit 3.17 shows how careful observation by Fleming gave rise to one of the most effective drugs.

Very often, drugs are discovered through persistent work and continual optimization and many tests and trials. This is exemplified by the history behind the discovery of sildenafil (Viagra, Pfizer), a drug for the treatment of erectile dysfunction, and the AIDS drug zidovidine (Retrovir, GlaxoSmith-Kline) (Exhibits 3.18 and 3.19).

3.8   CASE STUDY #3

Lipitor

Lipitor (atorvastatin calcium) is a synthetic lipid-lowering drug. The chemical name for Lipitor is [R-(R*, R*)]-2-(4-fluorophenyl)-β, δ-dihydroxy-5-(1-methylethyl)-3-phenyl-4-[(phenylamino)carbonyl]-1H-pyrrole-1-heptanoic acid, calcium salt (2:1) trihydrate. The molecular weight is 1209.42. It is a white

Exhibit 3.16 Omeprazole and Esomeprazole

AstraZeneca launched omeprazole in 1988. It is a safe and effective drug for acid reflux, functioning as a proton pump inhibitor. However, the patent has expired and AstraZeneca has to compete against generics. The company developed the active isomer and called it esomeprazole. It was approved by the Mutual Recognition process in Europe in July 2000, and by the US Food and Drug Administration in February 2001. The chemical formulas for omeprazole and esomeprazole are shown below.

It was reported that healing of reflux esophagitis with 40 mg/day of esomeprazole is effective in 78% of patients after 4 weeks of treatment and in 93% of patients after 8 weeks, compared to 65% and 87% of patients treated with 20 mg/day of omeprazole.

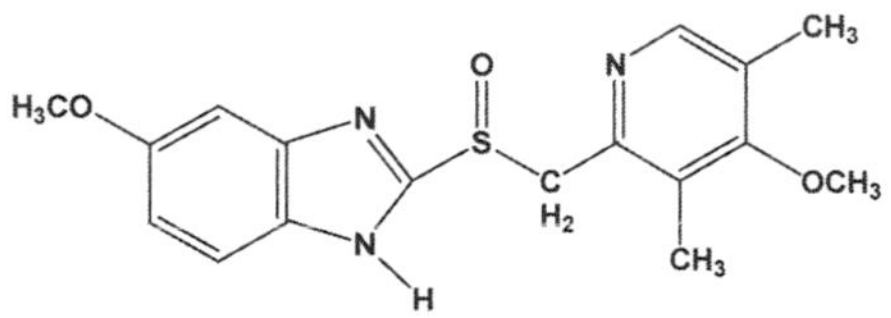

Omeprazole

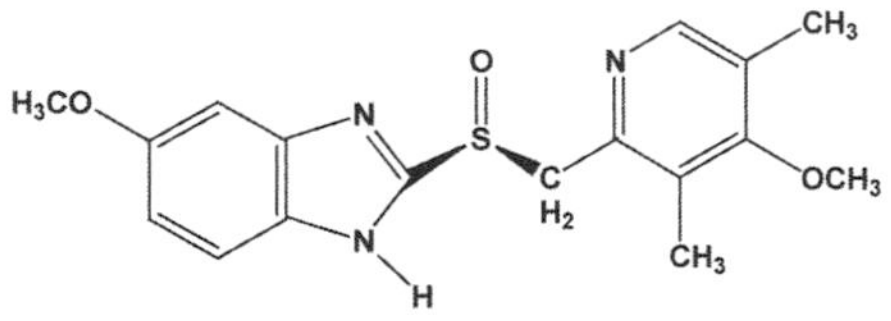

Esomeprazole

Omeprazole and esomeprazole. The two S atoms are chiral centers. (*Source*: Agranat I, Caner H, Caldwell J. Putting chirality to work: the strategy of chiral switches of drug molecules, *Nature Reviews Drug Discovery* 1:753–768 (2002). Used with permission.)

Exhibit 3.17 Importance of Observation in Drug Discovery

Alexander Fleming was studying bacteria. In 1928, he noticed that the bacterial cultures that he was growing were ruined when there was a mold present in the culture. The mold turned out to be *Penicillium*, which produces a substance called penicillin. This was found to be very effective in killing a variety of bacteria.

Exhibit 3.18 Discovery of Viagra

The long, tortuous path from drug discovery to commercialization is amply demonstrated by the sildenafil (Viagra) story.

Scientists at the Pfizer laboratory set out to discover an antihypertensive drug. The mechanism to lower blood pressure is the following:

- Atrial natriuretic peptide binds to the GPCR receptor.
- This binding activates the enzyme guanylate cyclase.
- Guanylate cyclase converts guanosine triphosphate (GTP) to cyclic guanosine monophosphate (cyclic GMP).
- Cyclic GMP lowers intracellular calcium, leading to (1) release of sodium in kidney cells or (2) relaxation of smooth muscle in blood vessels.

The enzyme phosphodiesterase (PDE) converts cyclic GMP to GMP. The Pfizer scientists wanted to develop a drug to inhibit PDE so that the level of cyclic GMP remains high, so that the last mechanism step can proceed.

Sildenafil was developed. However, there are different types of PDEs (nine are known today). As discussed previously, a potent drug has to be specific. Sildenafil inhibits PDE-5, which is absent in the kidney, although sildenafil's effect on smooth muscle relaxation was confirmed. The direction of the drug changed to treating angina instead, as sildenafil relaxes the vascular muscle of the heart.

At clinical trials, sildenafil did not work well as a treatment for angina. Instead, it was observed that it overcomes erectile dysfunction. Later, it was found that cyclic GMP also increases the level of nitric oxide, which is needed in penile erections.

Hence, we can see how the focus of treatment for sildenafil changed from antihypertensive to angina treatment to overcoming erectile dysfunction, giving rise to the drug called Viagra.

Exhibit 3.19   Retrovir, an AIDS Drug

Zidovudine (Retrovir, also known as AZT) was the first drug approved for the treatment of AIDS. The drug was first studied in 1964 as an anticancer drug, but it showed little promise. It was not until the 1980s, when desperate searches began for a way to treat victims of HIV, that scientists at Burroughs Wellcome Co. (Research Triangle Park, NC) took another look at zidovudine. After it showed very positive results in human testing, it was approved by the FDA in March 1987 for AIDS treatment.

to off-white powder. The tablet is formulated in 10, 20, 40, and 80 mg dosages. The excipients used are calcium carbonate, candelilla wax, croscarmellose sodium, hydroxypropyl cellulose, lactose monohydrate, magnesium stearate, microcrystalline cellulose, Opadry white YS-1-7040, polysorbate 80, and simethicone emulsion.

In our body cholesterol is manufactured from mevalonate **(1)**, which in turn is derived from (S)HMG-CoA **(2)**.

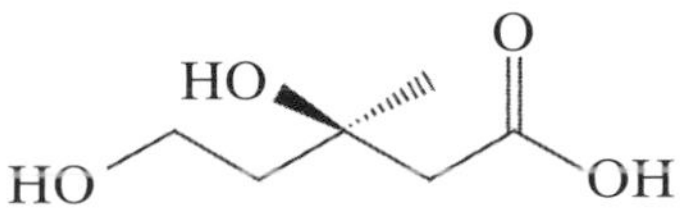

(1)

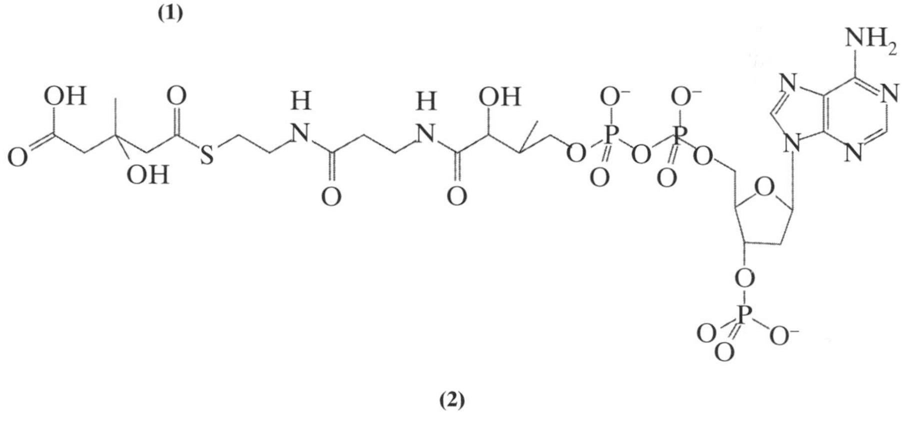

(2)

This cholesterol formation reaction is catalyzed by the enzyme HMG-CoA reductase. One means to stop or reduce the production of cholesterol is to interfere with the supply of mevalonate. This is the function of Lipitor, which acts as an inhibitor of HMG-CoA reductase.

Lipitor

Lipitor enters the active site of the enzyme, blocking the entry of HMG-CoA, and hence denies it from being reduced and converted to mevalonate. The diagram below depicts the Lipitor molecule at the active site.

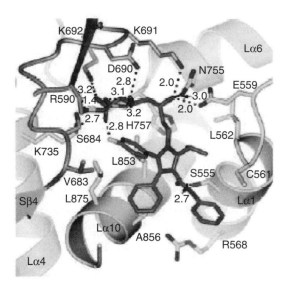

Lipitor molecule at active site. (*Source*: Istvan ES, Deisenhofer J. Structural mechanism for statin inhibition of HMG—CoA reductase, *Science* 292:1160 (2001). http://www.fda.gov/medwatch/SAFETY/2004/jul_PI/Lipitor_PI.pdf [accessed August 20, 2007]. Used with permission.)

3.9 SUMMARY OF IMPORTANT POINTS

1. The discovery of small molecule drugs can be separated into the irrational and rational approaches.

2. The irrational approach relies on the screening of many compounds in the hope of finding a "hit" with the disease target. Compounds screened include microorganisms, plants, and marine life forms. It is important to recognize conservation legislation and the sovereignty of the country of origin of these compounds. Extraction and purification are important steps to obtain the potential compound. High throughput screening is a necessary method to evaluate the potential use of the compounds in an efficient manner.

3. The rational approach commences with an understanding of the disease targets. The structures of the targets, including the active sites, are studied using X-ray crystallography and NMR.

4. Bioinformatics, genomics, and proteomics provide information about genes, proteins, and their functions on diseases. Combinatorial chemistry is used to generate different combinations of chemical compounds to test for their possible interactions with the putative disease targets. Structure–activity relationships of interactions are evaluated to find the potential drug candidates for further study. Metabolomics and systems biology are newer techniques in drug discovery.

5. Antisense and RNA interference techniques aim to utilize drug molecules to interfere with the transcription and translation process and stop diseases from progressing at the source.

6. Chiral drug development provides more effective drugs and extends the product life cycle of drugs for longer periods.

3.10 REVIEW QUESTIONS

1. Describe the irrational approach to drug discovery and provide examples of drugs discovered using this approach.

2. Explain the Access and Benefit-Sharing Agreement for biodiversity prospecting and discuss the pros and cons of this agreement.

3. Which are the techniques used for the rational approach to discover new drugs? Describe combinatorial chemistry and computational chemistry in drug discovery.

4. Give an example of a drug discovered under the rational approach and describe the process undertaken to optimize the drug effectiveness.

5. Discuss the pros and cons of X-ray crystallography and NMR for structural studies.

6. What are the reasons for the development of new methods, such as metabolomics and systems biology, to aid in the discovery of new drugs?

7. What are the mechanisms of action of antisense and RNA interference drugs in the treatment of diseases?

8. Describe chirality and why is the work on chiral drug important?
9. Describe the process for separating proteins in cells.

3.11   BRIEF ANSWERS AND EXPLANATIONS

1. Refer to Section 3.2 and associated exhibits.
2. Refer to Exhibit 3.2. The pros are that both parties—the prospectors and the source country—understand the obligations, whereas the cons are the negotiations may be protracted and delay prospecting activities that could yield beneficial compounds.
3. Refer to points 3 and 4 of Section 3.9. Computational chemistry and combinatorial chemistry are presented in Sections 3.3.4 and 3.3.5.
4. Exhibits 3.7 (Relenza and Tamiflu) and 3.11 (Gleevec) provide good examples of drugs discovered using the rational approach. The important criteria are (1) finding and validating the target, (2) determining the active site that can affect the disease pathway, and (3) designing drug candidates using computational chemistry. Once the drug candidates are designed, these compounds can be produced using combinatorial chemistry and tested with the high throughput system with tailored-made assays.
5. X-ray crystallography can provide very detailed information about the structure of target molecules but the technique requires good quality crystals to be grown and the structure determination process can be time consuming; the information provided is of a static nature. In contrast, NMR can provide dynamic information about the target interactions with drug candidates. However, the structural information is limited and the technique is not applicable for molecules >35 kDa. Hence, a combination of X-ray crystallography and NMR is needed to provide integrated information to enable more effective drug discovery.
6. Metabolomics and systems biology are new fields of study to better understand diseases and disease pathways. These new studies may help to discover and develop more effective drugs with fewer adverse reactions in shorter time spans.
7. Refer to Sections 3.4 and 3.5.
8. Refer to Section 3.6. Chiral drugs are more effective than racemic mixtures as they can better interact with active sites to alter disease progression. An important example is the case of omeprazole and esomeprazole (Exhibit 3.16). It is also strategically important for pharmaceutical companies to work on chiral drugs to further the product life cycle and compete with generics.
9. Refer to Exhibit 3.13.

3.12 FURTHER READING

Agranat I, Caner H, Caldwell J. Putting chirality to work: the strategy of chiral switches, *Nature Reviews Drug Discovery* 1:753–768 (2002).

Ambion. *siRNA Design Guidelines*. http://www.ambion.com/techlib/tb/tb_506.html [accessed April 17, 2007].

Andrews PR. *Drug–receptor interactions*, in *3D QSAR in Drug Design: Theory, Methods and Applications* (H. Kubinyi, ed.), Kluwer/Escom, The Netherlands, 2000.

Blundell TL, Jhoti H, Abell C. High-throughput crystallography for lead discovery in drug design, *Nature Reviews Drug Discovery* 1:45–54 (2002).

Bohm HJ, Schneider G, eds. *Virtual Screening for Bioactive Molecules*, Wiley-VCH, Weinheim, Germany, 2000.

Bolger R. High-throughput screening: new frontiers for the 21st century, *Drug Discovery Today* 4:251–253 (1999).

Brazil M. High throughput screening—molecular beacons for DNA binding, *Nature Reviews Drug Discovery* 1:98–99 (2002).

Bumcrot D, Manoharan M, Koteliansky V, Sah DWY. RNAi therapeutics: a potential new class of pharmaceutical drugs, *Nature Chemical Biology* 2:711–719 (2006).

Butcher EC, Berg EL, Kunkel EJ. Systems biology in drug discovery, *Nature Biotechnology* 22:1253–1259 (2004).

de Fougerolles A, Vornlocher HP, Maraganore J, Lieberman J. Interfering with disease: a progress report on siRNA-based therapeutics, *Nature Reviews Drug Discovery* 6:443–453 (2007).

Devlin JP, ed. *High Throughput Screening*, Marcel Dekker, New York, 1997.

Dove A. Antisense and sensibility, *Nature Biotechnology* 20:121–124 (2002).

Dunn D. The broader applications of uHTS, *Drug Discovery Today* 6:828 (2001).

Food and Drug Administration. *From Test Tube to Patient: New Drug Development in the United States*, 2nd ed., FDA, Rockville, MD, 1995. http://www.fda.gov/fdac/special/testtubetopatient/default.htm [accessed September 29, 2007].

Harvey AL, ed. *Advances in Drug Discovery Techniques*, Wiley, Hoboken, NJ, 1998.

Henry CM, Washington C. Systems biology, *CENEAR* 81:45–55 (2003).

Hood L, Perlmutter R. The impact of systems approaches on biological problems in drug discovery, *Nature Biotechnology* 22:1215–1217 (2004).

Howard K. The bioinformatics gold rush, *Scientific American* July:58–63 (2000).

Jhoti H, Leach A, eds. *Structure-Based Drug Discovery*, Springer, The Netherlands, 2007.

Kitchen DB, Decornez H, Furr JR, Bajorath J. Docking and scoring in virtual screening for drug discovery: methods and applications, *Nature Reviews Drug Discovery* 3:935–949 (2004).

Kitano H. A robustness-based approach to systems-oriented drug design, *Nature Reviews Drug Discovery* 6:202–210 (2007).

Larson RS, ed. *Bioinformatics and Drug Discovery*, Humana Press, Totowa, NJ, 2006.

Larvol BL, Wilkerson LJ. *In silico* drug discovery: tools for bridging the NCE gap, *Nature Biochemistry* 16(Suppl):33–34 (1998).

Liebman MN. Biomedical informatics: the future for drug development, *Drug Discovery Today* 7:s197–s203 (2002).

Loging W, Harland L, Williams-Jones B. High-throughput electronic biology: mining information for drug discovery, *Nature Reviews Drug Discovery* 6:220–230 (2007).

Lyne PD. Structure-based virtual screening: an overview, *Drug Discovery Today* 7:1047–1055 (2002).

NIH Guide. *Metabolomics Technology Development.htm, Request for Application-RM-04-002*. http://www.nih.gov [accessed August 3, 2007].

Opalinska JB, Gewirtz AM. Nucleic-acid therapeutics: basic principles and recent applications, *Nature Reviews Drug Discovery* 1:503–514 (2002).

Parrill AL, Reddy MR. *Rational Drug Design—Novel Methodology and Practical Applications*, American Chemical Society, Washington, DC, 1999.

Reid DG, ed. *Protein NMR Techniques*, Humana Press, Totowa, NJ, 1997.

Rupp B. *X-Ray 101—An Interactive Web Tutorial*. http://www-structure.llnl.gov/Xray/101index.html [accessed August 25, 2002].

Sittampalam SS, Kahl SD, Janzen WP. High-throughput screening: advances in assay technologies, *Current Opinion in Chemical Biology* 1:384–391 (1997).

Schneider G, Fechner U. Computer-based *de novo* design of drug-like molecules, *Nature Reviews Drug Discovery* 4:649–663 (2005).

Sneader W. *Drug Discovery: A History*, Wiley, Hoboken, NJ, 2005.

Terrett NK. *Combinatorial Chemistry*, Oxford University Press, Oxford, 1998.

Walters WP, Stahl MT, Murcko MA. Virtual screening—an overview, *Drug Discovery Today* 3:160–178 (1998).

Watt A, Morrison D. Strategic and technical challenges for drug discovery, *Drug Discovery Today* 6:290–292 (2001).

Wolke J, Ullmann D. Miniaturized HTS technologies—uHTS, *Drug Discovery Today* 6:637–646 (2001).

CHAPTER 4

DRUG DISCOVERY: LARGE MOLECULE DRUGS

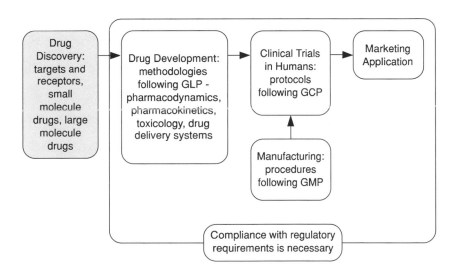

4.1	Introduction	94
4.2	Vaccines	95
4.3	Antibodies	106
4.4	Cytokines	113
4.5	Hormones	121
4.6	Gene Therapy	124
4.7	Stem Cells and Cell Therapy	126

4.8	Case Study #4	128
4.9	Summary of Important Points	131
4.10	Review Questions	132
4.11	Brief Answers and Explanations	133
4.12	Further Reading	134

4.1   INTRODUCTION

Unlike the small molecule drugs (pharmaceuticals) described in Chapter 3, large molecule drugs (biopharmaceuticals) are mainly protein based. Another distinction is that these protein-based drugs are, in the main, similar to natural biological compounds found in the human body or they are fragments that mimic the active part of natural compounds.

The discovery of pharmaceuticals commences with the scanning of hundreds of compounds, whether with actual materials (irrational approach) or virtual simulations (rational approach). To discover biopharmaceuticals, we have to examine the compounds within us, for example, hormones or other biological response modifiers, and determine how they affect the biological processes. In some cases, we study pathogens such as the influenza virus or bacteria to derive the vaccines. In other cases, we copy these biological response modifiers and use them as replacement therapy.

Pharmaceuticals are new chemical entities (NCEs) and they are produced (synthesized) in manufacturing plants using techniques based on chemical reactions of reactants. Biopharmaceuticals are made using totally different methods. These protein-based drugs are "manufactured" in biological systems such as living cells, producing the desired protein molecules in large reaction vessels as the living cells grow, or by extraction from animal serum.

Biopharmaceuticals are becoming increasingly important. The reason is that they are more potent and specific, as they are similar to the proteins within the body and hence are more effective in treating our diseases. There are three major areas in which biopharmaceuticals are used: as prophylactic (preventive, as in the case of vaccines), therapeutic (antibodies), and replacement (hormones, growth factors) therapy. Exhibit 4.1 presents selected statistics for biopharmaceuticals.

Another term that is used for protein-based drugs is biologics. The FDA gives the following definition for biologics:

> A biological product subject to licensure under the Public Health Service Act is any virus, therapeutic serum, toxin, antitoxin, vaccine, blood, blood component or derivative, allergenic product, or analogous product, applicable to the prevention, treatment or cure of diseases or injuries to humans. Biological products include, but are not limited to, bacterial and viral vaccines, human blood and plasma and their derivatives, and certain products produced by biotechnology, such as interferons and erythropoietins. Biologics encompass many different protein-based drugs, and include blood products such as clotting factors extracted from blood.

Exhibit 4.1   Biopharmaceuticals

There are at present about 165 biopharmaceutical drugs approved for marketing in the United States and the European Union. This market is worth more than US$50 billion, projected to reach US$70 billion by 2010. Approvals in the past 3 years are mainly for monoclonal antibodies, enzymes, and growth factors.

The annual R&D expenditure for biopharmaceuticals is around US$19–20 billion. There are estimated to be 2500 biopharmaceuticals in the discovery phase, 900 in preclinical trials, and 1600 in clinical trials. This represents 44% of all drugs in the development phase and 27% of all drugs in both preclinical and clinical trials. The most common target is cancer and monoclonal antibodies and vaccines have the largest amount of R&D activities.

Source: Walsh G. Biopharmaceutical benchmarks—2006, *Nature Biotechnology* 24:769–776 (2006).

Table 4.1 shows the major uses of therapeutic biologics, other than vaccines, in the treatment of patients suffering from various conditions.

In this chapter, we discuss the following topics, but exclude blood products:

- Vaccines
- Antibodies
- Cytokines
- Hormones
- Gene therapy
- Stem cells

We have included gene therapy and stem cells to present a more comprehensive perspective on medical treatments, although they are not drugs by our conventional definitions.

4.2   VACCINES

Most of us were vaccinated soon after we were born. As we grow up and go through different stages of life, we are further vaccinated against other diseases. The basis of vaccination is that administering a small quantity of a vaccine (antigen that has been treated) stimulates our immune system and causes antibodies to be secreted to react against the foreign antigen. Later in life, when we encounter the same antigen, our immune system will evoke a "memory" response and activate the defense mechanisms by generating antibodies to combat the invading antigen.

TABLE 4.1 Examples of Biologics and Their Uses

Type of Biologic	Treatment Use	Example
Cancer Antibodies	Metastatic cancer, lymphoma	Herceptin, Avastin, Rituxan, Erbitux, Procrit, Gleevec, Sutent
Erythropoietins	Anemia in kidney and cancer patients	Procrit, Eprex, Aranesp, Epogen, NeoRecormon
TNF blockers	Rheumatoid arthritis and psoriasis	Enbrel, Remicade, Humira, Rituxan
Insulin and insulin analogues	Diabetes	Novolin, Lantus, Lemevir, Novolog, Humalog
Interferon-α	Hepatitis B and C	Pegasys, Peg-Intron, Ribetron, Copegus
Interferon-β	Multiple sclerosis	Avonex, Rebif, Betaseron, Humira, Tysabri, Copaxone
G-CSF (granulocyte colony-stimulating factor)	Neutropenia (low level of neutrophils leading to susceptibility to infections)	Neulasta, Neupogen
Human growth hormone	Natural growth hormone deficiency	Genotropin, Norditropin, Humatrope, Nutropin, Saizen
Recombinant coagulation factors	Bleeding episodes or surgical bleeding in hemophiliacs	NovoSeven, Kogenate, Refacto, Benefix
Enzyme replacement	Gaucher's disease, Fabry disease, and mucopolysaccharidosis	Cerezyme, Fabrazyme, Aldurazyme
Antiviral antibodies	Preventing respiratory syncytial virus infections in premature infants	Synagis
Follicle stimulating hormones	Infertility	Gonal-f, Puregon
Others	Osteoporosis, asthma, cystic fibrosis, acute myocardial infarction, severe sepsis, psoriasis, non-Hodgkin's lymphoma	Forteo, Xolair, Pulmozyme, Activase/TNKase, Xigris, Raptiva, Zevalin

Source: Biologic Drug Report. http://www.biologicdrugreport.com/leading.htm [accessed May 25, 2007].

A vaccine formulation contains antigenic components that are obtained from or derived from the pathogen. These pathogens include mainly viruses, bacteria, parasites, and fungi. Research has shown that the part of the pathogen that causes disease, termed virulence, can be decoupled from the protective part, so-called immunity. Vaccine development focuses on means to reduce the virulence factor while retaining the immunity stimulation. Administration of vaccines may be oral or parenteral. After the initial vaccination, booster doses may be needed to maximize the immunological effects.

4.2.1   Traditional Vaccines

Traditionally, vaccines are prepared in a number of ways:

- Attenuated vaccines
- Killed or inactivated vaccines
- Toxoids

Attenuated Vaccines: The virulence of a pathogen can be reduced in a number of ways: by chemical treatment, by temperature adaptation, or by growing the pathogen in species other than the natural host (a process called "passaging").

The advantages of attenuated vaccines are (1) they have a low cost of preparation, (2) they elicit the desired immunological response, and (3) normally a single dose is sufficient. The disadvantages are (1) a potential to revert to virulence and (2) a limited shelf life.

Examples of attenuated vaccines are *Bacillus Calmette–Guerin* (BCG) for immunization against tuberculosis, Sabin vaccine for poliomyelitis, attenuated *Paramyxovirus parotitidus* against mumps, and attenuated measles virus against measles.

Killed or Inactivated Vaccines: Chemical and temperature treatment are normally used to kill or inactivate the pathogen. Formaldehyde treatment is one of the more common methods. Other chemicals used are phenol and acetone. Another method is to irradiate the pathogen to render it inactive.

The advantages are (1) nonreversal to virulence and (2) relatively stable shelf life. The disadvantages are (1) they have a higher cost of production, (2) more control is required for production to ensure reliable processes for complete inactivation, and (3) there is a possibility of reduced immunological response due to the treatment processes, so multiple booster vaccinations may be required.

Examples of killed or inactivated vaccines are cholera vaccine containing dead strains of *Vibrio cholerae*, hepatitis A vaccine with inactivated hepatitis A virus, pertussis vaccine with killed strains of *Bordetella pertussis,* typhoid vaccine with killed *Salmonella typhi*, and influenza vaccine with various strains of inactivated influenza viruses (see Exhibit 4.2 for a discussion of influenza viruses and vaccines and Exhibit 4.3 on avian influenza H5N1).

Toxoids: Toxoids are derived from the toxins secreted by a pathogen. The advantages and disadvantages are similar to those for killed or inactivated vaccines.

Exhibit 4.2 Influenza Viruses and Vaccines

Influenza is caused by the influenza viruses (orthomyxoviruses). There are three types of influenza viruses—A, B, and C (based on their protein matrix, influenza A and B have 8 RNA fragments, C has 7). Influenza A can infect humans and other animals, while influenza B and C mainly infect humans only. Unlike influenza A and B, influenza C viruses causes very mild illness and does not cause epidemics. Influenza A is categorized into subtypes based on its surface antigens: hemagglutinin and neuraminidase (see Exhibit 3.7). There are no subtype classifications for influenza B. Influenza viruses undergo frequent mutations as they replicate, with influenza A changing more rapidly than B, causing antigenic shifts in the hemagglutinin and neuraminidase.

Influenza virus. (*Source*: Cann AJ. Influenza virus haemagglutination. http://wwwmicro.msb.le.ac.uk/LabWork/haem/haem1.htm [accessed May 2, 2003]. Used with permission.)

The nomenclature for classifying influenza virus is as follows:

Type/Site Isolated/Isolate No./Year—for example, A/New Caledonia/20/99(H3N2), B/Hong Kong/330/2001

There have been three influenza pandemics in the 20th century: in 1918 (Spanish flu, H5N1), 1957 (Asian flu, H2N2), and 1968 (Hong Kong flu, H3N2). The pandemic in 1918 killed more than 20 million people worldwide, and the other two took the lives of 1.5 million people combined. In 1997, the first avian (chicken) flu was transmitted to humans in Hong Kong. It was caused by influenza A H5N1 (see Exhibit 4.3 for more information about H5N1).

Antibodies responsive to influenza antigens are specific to the subtype and strain. To have an effective influenza vaccine, it is a requirement that there is an accurate prediction of the subtypes and strains that are expected to circulate in the influenza season months before the season begins. When the antigenic match between vaccine and circulating viruses is close, influenza vaccine is 70–90% effective.

The World Health Organization (WHO) has a network of 110 centers worldwide that monitor influenza activity and ensure virus isolates and information are sent to the WHO for strain identification and action. Each

February (for the Northern Hemisphere winter) and September (for the Southern Hemisphere winter), the WHO provides advanced recommendations for the composition of the influenza vaccine to be manufactured. Similarly, the FDA CBER recommends trivalent influenza vaccine to be prepared for United States.

For the 2007–2008 winter season, the recommended trivalent vaccine by both the WHO and the FDA is:

A/Solomon Islands/3/2006 (H1N1)—like virus

A/Wisconsin/67/2005 (H3N2)—like virus

B/Malaysia/2506/2004—like virus

Exhibit 4.3   Avian Influenza H5N1

Avian influenza H5N1 is an infectious disease of birds. It can cause two distinct forms of disease: one is mild while the other is deadly. The virus is thought to be spread by migratory birds. Animals, especially farm poultry/animals, that lie under the migratory paths of the birds can become infected. To date, culling is the most effective means of controlling the spread of avian influenza in domestic poultry/animals.

There is concern that the virus can infect humans living in close proximity to the infected poultry/animals. As of May 31, 2007, there have been 309 cases of humans infected, with 187 fatalities. Most of the cases were in Indonesia, Vietnam, Egypt, Thailand, and China.

A further fear of this deadly infection is that it may cause a pandemic through two mechanisms: reassortment where the genetic material is exchanged between human and virus, or gradual adaptive mutation where the virus changes to a more potent form. The WHO and member countries through the IHR (see Case Study #7) are working together to control this infectious disease.

In April 2007, the FDA approved the H5N1 Influenza Virus Vaccine, which consists of the hemagglutinin (HA) of the virus strain A/Vietnam/1203/2004 (clade 1) in the presence of porcine gelatin (a stabilizer) and a mercury derivative, Thimerosal (a preservative).

The influenza vaccine induces the production of antibodies. These antibodies block the attachment of H5N1 virus to the human respiratory epithelial cells.

Source: World Health Organization. *Avian Influenza—Fact Sheet.* http://www.who.int/mediacentre/factsheets/avian_influenza/en/print.html [accessed June 5, 2007].

Examples are diphtheria and tetanus vaccines. Diphtheria vaccine is produced by formaldehyde treatment of the toxin secreted by *Corynebacterium diptheriae*. Similarly, tetanus vaccine is obtained from toxins of cultured *Clostridium tetani* that has been treated with formaldehyde.

4.2.2 New Vaccines

Advances in genomics, molecular biology, and recombinant technology have provided new directions for the discovery, development, and manufacture of vaccines. One of the current approaches is a minimalist strategy to decouple the virulence and immunity functions. The aim is to use only the immunity part to confer protection, so that the vaccine is safe to be administered. The approach can be divided into subunit, vector-based, DNA, and peptide vaccines.

Subunit Vaccines: Subunit vaccines use only a part of the bacteria or virus instead of the entire pathogen. Normally, the part is derived from the outside envelope protein of the pathogen. Discovery of the relevant envelope protein requires knowledge of the genome sequence of the pathogen by identifying open reading frames (ORFs, see Exhibit 4.4) that potentially encode novel antigenic surface proteins known as epitopes (Exhibit 4.4), which bind to antibodies. When identified, the ORFs are cloned to express protein epitopes using self-replicating plasmids (see Exhibit 10.10 and Appendix 4). The binding properties of the epitopes can be studied using enzyme-linked immunosorbent assay (ELISA, Exhibit 4.4) or a fluorescent activated cell sorter (FACS, Exhibit 4.4). After laboratory testing, the leading candidates of epitopes are injected into animals to determine whether they elicit any antibody response. Those that provoke a response are selected and optimized to become vaccine candidates with further tests before human clinical trials. Researchers are also working on multiple epitope subunit vaccines, which can provide different antigenic binding sites.

A new subunit recombinant vaccine is Gardasil; it is a tetravalent vaccine against human papillomavirus (HPV) implicated in cervical cancer. See Exhibit 4.5 for details.

Vector-Based Vaccines: Viruses and bacteria are detoxified and used as vehicles to carry vaccines. Subunit vaccines are delivered by carrier vehicles to elicit the immune response. An example is the use of canarypox (a virus that infects birds, but not humans) to carry envelope proteins for HIV treatment. Multiple types of envelope proteins can be delivered with this method. Clinical trials with this type of vector-based vaccines are being investigated.

DNA Vaccines: DNA vaccines are sometimes called nucleic vaccines or genetic immunization. The host (patient) is injected directly with selected viral

Exhibit 4.4 Important Concepts Related to Subunit Vaccines

An open reading frame (ORF) is a sequence of nucleotides in the RNA or DNA that has the potential to encode protein. The start triplet is ATG. It is followed by a string of triplets that code for amino acids. The stop triplet is TAA, TAG, or TGA (see Exhibit A2.2).

An epitope is an antigenic determinant of the pathogen. It consists of certain chemical groups that are antigenic, which means that it will elicit a specific immune response by binding to antibodies.

The enzyme linked immunosorbent assay (ELISA) is a method for determining the presence of antigen-specific antibodies. Antigens are first solubilized and coated onto solid support, such as 96-well plates. Test samples containing antibodies are added, and the antibodies bind to the antigens on the plate. Excess unbound sample is washed off, and the antigen–antibody complex is incubated with a second antibody linked to an enzyme (e.g., alkaline phosphatase, horseradish peroxidase). The labeling with the enzyme catalyzes certain biochemical reactions and provides a readout (color) to show the presence or absence of the specific antibodies. The process can also be used for detecting antigens. In this case, the antibodies are coated onto the substrate, followed by antigen attachment and conjugation to an enzyme.

A fluorescence-activated cell sorter (FACS) is a flow cytometry instrument used to separate and identify cells in a heterogeneous population. Cell mixtures to be sorted are first bound to fluorescent dyes such as fluorescein or phycoerythrin. The labeled cells are then pumped through the instrument and are excited by a laser beam. Cells that fluoresce are detected, and an electrostatic charge is applied. The charged cells are separated using voltage deflection.

Exhibit 4.5 Gardasil

Gardasil is a noninfectious recombinant vaccine consisting of capsid proteins from four different human papillomaviruses (HPVs) of types 6, 11, 16, and 18. HPV causes squamous cell cervical cancer and cervical adenocarcinoma, as well as 35–50% of vulvar and vaginal cancers.

The four antigens are produced in a fermentation process using the yeast *Saccharomyces cerevisiae* grown in chemically defined media. The purified antigens are formulated in aluminum-containing adjuvant in sterile liquid suspension.

In June 2006, the FDA approved the use of Gardasil to vaccinate females from ages 9 to 26.

Source: Food and Drug Administration. *Gardasil*. http://www.fda.gov/cber/label/hpvmer040307LB.pdf [accessed September 28, 2007].

genes, which contain engineered DNA sequences that code for antigens. The host's own cells take up these genes and express the antigens, which are then presented to the immune cells and activate the immune response.

Peptide Vaccines: Peptide vaccines are chemically synthesized and normally consist of 8–24 amino acids. In comparison with protein molecules, peptide vaccines are relatively small. They are also known as peptidomimetic vaccines, as they mimic the epitopes. Complex structures of cyclic components, branched chains, or other configurations can be built into the peptide chain. In this way, they possess conformations similar to the epitopes and can be recognized by immune cells. An *in silico* vaccine design approach has been used to find potential epitopes. A critical aspect of peptide vaccines is to produce 3D structures similar to the native epitopes of the pathogen.

4.2.3 Adjuvants

Very often, vaccines are formulated with certain substances to enhance the immune response. These substances are called adjuvants (from the Latin *adjuvare,* which means "to help"). The most common adjuvants for human use are aluminum hydroxide, aluminum phosphate, and calcium phosphate. Other adjuvants being used include bacteria and cholesterol. Mineral oil emulsions are normally the adjuvants used in animal studies. The adjuvant known as Freund's complete adjuvant consists of killed tubercle bacilli in water-in-mineral oil emulsion, and Freund's incomplete adjuvant is a water-in-oil emulsion. Both these adjuvants are effective in stimulating an immune response, but they cause unacceptable side effects in humans (see Table 4.2).

There are three basic mechanisms by which adjuvants assist in improving immune response. First, adjuvants help the immune response by forming reservoirs of antigens that provide a sustained release of antigens over a long period. Second, adjuvants act as nonspecific mediators of immune cell function by stimulating or modulating immune cells. Third, adjuvants can serve as vehicles to deliver the antigen to the spleen and/or lymph nodes, where immune response is initiated.

Edible food sources have been tested to deliver vaccines orally; for example, transgenic potato tuber-based vaccines have been developed. Other food sources, such as bananas, tomatoes, and corn, are being tested in laboratories (see Section 11.12). Mucosal vaccines, utilizing genetically modified enterotoxins, are delivered intranasally. Research in this area has to ensure the safety aspect of using enterotoxins.

4.2.4 Recent Vaccine Research and Clinical Activities

The field of vaccine research is very active. Exhibit 4.6 summarizes examples of some selected vaccines. Appendix 5 shows a table of the production methods for selected vaccines.

TABLE 4.2   Examples of Common Adjuvants and Their Mechanisms of Action

Adjuvant	Composition	Mechanism of Action
Alum (aluminum hydroxide or aluminum phosphate)	Aluminum hydroxide gel	Enhanced uptake of antigen by APC; delayed release of antigen
Alum with a mycobacterial-derived dipeptide	Aluminum hydroxide gel with muramyl dipeptide	Enhanced uptake of antigen by APC; delayed release of antigen; induction of costimulatory molecules on APCs
Alum with *Bordetella pertussis*	Aluminum hydroxide gel with killed *Bordetella pertussis*	Enhanced uptake of antigen by APC; delayed release of antigen; induction of costimulatory molecules on APCs
Freund's complete adjuvant	Oil in water with killed *tubercle bacilli*	Enhanced uptake of antigen by APC; delayed release of antigen; induction of costimulatory molecules on APCs
Freund's incomplete adjuvant	Oil in water	Enhanced uptake of antigen by APC; delayed release of antigen
Immune stimulatory complexes	Open cage-like structures containing cholesterol and a mixture of saponins	Delivery of antigen to cytosol, allowing induction of cytotoxic T cell responses

Source: Coico R, Sunshine G, Benjamini E. *Immunology*, 5th ed., Wiley-Liss, Hoboken, NJ, 2003.

Exhibit 4.6   Selected Vaccines

Cervical Cancer: See Exhibit 4.5.

Avian Influenza: See Exhibit 4.3.

Alzheimer's Disease: The vaccine being tested contains a small protein called β-amyloid (Aβ). This protein forms abnormal deposits, or "plaques," in the brains of people with Alzheimer's disease. Researchers believe that Aβ deposition causes loss of mental function by killing the brain neurons. The strategy of Aβ vaccination is to stimulate the immune system to clean up plaques and prevent further Aβ deposits. Although preclinical and Phase I studies showed the potential of the vaccine, the Phase II clinical trial was halted because 15 of 360 patients developed severe brain inflammation. Further studies showed that the Aβ did generate the desired antibody response. An acceptable vaccine may still be possible by modifying the epitope to reduce the inflammation effect.

Source: Frantz S. Alzheimer's disease vaccine revisited, *Nature Reviews Drug Discovery* 1:933 (2002).

Pneumococcal Disease: In October 2002, the FDA approved the use of Prevnar for immunization of infants and toddlers against otitis media—middle ear infection. Prevnar is a pneumococcal seven-valent conjugate vaccine. It is formulated with a sterile solution of saccharides conjugated to the antigen, *Streptococcus pneumoniae*.

Source: Center for Biologics Evaluation and Research, *Product Approval Information*, FDA, Rockville, MD, http://www.fda.gov/cber/approvltr/pneuled100102L.htm [accessed Oct 10, 2002].

Cancer: In cancer, the immune system does not recognize the changes in cancer cells. Cancer vaccines seek to mimic cancer-specific changes by using synthetic peptides to challenge the immune system. When these peptides are taken up by T cells, the immune system is activated. The T cells search for cancer cells with specific markers and proceed to kill them. Some vaccines being tested are (1) a peptide called β-defensin 2, which activates the immune system against tumor activity, and (2) an outer coat protein of the human papillomavirus to act as a vaccine against cervical cancer.

Source: National Institutes of Health. *Cancer Vaccine*. http://www.nih.gov [accessed October 28, 2002].

AIDS: (See Exhibit 2.12). AIDS is caused by HIV infection. HIV belongs to a large family of retroviruses, the Lentiviridae. The HIV genome is within the RNA. Following infection in humans, the RNA genome of HIV is reverse-transcribed into DNA and integrated within the human cell. HIV undergoes frequent mutation and therefore is highly variable. One technique for producing an AIDS vaccine is to reproduce, using recombinant technology, the surface proteins on the HIV. There are two particular envelope proteins being investigated: gp120 and gp41. gp120 is a trimeric protein and is held together by three transmembrane gp41 proteins. Laboratory studies have shown that vaccines based on these proteins can induce antibody responses to different strains of HIV. Other AIDS treatments are the use of (1) antiviral (AZT, a reverse transcriptase inhibitor) drugs, (2) drugs (indinavir) that target and inhibit the production of HIV protease, an enzyme required to assemble new virus particles, and (3) gene therapy—control of viral genome expression through the use of synthetic oligonucleotides.

Malaria: Malaria is a major disease in tropical countries. According to the WHO, 300–500 million individuals are infected with malaria. The death tolls are 1.5–3.5 million yearly. There are four species of malaria parasites that infect humans: *Plasmodium falciparum, P. vivax, P. ovale,* and *P. malariae.* Of these, *P. falciparum,* the predominant malarial parasite found in Africa, is the most virulent. There are four stages in the *P. falciparum* life cycle: (1) sporozoite (3–5 minutes when it is injected into the bloodstream by a mosquito); (2) liver stage (1–2 weeks after the parasite enters the liver, during which it matures; no symptoms are shown in stages 1 and 2); (3) blood stage (2 or more days/cycle during which red blood cells are invaded and parasites rupture out of red blood cells; fevers and chills are manifested); and (4) sexual stage (10–14 days during which parasites mature into the sexual form, ready to be picked up by a mosquito to infect the next person). Vaccine strategies are of three types: preerythrocytic (stages 1 and 2), blood stage (stage 3), and transmission blocking (stage 4).

Source: Dubovsky F. *Creating a Vaccine Against Malaria, Malaria Vaccine Initiative*, FDA, Rockville, MD, 2001. http://www.malariavaccine.org/files/Creating_a_Vaccine_against_Malaria.pdf, [accessed November 22, 2002].

A new malaria vaccine, RTS, S/AS02D, from GSK has shown very promising results in a Phase I/IIb double-blind randomized trial of 214 infants in Mozambique. For children in the 1–4 year-old age group, the most vulnerable group, the vaccine not only lowered the chances of infection by 65% over 3 months but also reduced episodes of clinical malaria by 35% in 6 months.

Source: Aponte, J. J. et al., Safety of the RTS,S/AS02D candidate malaria vaccine in infants living in a highly endemic area of Mozambique: a double blind randomised controlled phase I/IIb trial, *The Lancet* DOI:10.1016/S0140-6736(07)61542-6 (2007) [accessed Oct 19, 2007].

Chickenpox: Chickenpox is a highly contagious viral infection that causes rash-like blisters on the skin surface and mucous membranes. It is generally mild and not normally life-threatening. For adults, the symptoms are more serious and uncomfortable than for children. The disease can also be deadly for some people, such as pregnant women, people with leukemia, or immunosuppressed patients. Varivax (varicella virus vaccine live) from Merck & Co. was tested on about 11,000 children and adults and was approved by the FDA in March 1995 as a chickenpox vaccine.

Smallpox: Smallpox is a very contagious disease with a mortality rate as high as 30–35%. It is estimated that smallpox was responsible for 300–500 million deaths in the 20th century. Fortunately, it has been eradicated in 1979 through strict regimens of vaccination.

ACAM2000 is a smallpox vaccine using live vaccinia virus for active immunization of high-risk individuals. Vaccinia virus has the same taxonomic group (classification) as smallpox virus (variola) but it cannot cause smallpox to be developed. The vaccinia virus causes localized virus infection and stimulates production of neutralizing antibodies that cross-protect against smallpox virus. ACAM2000 is supplied in a lyophilized form and reconstituted into a liquid before vaccination.

Source: Food and Drug Administration. *ACAM2000*. http://www.fda.gov/cber/label/acam2000LB.pdf [accessed September 24, 2007].

Herpes Zoster (Shingles): Zostavax is a live attenuated varicella-zoster virus (VZV) vaccine for the prevention of herpes zoster in individuals 60 years or older. It is supplied in frozen lyophilized form and reconstituted before vaccination. The vaccine boosts VZV-specific immunity and protects individuals against zosters and its complications.

Source: Food and Drug Administration. *Zostavax*. http://www.fda.gov/cber/label/zostavaxLB.pdf [accessed September 24, 2007].

4.3 ANTIBODIES

Antibodies are produced by the B cells of the immune system (Exhibit 4.7). They are like weapons of our defense system and can be described as "homing devices" that target antigens and destroy them. Antibodies are immunoglobulins (proteins with immune functions) and are categorized into five different classes: immunoglobulin G and D (IgG and IgD, ~75%), immunoglobulin A (IgA ~ 15%), immunoglobulin M (IgM ~ 15%), and immunoglobulin E (IgE < 1%). They differ from each other in size, charge, carbohydrate content, and amino acid composition. Within each class, there are subclasses that show slight differences in structure and function from other members of the class.

4.3.1 Antibody Structure

The structure of an antibody is normally depicted as a capital letter Y configuration. IgG is the most predominant antibody. It is a tetrameric molecule consisting of two identical heavy (H) polypeptide chains of about 440 amino acids and two identical light (L) polypeptide chains of about 220 amino acids (Fig. 4.1). The four chains are held together by disulfide bonds and noncovalent interactions.

Within the light and heavy chains are domains, which consist of about 110 amino acids. The domains that have similar polypeptide sequence are termed constant domains. These are the C_H1, C_H2, and C_H3 domains of the heavy chain

Exhibit 4.7   Human Immune System

The human immune system is a remarkable system for combating against foreign substances that invade the body. It protects us from infections by pathogens such as viruses, bacteria, parasites, and fungi. An important aspect of the immune system is the self–nonself recognition function, by means of markers present on a protein called the major histocompatibility complex (MHC). Substances without such markers are discerned and targeted for destruction. Although in most cases the immune system functions properly, at times it breaks down. For some people, their immune systems lack the normal discrimination capability and revert to attack and destroy their own body cells as if they are foreign. This gives rise to autoimmune diseases such as rheumatoid arthritis, diabetes, multiple sclerosis, and systemic lupus erythematosus. There are also occasions when the immune system responds with undue sensitivities to innocuous substances such as airborne pollen, leading to allergies, as in the case of asthma and hay fever.

Immune responses are mediated through the lymphocytes called B cells and T cells. Lymphocytes are a particular type of white blood cell. White blood cells (leukocytes) are divided into granulocytes (neutrophils, 55–70%; eosinophils, 1–3%; and basophils, 0.5–1%) and agranulocytes (lymphocytes [B and T cells], 20–40%; and monocytes, 1–6%). There are 5000–10,000 white blood cells per milliliter of blood, compared with five million red blood cells in the same volume.

When pathogens enter the human body, cells called macrophages (meaning "big eaters") engulf and ingest the pathogens (antigens). The antigens are processed by the macrophages, and parts of the antigens are displayed on the surface in the form of short peptide chains bound to the MHC protein. These antigen-presenting cells (APCs) of macrophages and dendritic cells activate the immune response by sensitizing the B and T cells.

B cells are produced by the bone marrow. In response to activation of CD4$^+$ T helper cells (see below), B cells proliferate and produce antibodies. (The term CD stands for "cluster of differentiation." They are proteins coating cell surfaces. Altogether, there are more than 160 different types of CDs.) The antibodies produced by B cells circulate in the bloodstream and bind to antigens. Once bound, other cells are in turn activated to destroy the antigens.

T cells are lymphocytes produced by the thymus gland. There are two types of T cells involved in immune response: CD4$^+$ (CD positive, helper cells) and CD8$^+$ (CD positive, also called T killer, or suppressor, cells). When the APCs present the antigens to CD4$^+$ helper T cells, the secretory function is activated and growth factors such as cytokines are secreted to signal the proliferation of CD8$^+$ killer cells and B cells. When the CD8$^+$

cells are activated by the APCs, the CD8⁺ killer T cells directly kill those cells expressing the antigens. Activated B cells produce antibodies, as described above.

It is estimated that every B and T cell has about 100,000 protein molecules on the surface. There are many variations to these surface molecules, which act as receptors for antigens. As many as 10^{18} different surface receptors can be produced, thus giving rise to a vast probability for the B and T cells to recognize and bind to a vast array of antigens.

Note: CD4 is a receptor for HIV. Hence, people infected with HIV have suppressed immune response and develop AIDS because the CD4 cannot function normally.

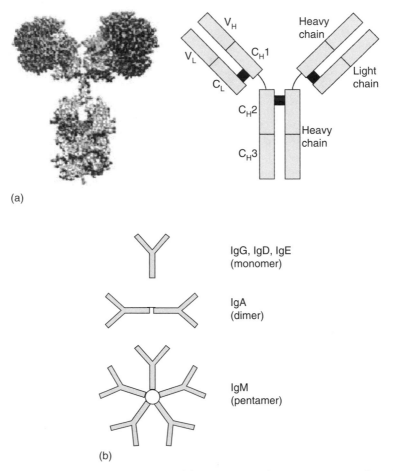

(a)

(b)

Figure 4.1 (a) IgG antibody molecule. (b) IgG, IgD, and IgE are monomeric antibodies. IgA and IgM are polymeric antibodies.

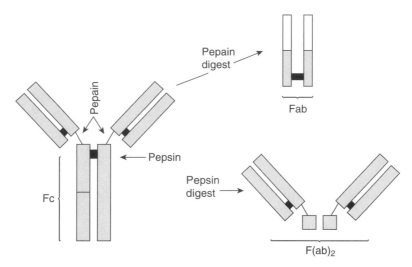

Figure 4.2 Different fragments of the antibody molecule.

and the C_L domain of the light chain. Where the sequence is variable, the domains are called variable domains, one each on the heavy and light chain: V_H and V_L. The variability is confined to particular regions of the variable domain, called the complementarity-determining regions. These regions have the appropriate 3D structure to bind to antigens.

An antibody can be cleaved by enzymes such as papain and pepsin into different fragments (Fig. 4.2).

These different fragments are the following:

- *Variable Fragment (Fv):* The tips of the two Y arms vary greatly from one antibody to another. They are the regions that bind to epitopes of antigens and bring them to the natural killer cells and macrophages for destruction.
- *Antigen-Binding Fragments (Fab), Fab′, and F(ab′)₂:* Various parts that contain the variable fragment.
- *Constant Fragment (Fc):* This is the stem of the letter Y. It is the part that is identical for all antibodies of the same class; for example, all IgGs have the same Fc. The Fc fragment is the part that links the antibody to other receptors and triggers immune response and antigen destruction.

4.3.2 Traditional Antibodies

Several decades ago, antibodies were obtained by extraction from blood samples of immunized animals or human donors. These are polyclonal antibodies, because several different types of antibodies are obtained through this method, although IgG is normally the predominant component. The steps for obtaining polyclonal antibodies are illustrated in Fig. 4.3.

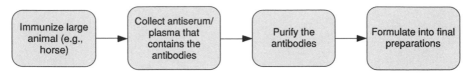

Figure 4.3   Production of polyclonal antibodies from horse antisera. (*Source:* Walsh G. *Biopharmaceuticals: Biochemistry and Biotechnology*, Wiley, Hoboken NJ, 1998.)

Examples of some of the polyclonal antibodies are the following:

- *Antibodies Derived from Horse Antisera:* Botulism antitoxin, diphtheria antitoxin, scorpion venom antisera, snake venom antisera, spider antivenins, and tetanus antitoxin.
- *Antibodies Derived from Human Donors:* Hepatitis A and B immunoglobulins, measles immunoglobulins, rabies immunoglobulin, and tetanus immunoglobulin.

Although polyclonal antibodies have been used for passive immunization and therapeutic treatments, there are cases when hypersensitivities are induced. The reason is that polyclonal antibodies contain not only the specific antibody that binds to the desired antigen but other antibodies, which our immune system will treat as foreign substances and act against.

4.3.3   Monoclonal Antibodies

The next development was the production of monoclonal antibodies (MAbs) in the mid-1970s. This uses hybridoma technology, which involves the fusion of antibody-producing B cells to immortal myeloma cells. Figure 4.4 shows the preparation of MAbs using hybridoma techniques. A more detailed discussion of biopharmaceuticals production is presented in Section 10.5.

MAbs are specific in binding to epitopes of antigens. Because MAbs are produced using murine (mouse) spleen cells, the human immune system can react against these murine MAbs. The allergic reaction is caused by human anti-mouse antibodies (HAMA) and they can neutralize the effect of the MAbs, or even induce rashes, swelling, and kidney problems; it may even be life threatening. In other cases, the murine MAbs may not be as effective as human antibodies because of their murine origin.

4.3.4   Humanization of Antibodies

As discussed previously, murine antibodies have limitations. The next phase of development is to make these murine MAbs more like human antibodies, by using genetic engineering techniques. A recent approach is to "humanize" the antibodies to reduce HAMA and improve the avidity of the MAbs (avidity

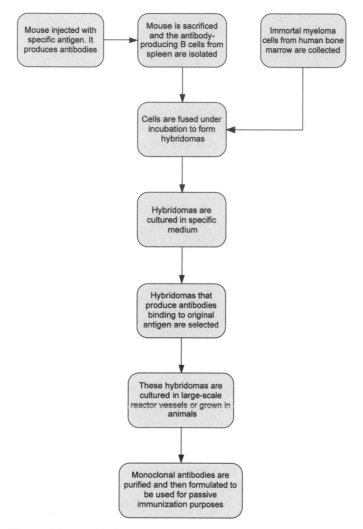

Figure 4.4   Production of MAbs using the hybridoma technique.

is a measure of the affinity or interaction of the binding of an antibody to an antigen). Several strategies have been adopted. They include replacing certain fragments of the antibodies.

Chimeric Antibodies: The first generation is the chimeric antibodies (chimeric comes from the word *Chimera*, a Greek mythology beast made of three animals: a lion, a snake, and a goat). This type of antibody consists of both murine and human parts. The murine Fv fragments are retained and linked to the Fc fragment of human IgG. An example of the chimeric antibody is ReoPro, which prevents blood clots by binding to a receptor on platelets.

Humanized Antibodies: To further improve the avidity and reduce antigenicity, only the specific antigen-binding region is derived from mouse, while the remainder of the antibody is constructed using human proteins. These are the humanized antibodies and include the breast cancer targeting MAb called Herceptin (see Case Study #4).

Full Human Antibody: Full human antibodies are the current engineered antibodies. Several techniques are used to construct these antibodies. One method is to fuse human B cells to myeloma cells. These hybridomas will produce fully human MAbs. Another method is to genetically alter mice in the laboratory to contain human antibody producing genes. In response to antigens, antibodies resembling the human antibodies are produced.

4.3.5 Conjugate Antibodies

Antibodies are also prepared to carry "payloads." Materials such as toxins, enzymes, or even radioisotopes can be fused to the antibodies (Fig. 4.5). The strategy here is to use antibodies as vehicles to deliver more effective treatment to specific target cells. Immunotoxins are fusion proteins consisting of a toxin connected to a MAb. Immunocytokines consist of a fusion of rDNA encoding the heavy chain of a MAb with the DNA encoding a cytokine. The aim is to obtain a high local concentration of cytokine to generate an antitumor response. Zevalin and Bexxar are two conjugate antibody drugs, which carry the radioisotopes yttrium (^{90}Y) and iodine (^{131}I) for the treatment of non-Hodgkin's lymphoma.

Another variation to conjugate antibodies is to use bispecific antibodies. These are produced using chemical means and recombinant techniques to fuse separate hybridomas into a hybrid hybridoma (Fig. 4.6). Bispecific antibodies use one arm of the Fv to target the antigen or tumor cell and the other arm carries the effector molecule of toxins, radioisotopes, or other drugs.

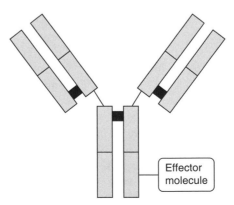

Effector molecule

Figure 4.5 Conjugate antibodies.

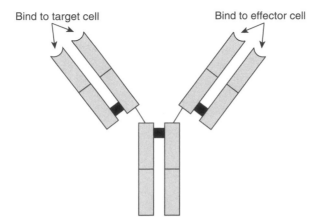

Figure 4.6 Bispecific antibody.

To date, there are 12 MAbs approved for the treatment of cancer around the world, as shown in Table 4.3.

4.4 CYTOKINES

Cytokines are produced mainly by the leukocytes (white blood cells). They are potent polypeptide molecules that regulate the immune and inflammation functions, as well as hematopoiesis (production of blood cells) and wound healing. There are two major classes of cytokines: (1) lymphokines and monokines and (2) growth factors.

4.4.1 Lymphokines and Monokines

Cytokines produced by lymphocytes are called lymphokines, and those produced by monocytes are termed monokines. Lymphocytes and monocytes are different types of white blood cells. The major lymphokines are interferons (IFNs) and some interleukins (ILs). Monokines include other interleukins and tumor necrosis factor (TNF).

Interferons: There are two types of interferons: Type I, which includes IFN-α and IFN-β, and Type II consisting of IFN-γ. IFN-α and IFN-β have about 30% homology in amino acid sequence. There are two more recently discovered Type I interferons; they are called IFN-ω and IFN-τ. IFN-α and IFN-β each have 166 amino acids, and IFN-γ has 143. Both IFN-α and IFN-β are of single chain structure and bind to the same type of cell surface receptors, whereas IFN-γ is a dimer of two identical chains and interacts with another type of receptor. All our cells can produce Type I interferons when infected by viruses, bacteria, and fungi. However, only T cells and natural killer cells can produce

TABLE 4.3 Monoclonal Antibodies Approved for Cancer Treatment

Generic Name	Trade Name	Description	First Approved Indication	Year and Country of First Approval
Edrecolomab	Panorex	Murine IgG2a	Postoperative colorectal cancer	1995 Germany
Rituximab	Rituxan	Chimeric IgG1κ anti-CD20	Relapsed or refractory low-grade non-Hodgkin's lymphoma	1997 United States
Trastuzumab	Herceptin	Humanized IgG1κ anti-HER2	HER2 overexpressing metastatic breast cancer	1998 United States
Gemtuzumab ozogamicin	Mylotarg	Humanized IgG4κ anti-CD33 immunotoxin (calicheamicin)	Relapsed acute myeloid leukemia	2000 United States
Alemtuzumab	Campath	Humanized IgG1κ anti-CD52	Chronic lymphocytic leukemia	2001 United States
Ibritumomab tiuxetan	Zevalin	Murine IgG1κ anti-CD20, radiolabeled Y-90	Relapsed or refractory low-grade, follicular transformed non-Hodgkin's lymphoma	2002 United States
I-131 ch-TNT	N/A	Chimeric IgG1κ anti-DNA associated antigens, radiolabeled I-131	Advanced lung cancer	2003 China
I-131 tositumomab	Bexxar	Murine IgG2aλ anti-CD20, radiolabeled I-131	Non-Hodgkin's lymphoma in rituximab-refractory patients	2003 United States
Cetuximab	Erbitux	Chimeric IgG1κ anti-EGF receptor	EGFR-expressing metastatic colorectal cancer	2003 Switzerland
Bevacizumab	Avastin	Humanized IgG1 anti-VEGF	Metastatic colorectal cancer	2004 United States
Nimotuzumab	TheraCIM	Humanized IgG1 anti-EGF receptor	Advanced head/neck epithelial cancer	2005 China
Panitumumab	Vectibix	Human IgG2κ anti-EGF receptor	Metastatic colorectal cancer	2006 United States

Source: Reichert JM, Valge-Archer WE. Development trends for monoclonal antibody cancer therapeutics, *Nature Reviews Drug Discovery* 6:340–356 (2007).

Type II interferon. Type I interferon binds to receptor, which in turn activates tyrosine kinase phosphorylation and the subsequent transcription pathway that induces viral resistance. Similarly, Type II interferon binds to another receptor and activates the immune response.

Because of its antiviral and anticancer effects, IFN-α is used in the treatment of hepatitis and various forms of cancer, such as Kaposi's sarcoma, non-Hodgkin's lymphoma, and hairy cell leukemia. Exhibit 4.8 describes the treatment of hepatitis C with IFN-α. IFN-β is used for treating multiple sclerosis, a chronic disease of the nervous system. The medical application of IFN-γ is for cancer, AIDS, and leprosy.

Interleukins: Interleukins are proteins produced mainly by leukocytes. There are many interleukins within this family (Table 4.4). Interleukins have a number of functions but are principally involved in mediating and directing immune cells to proliferate and differentiate. Each interleukin binds to a specific receptor and produces its response.

IL-2 is possibly the most-studied interleukin. It is also called T cell growth factor. IL-2 is a 15 kDa glycoprotein produced by CD4$^+$ T helper cells. It has 133 amino acids. There are four helical regions and a short β-sheet section (Fig. 4.7).

Exhibit 4.8   Hepatitis C and Interferon

Hepatitis C is caused by a virus contracted through contaminated blood. Most infected patients show no sign of hepatitis for a long time. Of those infected, about 15% will clear the virus, and 85% develop chronic hepatitis. Up to 30% of patients with chronic hepatitis C will develop cirrhosis within 20 years, and 5% will develop liver cancer. The WHO estimates that more than 170 million people worldwide are infected with hepatitis C.

Interferon is the approved treatment for hepatitis C. In general, there are four different treatments: (1) IFN-α, (2) combination therapy of IFN-α and another drug called ribavirin, (3) pegylated IFN-α, and (4) pegylated IFN-α with ribavirin. Pegylated interferon contains polyethylene glycol, which increases the half-life (see Section 5.3.5) from 6 hours to 45 hours and slows down the body's absorption of interferon. In this way, a more controlled release of interferon is achieved to prolong absorption. Instead of a subcutaneous injection of three times weekly, the frequency can be reduced to once weekly.

Interferons were extracted and purified from human blood supplies up until the 1980s. The amount produced was very low. Since then, interferons have been produced using recombinant technology from a variety of cells: *Escherichia coli*, fungus, yeast, and mammalian.

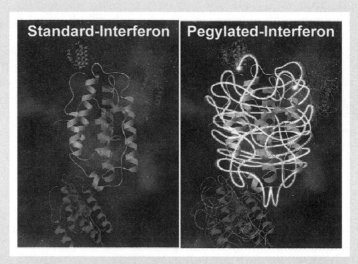

Interferon. (*Source*: Roche. *Pegasys improves things for patients with chronic hepatitis C*. http://www.roche.com/pages/facets/10/pegasyse.htm [accessed November 2, 2007]. Used with permission.)

Patients receiving IFN experience side effects similar to influenza symptoms: headache, nausea and tiredness. IFN also decreases red blood cells, white blood cells and platelet counts. A measure of the effectiveness of IFN treatment is the marker called alanine aminotransferase in blood. The normal range is 10–70 U/L.

TABLE 4.4 Selected Interleukins

Cytokine	Origin	Target cell	Effect on Immune Response
IL-2	T cell	T cell, NK cell	Proliferation of antigen-specific cells and other immune cells
IL-3	T cell	Bone marrow cells	Growth/differentiation of all cell types
IL-4	T cell	B cell	Stimulation of immunoglobulin (Ig) heavy chain switching to IgE
IL-5	T cell	B cell, eosinophil	Growth /differentiation of eosinophils
IL-7	Bone marrow stroma cell	Lymphocyte	Growth factor
IL-10	Macrophage, T cell	Macrophage, T cell	Inhibition of macrophage function, control of immune response
IL-15	Macrophage	NK cell, T cell	Proliferation

Source: Adapted from Zane HD. *Immunology—Theoretical and Practical Concepts in Laboratory Medicine*, Saunders Philadelphia, 2001.

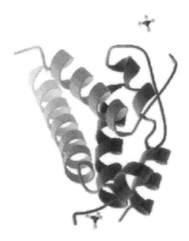

Figure 4.7 Interleukin 2 (IL-2) molecule. (*Source*: Protein Date Bank, PDB ID: 1M47. http://www.rcsb.org/pdb/cgi/explore.cgi?job=summary&pdbId=1M47&page=. Arkin MM, Randal M, Delano WL, et al. Binding of small molecules to an adaptive protein–protein interface, *Proceedings of the National Academy of Sciences USA* 100:1603 (2003). Used with permission.)

Exhibit 4.9   Proleukin and Ontak

Proleukin is a recombinant form of IL-2. It is approved for the treatment of malignant melanoma and renal cell cancer. Ontak (denileukin diftitox) is a fusion protein for the treatment of persistent or recurrent T-cell lymphoma. Activated T cells express IL-2 receptors. Ontak has a fragment that binds to the IL-2 receptor while the other part presents a diphtheria toxin to kill the activated T cell.

Sources: (1) Food and Drug Administration. *List of Orphan Products Designations and Approvals*. http://www.fda.gov/ohrms/dockets/dailys/00/mar00/030100/lst0094.pdf [accessed September 21, 2007]. (2) Food and Drug Administration. *Ontak*. http://www.fda.gov/ medwatch/SAFETY/2006/Mar%20PIs/Ontak_PI.pdf [accessed September 21, 2007].

IL-2 promotes the growth of B cells for antibody production and induces the release of IFN-γ and TNF (see below). It has been approved by the FDA for the treatment of different types of cancer, including metastatic melanoma and metastatic renal carcinoma. Examples of IL-2 for the treatment of malignant melanoma and a protein that targets IL-2 receptor in T-cell lymphoma are given in Exhibit 4.9.

Although IL-2 has not been approved to treat HIV/AIDS, many clinical trials using IL-2 are being conducted. The strategy is to complement the anti-HIV therapy by boosting the immune system with IL-2. The replacement

Exhibit 4.10   Rheumatoid Arthritis and TNF-α

Rheumatoid arthritis is an autoimmune disease of the synovial lining of joints. Typically, the joints affected are those in the extremities: fingers, wrists, toes, and ankles. It is a debilitating disease in which ligaments may be damaged and joints deformed.

In the late 1980s, scientists found that TNF-α is involved in causing arthritis. Standard drug treatment for rheumatoid arthritis used to be methotrexate, steroids, and nonsteroidal anti-inflammatory drugs (NSAIDs). These drugs are nonspecific and their effectiveness is variable. The new set of drugs in the late 1990s was designed to specifically target TNF-α. Two drugs are especially effective: infliximab (Remicade, Centocor) (a chimeric antibody that targets the TNF-α) and etanercept (Enbrel, Wyeth) (a soluble protein receptor for TNF-α that neutralizes its effect).

The success of these two drugs provides impetus for the development of other anti-inflammatory drugs aiming at specific inflammatory agents.

therapy of IL-2 administered to AIDS patients increases production of CD4$^+$ T cells and the activities of natural killer cells to combat HIV. Therapeutic IL-2 is manufactured using recombinant technology.

Tumor Necrosis Factor: There are two types of tumor necrosis factor: TNF-α and TNF-β. Of the two, TNF-α has been studied in more detail. TNF-α is a 157 amino acid polypeptide. It is a mediator of immune regulation, including the activation of macrophages and induction of the proliferation of T cells. Another TNF-α function is its cytotoxic effects on a number of tumor cells. Recent research, however, concentrates on its property in the stimulation of inflammation, particularly in the case of rheumatoid arthritis. Clinical trials are being conducted with drugs to block TNF-α with anti-TNF-α monoclonal antibodies. These antibodies target the excessive levels of TNF-α in the synovial fluid of joints and provide relief to sufferers of rheumatoid arthritis (Exhibit 4.10).

4.4.2   Growth Factors

As the name implies, growth factors stimulate cell growth and maintenance. We will discuss the following growth factors:

- Erythropoietin
- Colony stimulating growth factors
- Vascular endothelial growth factors

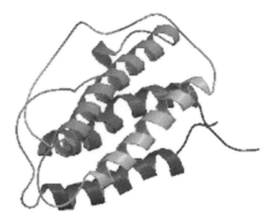

Figure 4.8   Erythropoietin. (*Source*: Cheetham JC, Smith DM, Aoki KH, et al. NMR structure of human erythropoietin and a comparison with its receptor bound conformation, *Nature Structural Biology* 5:861–866 (1998).)

Erythropoietin: Erythropoietin (EPO) (Fig. 4.8) is a glycoprotein produced by specialized cells in the kidneys. It has 166 amino acids and a molecular weight of approximately 36 kDa. EPO stimulates the stem cells of bone marrow to produce red blood cells. It is used to treat anemia and chronic infections such as HIV and cancer treatment with chemotherapy where anemia is induced. Patients feel tired and breathless owing to the low level of red blood cells. EPO can be prescribed instead of blood transfusion.

Biopharmaceutical quantities of EPO are produced with recombinant cells. This is achieved through the isolation of the human gene that codes for EPO and transfection of the gene into cell lines such as Chinese hamster ovary cells (see Section 10.5). The product is called rhEPO—recombinant human EPO. EPO is normally administered subcutaneously and is generally well tolerated by patients.

EPO is considered a banned performance-enhancing drug in the sports arena, where athletes use EPO to boost their red blood cells with the expectation of boosting performance (see Exhibit 4.11 for a brief review of performance-enhancing drugs and Section 11.10 on banned drugs in sports).

Colony Stimulating Growth Factors: Growth factors such as granulocyte macrophage colony stimulating factor (GM-CSF) and macrophage colony stimulating factor (M-CSF) are involved in the regulation of the immune and inflammatory responses. GM-CSF is a glycoprotein with 127 amino acids and a molecular weight of about 22 kDa. It is produced by macrophages and T cells.

Clinically, GM-CSF is used to stimulate production of blood cells, in particular, in patients who have received chemotherapy. M-CSF is a glycoprotein

Exhibit 4.11   Performance-Enhancing Drugs

To help them excel in sports, some athletes use drugs to boost their performance. There are several reasons why drugs are used by athletes:

- To increase oxygen delivery
- To build muscle and bone
- To mask pain
- To mask use of other drugs
- As stimulants

EPO is used in blood doping to generate more red blood cells for carrying oxygen. It is particularly favored by endurance athletes to enhance their performance. Human growth hormone (hGH, see description in Section 4.5.2) is used to build up muscle and bone strength. Both EPO and hGH are banned in sports.

The recombinant EPO and hGH produced are almost replicas of those that occur naturally in our body. Hence, it is very difficult to detect these banned substances if taken by athletes. Another difficulty is the need to develop reliable and sensitive test methods that take into account differences of these substances in athletes of different racial groups.

Sources: (1) Zorpette G. All doped up—and going for gold, *Scientific American* May: 20–22 (2000). (2) Freudenrich C. *How Performance-Enhancing Drugs Work*. http://entertainment. howstuffworks.com/athletic-drug-test1.htm [accessed September 24, 2002].

that can exist in different forms. The number of amino acids ranges from just over 200 to about 500, and molecular weight varies between 45 and 90 kDa. M-CSF is being evaluated clinically for its antitumor activity.

Vascular Endothelial Growth Factor: Vascular endothelial growth factor (VEGF) is a homodimeric glycoprotein with a molecular weight of 45 kDa. It is a major regulator of tumor angiogenesis (growth of blood vessels).

For cells to grow, a supply of oxygen is required. Our blood delivers oxygen to the cells, which are within a tenth of a millimeter from blood capillaries. For tumor cells, if they grow larger than a millimeter, they will be denied oxygen if there are no new vessels formed. VEGF is used by tumor cells to form new blood vessels. It binds to receptors on the surface of endothelial cells and signals them to form new vessels. This will promote further tumor growth and metastasis, leading to the spread of tumor to other parts of the body.

Two recent MAbs have been approved by the FDA to treat prostate cancer and macular degeneration through the use of the antibodies to target VEGF and stop angiogenesis (Exhibit 4.12).

Exhibit 4.12   Avastin and Lucentis

Avastin: Bevacizumab (Avastin) is a monoclonal IgG₁ antibody. It binds to vascular endothelial growth factor (VEGF) and prevents VEGF from interacting with its receptors (Flt-1 and KDR) on the surface of endothelial cells. Thus, it inhibits endothelial cell proliferation and new blood vessel formation, leading to reduction of microvascular growth and inhibition of metastatic disease progression.

Bevacizumab is produced using CHO expression in a nutrient medium with gentamicin antibiotic. It has a molecular weight of 149 kDa. The antibody is humanized, with human framework and murine complementarity-determining regions.

Source: Food and Drug Administration. *Avastin.* http://www.fda.gov/medwatch/SAFETY/2005/Jan_PI/Avastin_PI.pdf [accessed September 17, 2007].

Lucentis: Ranibizumab (Lucentis) is a monoclonal IgG₁ antibody. It is designed for intraocular use for the treatment of age-related macular degeneration (AMD)—thinning of the retina—which affects the central vision of the elderly. The antibody binds to the vascular endothelial growth factor A (VEGF-A) and inhibits the biological activity.

Ranibizumab is produced in *E. coli* expression system in a nutrient medium containing the tetracycline antibiotic. It has a molecular weight of 48 kDa.

Source: Food and Drug Administration. *Lucentis.* http://www.fda.gov/cder/foi/label/2006/125156lbl.pdf [accessed September 17, 2007].

4.5   HORMONES

Hormones are intercellular messengers. They are typically (1) steroids (e.g., estrogens, androgens, and mineral corticoids, which control the level of water and salts excreted by the kidney), (2) polypeptides (e.g., insulin and endorphins), and (3) amino acid derivatives (e.g., epinephrine, or adrenaline, and norepinephrine, or noradrenaline). Hormones maintain homeostasis—the balance of biological activities in the body; for example, insulin controls the blood glucose level, epinephrine and norepinephrine mediate the response to the external environment, and growth hormone promotes normal healthy growth and development.

4.5.1   Insulin

Insulin is produced in the pancreas by β cells in the region called the islets of Langerhans. It is a polypeptide hormone consisting of two chains: an A chain

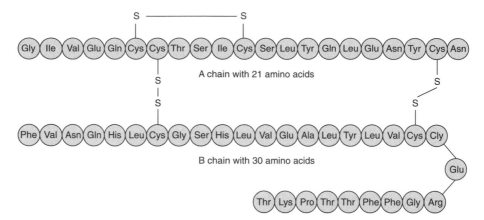

Figure 4.9   Human insulin molecule. Refer to Table A2.1 for the names of amino acids.

with 21 amino acids with an internal disulfide bond, and a B chain with 30 amino acids. There are two disulfide bonds joining these two chains together (Fig. 4.9). The molecular weight is around 6.8 kDa. Insulin regulates the blood glucose level to within a narrow range of 3.5–8.0 mmol/L of blood.

Insulin was originally (since the 1930s) obtained from porcine and bovine extracts. Bovine insulin differs from human insulin by three amino acids, and it can elicit an antibody response that reduces its effectiveness. Porcine insulin, however, differs in only one amino acid. An enzymatic process can yield insulin identical to the human form. Currently, insulin is produced via the rDNA process; it was the first recombinant biopharmaceutical approved by the FDA in 1982. The recombinant insulin removes the reliance on animal sources of insulin and ensures that reliable and consistent insulin is manufactured under controlled manufacturing processes. A description of diabetes mellitus and insulin is presented in Exhibit 4.13.

In January 2006, the FDA approved the inhalable insulin Exubera for type I and type II diabetes. Details are presented in Exhibit 4.14.

4.5.2   Human Growth Hormone

Human growth hormone (hGH) is a polypeptide with 191 amino acids. It is secreted by the pituitary gland. This hormone stimulates the production of insulin-like growth factor-1 (IGF-1) from the liver. Most of the positive effects of hGH are mediated by the IGF-1 system, which also includes specific binding proteins.

A major function of hGH is the promotion of anabolic activity, that is, bone and tissue growth due to increase in metabolic processes. Other biological effects of hGH are stimulation of protein synthesis, elevation of blood glucose level, and improvement of liver function.

Exhibit 4.13   Diabetes Mellitus and Insulin

Diabetes mellitus occurs when the human body does not produce enough insulin. This form of diabetes is called insulin-dependent diabetes mellitus (IDDM, or juvenile diabetes, or type I diabetes). IDDM is an autoimmune disease (see Exhibit 4.7) in which the β cells are targeted by the body's own immune system and progressively destroyed. Once destroyed, they are unable to produce insulin.

Production of insulin is triggered when there is a rise in blood sugar, for example, after a meal. Most of our body cells have insulin receptors, which bind to the insulin secreted. When the insulin binds to the receptor, other receptors on the cell are activated to absorb sugar (glucose) from the bloodstream into the cell.

When there is insufficient insulin to bind to receptors, the cells are starved because sugar cannot reach the interior to provide energy for vital biological processes. Patients with IDDM become unwell when this happens. They depend on insulin injection for survival.

Another form of diabetes is non-insulin-dependent diabetes mellitus (NIDDM, or adult diabetes, or type II diabetes). In this case, insulin is produced and a normal insulin level is detected in blood. But for various reasons its effect is reduced. This may be caused by a reduced number of insulin receptors on cells, or reduced effectiveness in binding to these receptors. The cause is complex and may involve genetic make-up, changes in lifestyle, nutritional habits, and environmental factors.

Exhibit 4.14   Inhalable Insulin

Exubera is an inhaled insulin. It represents a major step forward since the first insulin injection was approved in the 1920s. The insulin particles are formulated to a certain micron size for deep lung delivery. An inhaler is used to achieve the delivery. The large surface area of the thin alveolar walls in the lungs allows for fast absorption of the insulin into the bloodstream.

A number of side effects, however, have been reported, such as coughing, shortness of breath, sore throat, and dry mouth. There is also concern over the prolonged delivery of insulin into the lungs.

It was expected to be a US$2 billion drug but in October 2007, Exubera was withdrawn from the market due to low demand by patients. The problem stems from the higher cost of the medication, the cumbersome inhaler, confusing dosage calculations, and possible effects on pulmonary function.

Sources: 1. Walsh J. *Insulin—Diabetes Mall*. http://www.diabetesnet.com/diabetes_treatments/insulin_inhaled.php [accessed August 21, 2007]. (2) Exubera Official Site. http://www.exubera.com/content/con_index.jsp?printFriendl=true [accessed November 13, 2007].

Overproduction of hGH during puberty leads to gigantism, and deficiency during this period results in dwarfism. The current main therapeutic use of hGH is for the treatment of short stature. As discussed in Exhibit 4.11, hGH is used by athletes illegally to enhance their performance. This hormone is also sold without prescription with claims of improvement to body composition (lean body mass, fat mass, fluid volume), bone strength, immune function, youthful vigor, and general well-being.

4.6 GENE THERAPY

In essence, gene therapy can be described as "good genes for bad genes." The technology involves the transfer of normal, functional genes to replace genetically faulty ones so that proper control of protein expression and biochemical processes can take place. Although this seems straightforward, the major question is: How do we get the normal genes to the intended location?

This question revolves around the delivery tools for the genes. The transport system or vehicles used are called vectors (gene carriers). There are two basic gene therapy techniques: *in vitro* and *in situ* methods.

For the *in vitro* method, some of the patient's tissues, which have the genetic fault, are removed. Cells are selected from these tissues and normal genes are loaded into the cells with vectors. The modified cells are then returned to the patient to correct the genetic fault. With the *in situ* method, genes encapsulated by the vectors are injected directly into the tissues to be treated. Figure 4.10 shows the basis for gene therapy.

Whether using the *in vitro* or *in situ* method, genes are first loaded onto the vectors. A number of vectors are used (Table 4.5).

The most common vectors used today are viruses, with retroviruses being the preferred candidates, as they are efficient vectors for entering humans and

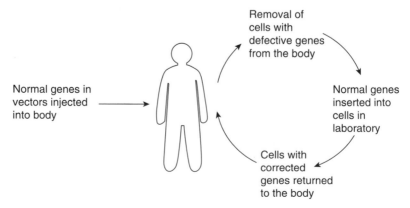

Figure 4.10 Basis of gene therapy.

TABLE 4.5   Vectors for Gene Therapy

Retrovirus	Adenovirus	Adeno-Associated Virus	Liposomes	Naked DNA
		Advantages		
Integrates genes to host chromosomes, chance of long-term stability	Large capacity for carrying foreign genes	Integrates genes to host chromosomes	Do not have viral genes, so do not cause disease	Do not have viral genes, so do not cause disease
		Disadvantages		
Integration is random, mostly on dividing cells	Transient function of genes	Small capacity for foreign genes	Less efficient than viruses	Inefficient gene transfer

Source: Friedman T. Overcoming the obstacles to gene therapy, *Scientific American* June:96–101 (1997).

replicating their genes within human cells. Scientists take advantage of this biological function. Disease-causing genes from the viruses are removed, and the therapeutic genes are inserted. Retroviruses carrying the desired therapeutic genes are placed in the patient's body. When the viruses invade the cells, they "infect" these host cells and the therapeutic genes are added to the host DNA. In this way, the new genes function and take over from the original faulty genes.

Theoretically, this appears to be a fitting solution to gene problems. However, there are problems, such as immune and inflammation responses, toxicity, and means to target the intended cells. Nonviral vectors may overcome the problems with viral delivery agents. Lipids, in the form of liposomes and other lipid complexes, are being studied. Injection of DNA directly into a patient's muscle cells is another avenue being researched.

Another hurdle surrounding gene therapy is the identification of genes causing the disease. Effective cures can only arise when there is a good understanding of the roles of particular genes in diseases. Some of the diseases to which gene therapy may be applicable are cancer, hemophilia, sickle cell anemia, cystic fibrosis, insulin-dependent diabetes mellitus (see Exhibit 4.13), emphysema, Alzheimer's disease, Huntington's disease, and severe combined immune deficiency (SCID). To date, the FDA has not approved any gene therapy product. Numerous clinical trials are in progress (Exhibit 4.15). Gene therapy also involves ethical considerations. This issue is discussed in Section 11.6.

Exhibit 4.15   Gene Therapy Trials

The first gene therapy trial was conducted in September 1990. A 4-year-old girl with SCID (an inherited immune disorder disease, otherwise known as the "bubble boy" syndrome) was treated in Cleveland, Ohio. She is doing well some 10 years after the treatment. A second girl with the same disorder underwent gene therapy and she too continues to do well.

These are the successes; there are many failures as well. More than 400 gene therapy clinical trials have been conducted, mainly on cancer, but not many cases worked. In 1999, an 18-year-old boy in Pennsylvania unexpectedly died from a reaction to gene therapy when he was treated for a metabolic disease. This trial raised many issues, and many trials with discrepancies and unreported adverse events were suspended by the FDA. The FDA has since introduced tighter controls for gene therapy trials.

In April 2007 British doctors performed gene therapy on some young adults with a type of childhood blindness, Leber's congenital amaurosis. The blindness is caused by a faulty RPE65 gene. The doctors inserted normal *RPE65* genes into the retina using a virus vector.

On July 24, 2007 the US Food and Drug Administration (FDA) was informed by Targeted Genetics Corporation of Seattle about the death of a patient who received an investigational gene therapy product in a clinical trial for the treatment of active inflammatory arthritis. The product that was being studied uses a recombinant adeno-associated virus (AAV) derived vector to deliver the gene to the TNF receptor, with the intent to inhibit a key mediator of inflammation. In the study, the gene therapy was administered into the joint affected by the disease to reduce inflammation and disease. More than 100 subjects were enrolled in the trial. One patient unfortunately died following a second injection of the product. The trial has since been put on hold pending further investigation by the FDA and the company.

4.7   STEM CELLS AND CELL THERAPY

Stem cells are divided into three different categories: totipotent, pluripotent, and multipotent. A description of the genesis of stem cells is shown in Fig. 4.11.

Totipotent stem cells are obtained from embryos that are less than 5 days old. These cells have the full potential to develop into another individual and every cell type.

After about 5 days and several cycles of cell division, the totipotent cells form a hollow sphere of cells called a blastocyst. The blastocyst has an outer layer of cells surrounding clusters of cells. Those cells on the outside continue

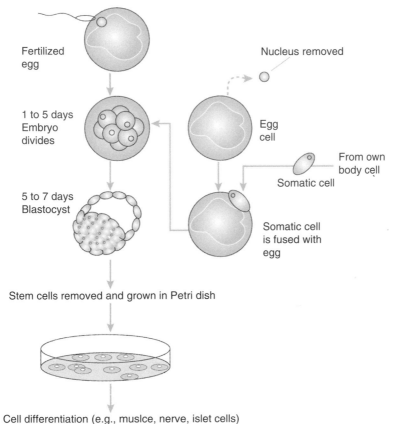

Fertilized
egg

Nucleus removed

1 to 5 days
Embryo
divides

Egg
cell

From own
body cell

Somatic cell

5 to 7 days
Blastocyst

Somatic cell
is fused with
egg

Stem cells removed and grown in Petri dish

Cell differentiation (e.g., muslce, nerve, islet cells)

Figure 4.11   Genesis of different stem cells.

to divide and grow into the placenta and supporting tissues. The clusters of cells on the inside divide and form virtually all the cell types, except the placenta and supporting tissues, which give rise to a human being. These are the pluripotent stem cells, and they give rise to many different types of cells, but not a new individual.

Pluripotent cells continue to develop, differentiate, and specialize into different cells. They become the specialized stem cells, such as blood, skin, and nerve stem cells. These differentiated stem cells are multipotent; that is, they have the potential to produce specialized cells. For example, blood stem cells in bone marrow produce red blood cells, white blood cells, and platelets, but not other types of cells.

There are two general avenues for stem cell research: pluripotent and multipotent stem cells. Pluripotent stem cells are obtained by two methods. One method is to harvest the clusters of cells from the blastocysts of human embryos. Another method is the isolation of pluripotent cells from fetuses in terminated pregnancies. Multipotent stem cells are derived from umbilical cords or adult stem cells. However, because of the specialization of these cells, their potential to develop into a myriad of different cells is limited.

A burning issue is the ethics of obtaining pluripotent stem cells from embryos and fetuses. The US government has acted on this issue and declared that federal funds for stem cell research have to meet certain criteria. It requires that funding will only be provided to research with stem cells obtained before August 9, 2001, as a cut-off date to limit research to preexisting stem cells. Refer to Section 11.7 for an ethical debate on stem cells.

The potential contribution of stem cells to medical treatment lies in their capability to differentiate and grow into normal, healthy cells. Using pluripotent stem cells, scientists are devising means to culture them in the laboratories and coax them to grow into various specialized cells. Rather than gene therapy, with stem cells we have the potential of cell therapy to repair our diseased tissues and organs. This will circumvent the lack of donor organs. Stem cells also provide the possibility for healthy cells to cure disabilities such as strokes, Parkinson's disease, and diabetes.

A drawback for stem cell therapy is the problem of cell rejection due to the host's immune system recognizing the cells as foreign. This rejection issue has to be overcome to ensure stem cell therapy is a viable treatment. Recently, French scientists reported on research progress in stem cell transplants for curing children with sickle cell anemia. A mix of antirejection drugs was used to suppress rejection of the new stem cells.

Although research into stem cells is new, the use of stem cells for therapy has been with us for some time. Most of us are familiar with bone marrow transplant for patients with leukemia. This procedure involves finding a matching donor to harvest bone marrow stem cells and transfuse them into the patient with leukemia (see Exhibit 4.16 for details).

Another cell therapy method includes the excision of cells from the body. These cells are then modified and returned to the host body. Provenge, a cancer vaccine using cell therapy, has completed Phase III trial and is being reviewed by the FDA. The technique for this therapy is given in Exhibit 4.17.

4.8   CASE STUDY #4

Herceptin and Tykerb (See Also Exhibit 8.3)*

Studies found that in about 25–30% of early stage cancer, the cancer cells overexpress the HER2/neu receptors due to *HER2/neu* gene amplification. The name HER2/neu is derived from its structural similarity to human epidermal growth factor, HER1, and neu is a derivative of the oncogene from a neuroglioblastoma cell line.

Sources: (1) Jarvis LM. Battling breast cancer, *Chemical & Engineering News* 84:21–27 (2007). (2) FDA News, March 2007. http://www.fda.gov/bbs/topics/NEWS/2007/NEW01586.html, http://www.fda.gov/cder/foi/label/2007/022059lbl.pdf [accessed September 20, 2007]. (3) Moy B, et al. Lapatinib, *Nature Reviews Drug Discovery* 6:431–432 (2007).

Exhibit 4.16   Bone Marrow Transplant

Bone marrow is the spongy tissue inside the cavities of our bones. Bone marrow stem cells grow and divide into the various types of blood cells: white blood cells (leukocytes) that fight infection, red blood cells (erythrocytes) that transport oxygen, and platelets that are the agents for clotting.

Patients with leukemia have a condition in which the stem cells in the bone marrow malfunction and produce an excessive number of immature white blood cells, which interfere with normal blood cell production.

The aim of a bone marrow transplant is to replace the abnormal bone marrow stem cells with healthy stem cells from a donor. Healthy stem cells are normally harvested using a syringe to withdraw bone marrow from the rear hip bone of the donor. They are then infused into the patient via a catheter in the chest area. Before the infusion, the patient receives chemotherapy or radiotherapy to destroy the diseased bone marrow stem cells so that the infused stem cells have a chance to grow free of complications from diseased cells.

There are a number of terms used in the transplant procedure:

Allogeneic Transplant: The person giving the bone marrow or stem cells is a genetically matched family member (usually a brother or sister).

Unrelated Allogeneic Transplant: The person donating marrow is unrelated to the patient.

Syngeneic Transplant: The person donating the bone marrow or stem cell is an identical twin.

Autologous Transplant: The patient donates his or her own bone marrow or stem cells before treatment, for reinfusion later. This happens when a patient is receiving radiotherapy or chemotherapy in such a high dose that the bone marrow is destroyed. The bone marrow stem cells collected previously are reinjected into the patient to reinforce the immune system.

Source: Bone Marrow and Stem Cells Transplant Support, *Bone Marrow Transplant Overview*. http://www.bmtsupport.ie/bmtoverview.html [accessed September 2, 2002].

Trastuzumab (Herceptin): Herceptin is a humanized MAb that targets the HER2/neu receptor. It is an IgG_1 kappa antibody with a human framework and murine complementarity-determining regions (4D5), which bind to HER2/neu growth factor receptor. Women are selected for Herceptin treatment based on immunohistochemistry (IHC) and fluorescence *in situ* hybridization (FISH) diagnostic tests, which check for HER2 overexpression and gene

Exhibit 4.17   Cell Therapy—Provenge

Provenge is a cancer vaccine using cell therapy technique. Dendritic cells are removed from patients. These cells are treated with the prostate-specific antigen prostatic acid phosphatase (PAP), which is present in 95% of prostate cancer cases. The activated dendritic cells are returned to the patients and they stimulate the T cells to destroy cancer cells expressing the PAP, thus treating the tumor.

Source: Jones D. Cancer vaccines on the horizon, *Nature Reviews Drug Discovery* 6:333–334 (2007).

amplification, respectively. Herceptin is supplied in sterile lyophilized form with 440 mg Trastuzumab, 400 mg α,α-trehalose dehydrate, 9.9 mg L-histidine HCl, 6.4 mg L-histidine, and 1.8 mg polysorbate 20.

Herceptin attaches to the HER2/neu receptor and activates the complement system (a series of serum and cell-associated proteins involved in immune response) to destroy those cells expressing such receptors. Through this action, Herceptin disrupts the signaling pathway for breast cancer cell proliferation (refer to diagram below).

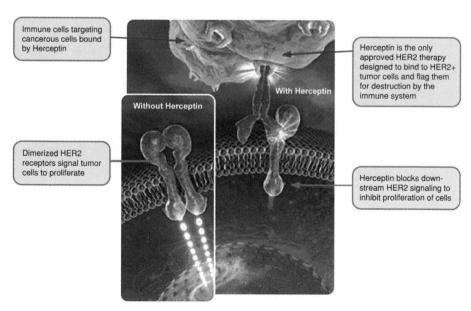

Mechanism of action of herceptin. (*Source*: Genentech website, http://www.gene.com/gene/products/information/oncology/herceptin/factsheet.html [accessed January 23, 2008].)

Lapatinib (Tykerb): Tykerb is a small molecule drug that acts as a kinase inhibitor. It was approved in 2007 by the FDA as a combination therapy with capectabine (Xeloda) for the treatment of advanced HER2 positive breast cancer. The chemical name of the active ingredient is *N*-(3-chloro-4-{[(3-fluorophenyl)methyl]oxy}phenyl)-6-[5-({[2-(methylsulfonyl)ethyl]amino}methyl)-2-furanyl]-4-quinazolinamine-bis(4-methylbezenesulfonate)monohydrate, $C_{29}H_{26}ClFN_4O_4S(C_7H_8O_3S)_2H_2O$, with a molecular weight of 943.5 and the following chemical structure:

It is a yellow solid with a solubility in water of 0.007 mg/mL. The Tykerb tablet contains 405 mg of the active ingredient, and the inactives are magnesium stearate, microcrystalline cellulose, povidone, and sodium starch glycolate.

Tykerb attaches to the HER1 and HER2 receptors and block the tyrosine kinase reactions, hence turning off the growth of breast cancer cells. By widening the targets to beyond HER2 which Herceptin attaches, Tykerb may help in those cases where Herceptin has failed. It is postulated that multikinase inhibitors such as Tykerb may be able to interfere with more biochemical signaling pathways to block the functions of the HER family type of receptors. Furthermore Tykerb may also help in cases where the metastases had spread from the breast to the brain, and Herceptin may not be able to cross the blood-brain barrier (refer to Exhibit 5.4) with its large molecular size.

4.9 SUMMARY OF IMPORTANT POINTS

1. Biopharmaceuticals are mainly protein-based molecules, which are copies of natural biological compounds. The aim of these biopharmaceuticals is to modify or alter the undesirable biological responses in our body in the disease states. Biopharmaceuticals have molecular weights of tens of thousands of daltons, unlike the small molecule drugs of mostly less than 500 Da.

2. Vaccines are derived from whole or fragments of pathogens, either through inactivation or attenuation of the pathogen or by generating the

requisite molecules using recombinant technology. Adjuvants are added to enhance the efficacy of the vaccines.

3. Antibodies are proteins that mimic the natural antibodies, especially designed to interact with endogenous or exogenous protein molecules. They are Y-shaped molecules with the tips of the two arms binding to antigens while the stem part is used to elicit the immune response to destroy the antigens. The variability of the tips means that they can bind to many different antigens. Parts of the tip can be replaced by toxins or radioactive elements to help destroy the antigens.

4. Cytokines such as interferons are used for the treatment of hepatitis, cancer, and lymphoma; interleukins are used to enhance immune response and growth factors are used for the treatment of anemia and regulation of tumor angiogenesis.

5. Hormones are intercellular messengers: insulin is used to treat diabetes, while growth hormone promotes bone and tissue growth.

6. Gene therapy is a techniques to deliver genes into the body to replace faulty genes or insert new genes if they are missing in the body. To date, the FDA has not approved any gene therapy product.

7. Stem cells and cell therapy is the use of pluripotent and multipotent cells to generate healthy cells and tissues to replace the faulty ones in disease conditions. The main ethical questions are the source of the cells and the possibility of cloning humans.

4.10 REVIEW QUESTIONS

1. Describe the different types of vaccines: give examples of those produced by traditional methods and current techniques. Explain how the use of adjuvants can help to improve the efficacy of vaccines.

2. Discuss the structure and naming convention for the influenza virus. Provide reasons for the variations in the yearly compositions of the influenza vaccines.

3. What are the characteristics that make the influenza virus, for example, avian influenza, a potential pandemic agent?

4. Compare and contrast the different types of antibody immunoglobulins. Provide a detailed description of the structure of the IgG antibody with particular reference to how it binds to antigens.

5. Explain why the humanization of antibodies is important and, through the use of examples, demonstrate the progress made to the modification of antibodies as technology advances.

6. Provide examples for conjugating antibodies with toxins and radioactive elements.

7. Describe how the human immune system works, focusing on the B and T cells.

8. What is the mechanism of action for drugs that target growth factors? Describe the mechanism of action for some of the latest drugs, such as Avastin.

9. Distinguish the technologies based on gene therapy and cell therapy. Describe the use of vectors for gene delivery. Describe how cell rejection can be overcome in cell therapy.

10. Demonstrate by citing examples the current treatment of breast cancer using small and large molecule drugs.

4.11 BRIEF ANSWERS AND EXPLANATIONS

1. Refer to Sections 4.2.1 and 4.2.2 on traditional and new vaccines. The mechanisms of vaccine efficacy enhancement are explained in Section 4.2.3.

2. Refer to Exhibits 3.7 and 4.2 for an explanation of the roles of hemagglutinin and neuraminidase, followed by the nomenclature for classifying influenza virus and the procedure that the FDA and WHO recommend for the preparation of multivalent vaccines.

3. Avian influenza is extremely deadly, with a 60% fatality rate for infected human cases to date. The virus may become even more deadly through the process of reassortment and gradual adaptive mutation.

4. Refer to Section 4.3 and Fig. 4.1b for a comparison and contrast of the structures of IgG, IgD, IgE, IgA, and IgM, and an explanation of the heavy and light chain structure of IgG and the complementarity-determining regions that bind to antigens.

5. The problem is the neutralizing or allergic reactions caused by the production of human anti-mouse antibodies because the body treats the MAbs as foreign (refer to Section 4.3.4). The humanization of antibodies, through chimeric to humanized and full human types, helps to address this problem.

6. Refer to Section 4.3.5.

7. Refer to Exhibit 4.7.

8. Section 4.4.2 shows the actions for EPO, CSFs, and VEGFs. The anti-angiogenesis mechanism of Avastin is explained in Exhibit 4.12.

9. Refer to Sections 4.6 and 4.7. The vectors for gene therapy are tabulated in Table 4.5. In cell therapy, antirejection drugs are used to suppress the rejection of transplanted cells.

10. Herceptin and Tykerb are examples of large and small molecule drugs prescribed for the treatment of breast cancer. Refer to Section 4.8 for details.

4.12 FURTHER READING

Aagaard L, Rossi JJ. RNAi therapeutics: principles, prospects and challenges, *Advanced Drug Delivery Reviews* 59:75–86 (2007).

Austen FA, Burakoff SJ, Rosen FS, Strom TB. *Therapeutic Immunology*, 2nd ed., Blackwell Science, Malden, 2001.

Atun RA, Sheridan D, eds. *Innovation in the Biopharmaceutical Industry*, World Scientific, Singapore, 2007.

Dubel S. Recombinant therapeutic antibodies, *Applied Microbiology and Biotechnology* 74:723–729 (2007).

Ezzell C. Magic bullets fly again, *Scientific American* October:34–41 (2001).

Felgner PL. Nonviral strategies for gene therapy, *Scientific American* June:103–106 (1997).

Funaro A, Horenstein AL, Santoro P, et al. Monoclonal antibodies and therapy of human cancers, *Biotechnology Advances* 18:385–401 (2000).

George AJT, Urch CE, eds. *Diagnostic and Therapeutic Antibodies*, Humana Press, Totowa, NJ, 2000.

Green BA, Baker SM. Recent advances and novel strategies in vaccine development, *Current Opinion in Microbiology* 5:483–488 (2002).

Groves MJ, ed. *Pharmaceutical Biotechnology*, 2nd ed., Taylor & Francis, Philadelphia, 2006.

Hanly WC, Bennett BT, Artwohl JE. *Overview of Adjuvants*, Biologic Resources Laboratory, University of Illinois, Chicago. http://www.nal.usda.gov/awic/pubs/antibody/overview.htm [accessed November 21, 2002].

Haseltine WA. Beyond chicken soup, *Scientific American* November:56–63 (2001).

Ho RJY, Gibaldi M. *Biotechnology and Biopharmaceuticals: Transforming Proteins and Genes into Drugs*, Wiley-Liss, Hoboken, NJ, 2003.

Hudson PJ. Recombinant antibody constructs in cancer therapy, *Current Opinion in Immunology* 11:548–557 (1999).

Laver WG, Bischofberger N, Webster RG. Disarming flu viruses, *Scientific American* January:78–87 (1999).

Lubiniecki AS. Monoclonal antibody products: achievement and prospects, *Bioprocessing Journal* 2(Mar/Apr):21–26 (2003).

Mandavilli A. Gene-therapy trials for hemophilia make comeback, *BioMedNet*, October 1, 2002. http://news.bmn.com/news/story?day=021001&story=1 [accessed November 27, 2002].

McKay D. Alzheimer's vaccine? *Trends in Biotechnology* 19:379–380 (2001).

Medline Plus. *Malaria.* http://www.nlm.nih.gov/medlineplus/malaria.html [accessed November 29, 2007].

National Institutes of Health. *Stem Cell Research*, NIH, Bethesda, MD, 2000. http://www.nih.gov/ [accessed November 25, 2002].

National Institutes of Health. *Stem Cells: A Primer*, NIH, Bethesda, MD, 2002. http://www.nih.gov/ [accessed November 25, 2002].

Penichet ML, Morrison SL. Design and engineering human forms of monoclonal antibodies, *Drug Development Research* 61:121–136 (2004).

Pizzi RA. The science and politics of stem cells, *Modern Drug Discovery* 5:32–37 (2002).

Robinson MK, Weiner LM, Adams GP. Improving monoclonal antibodies for cancer therapy, *Drug Development Research* 61:172–187 (2004).

Sayers JR. Acres of antibodies: the future of recombinant biomolecule production? *Trends in Biotechnology* 19:429–430 (2001).

Scheibner V. Adjuvants, preservatives and tissue fixatives in vaccines, *Nexus* 8 (Dec & Feb) (2000–2001). http://www.whale.to/vaccine/adjuvants.html [accessed November 21, 2002].

Sinclair M. Surface vaccine combo, *Nature Biotechnology* 18:586 (2000).

Singh M, O'Hagan D. Advances in vaccine adjuvants, *Nature Biotechnology* 17:1075–1081 (1999).

Walsh G. *Biopharmaceuticals: Biochemistry and Biotechnology*, Wiley, Chichester, UK, 1998.

Walsh G, Jefferis R. Post-translational modifications in the context of the therapeutic proteins, *Nature Biotechnology* 24:1241–1252 (2006).

Wink M, ed. *An Introduction to Molecular Biotechnology*, Wiley-VCH, Weinheim, Germany, 2006.

CHAPTER 5

DRUG DEVELOPMENT AND PRECLINICAL STUDIES

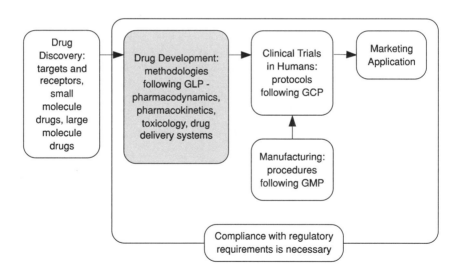

5.1	Introduction	137
5.2	Pharmacodynamics	139
5.3	Pharmacokinetics	143
5.4	Toxicology	155
5.5	Animal Tests, *In Vitro* Assays, and *In Silico* Methods	158

5.6	Formulations and Delivery Systems	161
5.7	Nanotechnology	168
5.8	Case Study #5	169
5.9	Summary of Important Points	171
5.10	Review Questions	172
5.11	Brief Answers and Explanations	173
5.12	Further Reading	174

5.1 INTRODUCTION

In earlier chapters, we discussed the discovery of new drugs. After a lead compound has been identified, it is subjected to a development process to optimize its properties. The development process includes pharmacological studies of the lead compound and its effects on toxicity, carcinogenicity, mutagenicity, and reproductive development. These data are important for determining the safety and effectiveness of the lead compound as a potential drug.

An ideal drug is potent, efficacious, and specific; that is, it must have strong effects on a specific targeted biological pathway and minimal effects on all other pathways, to reduce side effects. In reality, no drugs are perfectly effective and absolutely safe. The aim of pharmacological studies is to obtain data on the safety and effectiveness of the lead compound. Many iterations of optimization of the lead compound may be necessary to yield a potential drug candidate for clinical trial.

The potency, efficacy, and safety of a drug depend on the chemical and structural specificity of drug–target interaction. In pharmacology, we are concerned with pharmacodynamics (PD), pharmacokinetics (PK), and toxicity. In simplified terms, PD deals with the actions of the drug on the target, whereas PK is about the actions of the body on the drug. Toxicity information in preclinical studies provides us with confidence in the safety aspect of the potential drug. These data for PD, PK, and toxicity enable the dose and dosing regimen to be set for the clinical trials.

Although pharmaceutical firms are increasingly using *in vitro* methods to evaluate pharmacological responses, some aspects of pharmacological development have no alternatives but to use *in vivo* tests in animals to study the effects of a potential drug in living systems. Pharmacological and toxicity studies using animals are regulated under Good Laboratory Practice with strict guidelines, requiring scientists to follow established protocols. Readers are referred to FDA 21 CFR Part 58 *Good Laboratory Practice for Nonclinical Laboratory Studies*. This regulation details the requirements for the conduct of nonclinical laboratory studies intended to support applications for clinical trials and marketing approvals (Investigational New Drug (IND), New Drug Application (NDA), and Biologics License Application (BLA); see Chapter 8). The contents list for this guideline is presented in Exhibit 5.1.

Exhibit 5.1   FDA 21 CFR Part 58, *Good Laboratory Practice for Nonclinical Laboratory Studies*: **Table of Contents**

Scope
Definitions
Applicability to studies performed under grants and contracts
Inspection of a testing facility
Personnel
Testing facility management
Study director
Quality assurance unit
General
Animal care facilities
Facilities for handling test and control articles
Laboratory operation areas
Specimen and data storage facilities
Equipment design
Maintenance and calibration of equipment
Standard operating procedures
Reagents and solutions
Animal care
Test and control article characterization
Test and control article handling
Mixtures of articles with carriers
Protocol
Conduct of a nonclinical laboratory study
Reporting of nonclinical laboratory study results
Storage and retrieval of records and data
Retention of records
Purpose
Grounds for disqualification
Notice of and opportunity for hearing on proposed disqualification
Final order on disqualification
Actions upon disqualification
Public disclosure of information regarding disqualification
Alternative or additional actions to disqualification
Suspension or termination of a testing facility by a sponsor
Reinstatement of a disqualified testing facility

Examples of some of these requirements are the following:

- Personnel must have the education, training, and experience to conduct the nonclinical studies.
- A quality assurance unit should be set up.
- Materials for the studies must be appropriately tested for identity, strength, purity, stability, and uniformity.
- Appropriate personnel, resources, facilities, equipment, materials, and methodologies must be available.
- The studies must be conducted under specifically designed protocols with an appropriate quality system established to handle data, deviations, and reporting.
- Animals must be isolated, their health status checked, and they must be given the appropriate welfare.
- Nonclinical laboratory studies must be conducted in accordance with the protocols.

Drug development also extends to formulation and delivery. Most drugs that are administered to patients contain more than just the active pharmaceutical ingredients (APIs, the drug molecules that interact with the receptors or enzymes). Other chemical components are often added to improve manufacturing processing or the stability and bioavailability of drugs, such as the use of adjuvants for vaccines. Effective delivery of drugs to target sites is an important factor to optimize efficacy and reduce side effects. The development process also includes the design and development of new manufacturing and testing methodologies for cost-effective production of drugs in compliance with regulatory requirements. Drugs are manufactured under Good Manufacturing Practice, which is discussed in Chapters 9 and 10.

5.2 PHARMACODYNAMICS

The chemical and structural aspects of pharmacodynamics (PD) are discussed in Chapters 2–4, where we considered drug–target interactions. When a drug binds to a target, it may regulate the receptor as an agonist or antagonist, or act as an inducer or inhibitor in the case of an enzyme. The lock and key chemical and structural interaction is *a priori* to achieving a potent and safe drug. In this chapter, we focus on the quantitative mathematical relationships of drug–target interactions, in addition to the chemical and structural aspects covered in Chapters 2–4.

PD is the study to determine dose–response effects. We are interested in finding out the effects of a drug on some particular response, such as heart rate, enzyme levels, antibody production, or muscle relaxation or contraction. When a drug binds to a receptor, the ensuing response is complex. The following example is an idealized case, which illustrates the drug–receptor interaction:

$$D + R \rightleftarrows D^*R \Longrightarrow Response$$

where D is the drug, R is the receptor, and D*R is the drug–receptor complex.

The response may be local or via a signal transduction process. The rate for the forward reaction of drug binding to receptor is proportional to the concentrations of both the drug and target. Conversely, the rate for the reverse reaction (i.e., dissociation of the drug–receptor complex) is proportional to the concentration of the drug–receptor complex. At equilibrium, both forward and reverse reactions are equal. Mathematically, we have

$$k_1[D][R] = k_{-1}[D^*R] \tag{5.1}$$

where k_1 is the forward reaction rate constant, k_{-1} is the reverse reaction rate constant, and [D], [R], and [D*R] are the concentrations of the drug, receptor, and drug–receptor complex, respectively.

Rearranging Eq. (5.1), we obtain

$$\frac{[D][R]}{[D^*R]} = \frac{k_{-1}}{k_1} = K_D \tag{5.2}$$

where K_D is the equilibrium dissociation constant.

When half the receptors are bound, we have [R] = [D*R]. Substituting into Eq. (5.2), K_D is equal to [D]. This means that K_D is the concentration of the drug that, at equilibrium, will bind to half the number of receptors.

If we consider all the available receptors as 100% and [D*R] are the occupied receptors with drug at the binding sites, then [R], which is the percentage of free, unoccupied receptors, can be substituted with 100 − [D*R]. Equation (5.2) can be rewritten as

$$[D] = \frac{K_D[D^*R]}{100 - [D^*R]} \tag{5.3}$$

Equation (5.3) is a hyperbolic function showing the relationship between dose of drug, [D], and its effects resulting from drug–receptor interaction, [D*R]. A graphical representation of this dose–effect, or dose–response curve, is shown in Fig. 5.1. The graph shows that, at low doses, the effects are approximately linear in proportion to the doses. However, as the dose increases, there is gradually a diminishing return in effects. A maximum is reached and at this point all available receptors are bound with drug molecules. Further increase in dose does not generate any increase in effects. The point E_{max} is the maximum effect, and EC_{50} is the concentration of the drug that produces 50% of the maximum effect.

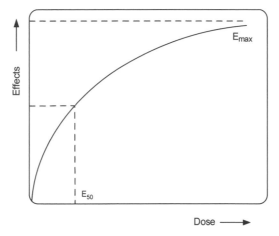

Figure 5.1   Dose–effect curve.

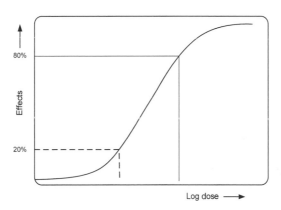

Figure 5.2   Dose–effect curve with logarithmic scale for dose.

Very often, the dose–effect curve is redrawn using a logarithmic scale for the dose. This gives rise to a sigmoid curve, as shown in Fig. 5.2. It is a mathematical transformation, which shows an approximate linear portion for the 20–80% maximal effect scale, which is usually the dose level for a therapeutic drug. Doses above 80% provide very little increase in therapeutic effects but with a concomitant rise in the risk of adverse reactions.

Scientists also study the potency, effectiveness, safety margin, and therapeutic index of a drug. These terms are described next with reference to Figs. 5.3 and 5.4.

Potency: This is the dose required to generate an effect. A potent drug elicits an effect at a low dose.

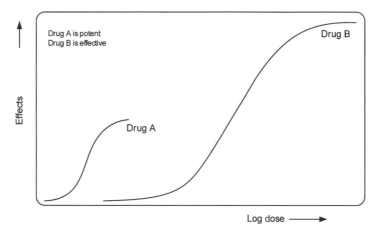

Figure 5.3   Potency and effectiveness.

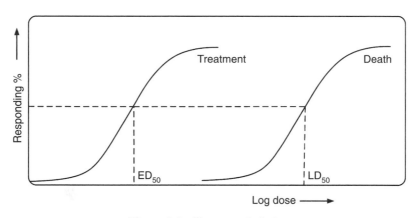

Figure 5.4   Therapeutic index.

Effectiveness: This is the intensity of the effect or response. It is a measure of the affinity of the drug for the receptor. An effective drug is one that can achieve effects in the vicinity of E_{max}.

Therapeutic Index: The index is given by the ratio

$$\frac{LD_{50}}{ED_{50}} \tag{5.4}$$

where LD_{50} is the lethal dose for 50% of the population, and ED_{50} is the effective dose for 50% of the population.

When there is a large difference in dose between ED_{50} and LD_{50} the therapeutic index of the drug is high.

Safety Margin: This is the separation of two doses: one that produces therapeutic effects and one that elicits adverse reaction. The standard safety margin (SSM) is given by

$$SSM = \frac{LD_1 - ED_{99}}{ED_{99}} \times 100 \qquad (5.5)$$

where LD_1 is the lethal dose for 1% of the population, and ED_{99} is the effective dose for 99% of the population.

A large safety margin is achieved when there is a significant difference between the ED_{99} and LD_1 doses.

5.3 PHARMACOKINETICS

For a drug to interact with a target, it has to be present in sufficient concentration in the fluid medium surrounding the cells with receptors. Pharmacokinetics (PK) is the study of the kinetics of absorption, distribution, metabolism, and excretion (ADME) of drugs. It analyzes the way the human body deals with a drug after it has been administered, and the transportation of the drug to the specific site for drug–receptor interaction. For example, a person has a headache and takes an aspirin to abate the pain. How does the aspirin travel from our mouth to reach the site in the brain where the headache is and act to reduce the pain?

There are several ways to administer a drug. They include the following:

- Intravenous
- Oral
- Buccal
- Sublingual
- Rectal
- Subcutaneous
- Intramuscular
- Transdermal
- Topical
- Inhalational

With the exception of intravenous administration, where a drug is injected directly into the bloodstream, all the routes of administration require the drug to be absorbed before it can enter the bloodstream for distribution to target sites. Metabolism may precede distribution to the site of action, for example, in the case of oral administration. The human body also has a clearance process to eliminate drugs through excretion. We will now consider absorption, distribution, metabolism, and excretion with reference to Fig. 5.5.

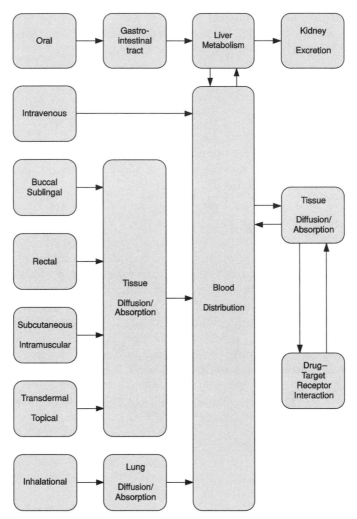

Figure 5.5   Schematic representation of drug absorption, distribution, metabolism, and excretion.

5.3.1   Transport Mechanism

Except for intravenous injection, drug molecules have to cross cell membranes to reach target sites. There are four basic transport mechanisms:

- Passive diffusion
- Facilitated diffusion
- Active transport
- Pinocytosis

Passive Diffusion: Diffusion is the random movement of molecules in fluid. If a fluid is separated by a semipermeable membrane, more dissolved molecules will diffuse across the membrane from the higher concentration side to the lower concentration side than in the reverse direction. This process will continue until equilibrium is achieved, whereby both sides have the same concentration. When equilibrium is reached, there are equal numbers of molecules crossing the membrane in both directions.

Drug molecules are transported across cell membranes. Because of the lipid bilayer construction of the membrane (Appendix 2), nonpolar (lipid-soluble) molecules are able to diffuse and penetrate the cell membrane. Polar molecules, however, cannot penetrate the cell membrane readily via passive diffusion and rely on other transport mechanisms.

Lipid solubility determines the readiness of drug molecules to cross the gastrointestinal tract, blood–brain barrier, and other tissues. Molecular size is another factor that determines the diffusion of drugs across the membrane, with the smaller molecules able to diffuse more readily. Exhibit 5.2 describes the kinetics for diffusion of drug molecules across the cell membrane.

Facilitated Diffusion: Polar drug molecules have been observed to cross cell membranes. The transport mechanism is via carrier systems. Transmembrane carriers, such as proteins, are similar to receptors and bind to polar and nonpolar drug molecules. They facilitate the diffusion of drugs across the cell membrane. The facilitated diffusion rate is faster than passive diffusion and may be controlled by enzymes or hormones. Facilitated diffusion is from a region of high concentration to one of low concentration. However, these carriers, or transporters, may become saturated at high drug concentration. In this case, the transportation rate plateaus until the carriers are cleared of the drug in preparation for another cycle of transportation.

Active Transport: The active transport mechanism requires energy to drive the transportation of drugs against the concentration gradient, from low to high. The transportation rate is dependent on the availability of carriers and energy supply via a number of biological pathways.

Pinocytosis: Pinocytosis involves the engulfing of fluids by a cell. The process commences with the infolding of cell membrane around fluids containing the drug. The membrane then fuses and forms a vesicle with the fluid core. In this way, the drug is taken into the cell interior within the vesicle.

5.3.2   Absorption

Oral Administration: The oral route is the most common way of administering a drug. For a drug to be absorbed into the bloodstream, it has to be soluble in the fluids of our gastrointestinal tract. Drugs are often formulated with excipients (components other than the active drug) to improve manufacturing and dissolution processes (see Section 5.6).

Exhibit 5.2 Diffusion of Drugs

Most drugs are weak acids or bases. Under different pH conditions, they become ionized and cannot diffuse through the cell membrane. This ionization process is illustrated below:

$$\text{Weak acid:} \quad AH \leftrightarrow A^- + H^+$$

$$pK_a = pH + \log_{10} \frac{[AH]}{[A^-]} \tag{1}$$

$$\text{Weak base:} \quad BH^+ \leftrightarrow B + H^+$$

$$pK_a = pH + \log_{10} \frac{[BH^+]}{[B]} \tag{2}$$

AH and B are the un-ionized acid and base, respectively, and A^- and BH^+ are the ionized forms. The lipid solubility of AH and B are dependent on the chemical structure of the drugs. In most instances, they are of sufficient solubility to diffuse across the cell membranes. However, as the equations show, the pH environment affects the ionization of a drug. We illustrate this with aspirin (a weak acid drug, $pK_a = 3.5$) as an example and apply Eq. (1):

Blood: High pH (7.4) Environment	Stomach: Low pH (3.0) Environment
$3.5 = 7.4 + \log_{10} \dfrac{[AH]}{[A^-]}$	$3.5 = 3.0 + \log_{10} \dfrac{[AH]}{[A^-]}$
$\log_{10} \dfrac{[AH]}{[A^-]} = -3.9$	$\log_{10} \dfrac{[AH]}{[A^-]} = 0.5$
$\dfrac{[AH]}{[A^-]} = 0.00126$	$\dfrac{[AH]}{[A^-]} = 3.16$
The ionized form is dominant: therefore, less lipid soluble.	There is more of the un-ionized form: therefore, more lipid soluble.

A similar method is used to calculate the un-ionized to ionized forms for basic drugs using Eq. (2).

Source: Rang HP, Dale MM, Ritter JM, Gardner R. *Pharmacology*, 3rd ed., Churchill Livingstone, New York, 1995, p. 70. Adapted with permission.

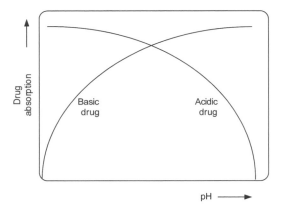

Figure 5.6 Absorption of drugs in different pH environments.

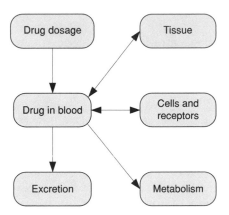

Figure 5.7 Process of drug in the body.

Our gastrointestinal tract is lined with epithelial cells, and drugs have to cross the cell membrane (see Exhibit 5.2). In the stomach, where pH is low, drugs that are weak acids are absorbed faster. In the intestine, where pH is high, weak basic drugs are absorbed preferentially. Figure 5.6 shows the absorption of drugs under different pH environments.

In reality, there is more than just passive diffusion at work for drugs to traverse the cell membrane. Most drugs are absorbed in the intestine. Often, if an oral drug is taken and a fast response is desired, the drug is taken on an empty stomach to ensure a quick passage through the stomach for absorption in the intestine to take place.

Drugs absorbed through the gastrointestinal tract pass into the hepatic portal vein, which drains into the liver. The liver metabolizes the drug, which leads to reduction in the availability of the drug for interaction with receptors. This is called first pass metabolism. A schematic representation of the process of drug in the body is given in Fig. 5.7.

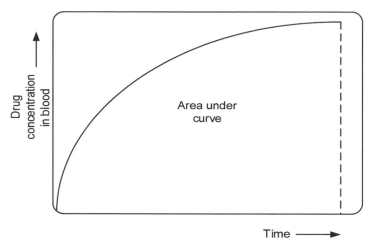

Figure 5.8 Drug concentration in the bloodstream versus time for a single dose.

A plot of the drug concentration in the bloodstream over time for a single dose is shown in Fig. 5.8. At a certain time after administration, the rate of drug absorption equals the rate of clearance. This is an equilibrium condition called "steady state".

The area under the curve (AUC) represents the total amount of drug in the blood. It is a measurement of the bioavailability of the drug. Comparison of drug concentrations in the bloodstream administered via intravenous injection and the oral route provides information for the bioavailability of the oral drug. This is because the oral drug is metabolized in the liver before reaching the general blood circulation (see Section 5.3.4), whereas for intravenous injection the total amount of drug is injected directly into the bloodstream. In general, oral doses are higher than intravenous doses to take into account the effects of first pass metabolism.

Buccal and Sublingual Administration: Drugs can be absorbed through the oral cavity. Buccal (between the gums and cheek) and sublingual (underneath the tongue) can be effective means of drug administration. In both cases, drugs can enter the blood circulation without first pass metabolism in the liver.

Rectal Administration: Rectal administration of a drug may be applied when the patient is unable to take the drug orally and some other routes are impractical. The drug administered via the rectum is absorbed and partially bypasses the liver. However, the absorption of drugs may be unreliable in certain cases.

Subcutaneous and Intramuscular Administration: Subcutaneous and intramuscular administration can be used to deliver protein-based drugs. The absorption of drug is faster than with the oral route. The rate of absorption is determined by the blood flow pattern and diffusion of drug molecules in tissues.

Transdermal and Topical Administration: Transdermal administration is used to apply the drug on the skin surface. The drug is absorbed and transported by blood to receptors, which may be remote from the part of the skin where the transdermal patch is. The first pass metabolism is circumvented. Topical administration is used to apply the drug for local effects. The typical areas for topical application are the skin, eyes, throat, nose, and vagina.

Inhalation Administration: Aerosol particles of drug can be inhaled into the lungs. Because of the large surface area of the alveoli, absorption is rapid and effective. As the lungs are richly supplied with capillaries, distribution of inhalational drugs is very quick (refer to Exhibit 4.14 on inhalable insulin).

Intravenous Administration: When a drug is injected, the entire dose can be considered as being available in the bloodstream to be distributed to the target site. Hence, the dosage can be controlled, unlike with other routes of administration, where the bioavailability of the drug may be unpredictable because of diffusion processes. Intravenous injection is the normal route for administration of protein-based drugs, as they are likely to be destroyed if taken orally because of the pH conditions in the gastrointestinal tract.

The onset of drug action with intravenous injection is quick, and this method is especially useful for emergency cases. However, intravenous injection is potentially the most dangerous. Once a drug is injected, there is no means to stop it from circulating throughout the body. The complete circulation of blood in the body takes about a minute, and hence an adverse reaction can occur almost instantaneously.

5.3.3 Distribution

When a drug is in the bloodstream, it is distributed to various tissues. The distribution pattern depends on a number of factors:

- Vascular nature of the tissue
- Binding of the drug to protein molecules in blood plasma
- Diffusion of the drug

When a tissue is perfused with blood supply, drug molecules in the blood are transported to the tissue rapidly until equilibrium is reached. On the other hand, the drug may bind to albumin and proteins in the blood, rendering less of it available for distribution to tissues. In general, acid drugs bind to albumins and basic drugs to glycoproteins. The third factor for drug distribution is passive diffusion. Lipid-soluble drugs can cross the cell membrane more readily than polar drugs and move into the tissues to interact with receptors.

Exhibit 5.3   Barriers to Drug Distribution

Blood–Brain Barrier (BBB): Distribution of drugs to the brain tissue is restricted for some types of drugs. The reason is that the brain has a sheath of connective tissue cells, the astrocytes, surrounding it, forming a barrier to passive diffusion for polar drugs. In addition, the endothelial cells of the brain capillaries are joined more tightly together, further curtailing the diffusion of polar drugs to the brain. Lipid-soluble drugs, however, can diffuse into the brain more readily and bring forth their effects.

For neuropharmaceuticals that target the brain, as in the cases of neurodegenerative disorders (Alzheimer's, multiple sclerosis), psychiatric or psychotherapy, stroke, and infectious diseases, drug candidates are tested using *in vivo* and *in vitro* models to assess the transfer of the drug compound across the BBB.

Placental Barrier: The placental barrier consists of several layers of cells between the maternal and fetal circulatory systems. Diffusion of polar drugs is limited. However, lipid-soluble drugs can pass through the barrier. Fetuses are rich in lipids and may form a reservoir for sequestering lipid-soluble drugs.

The volume of distribution (V_d) is an important parameter. It is represented by the following equation:

$$V_d = \frac{\text{Dose}}{C_b} \tag{5.6}$$

V_d is a hypothetical volume and C_b is the concentration of drug in blood. When C_b is low, V_d may turn out to be a large value, many times more than the volume of a person of around 60–70 liters. Highly lipid-soluble drugs have a very high volume of distribution. Lipid-insoluble drugs, which remain in the blood, have a low V_d.

For example, obesity affects V_d because lipid-soluble drugs diffuse into the adipose tissues of the obese person. V_d is a useful parameter for determining the loading dose for a drug to attain equilibrium after the drug is administered.

Distribution of drugs is restricted in two areas: the brain and the placenta. Refer to Exhibit 5.3 for a brief description on how drugs cross these barriers. Exhibit 5.4 presents a potential new method to deliver drugs across the blood–brain barrier.

5.3.4   Metabolism

Many drugs are metabolized in the body; their chemical structures are altered and pharmacological activity reduced. The liver is the major organ for

Exhibit 5.4 Drug Delivery Across the Blood–Brain Barrier

Using the knowledge that rabies virus can spread into the brain neurons, scientists mimic its delivery system. A short, 29 amino acid peptide chain is derived from the rabies virus glycoprotein (RVG). The RVG binds to the acetylcholine receptor on the neurons and the endothelium cells of the blood–brain barrier. Through this interaction, transvascular delivery is enabled.

The drug in this case, a siRNA, was coupled to the RVG peptide and successfully delivered to the neurons in mice. This work opens up the possibility of using the RVG as delivery tools for drugs designed for interaction with brain neurons.

Source: Kumar P, et al. Transvascular delivery of small interfering RNA to the central nervous system, *Nature* 448:39–43 (2007).

metabolizing drugs, a secondary role is played by the kidneys. Some drugs are metabolized in tissue systems.

Two types of biochemical metabolism reactions take place in the liver: Phase I and Phase II reactions. Phase I reactions include oxidation, reduction, and hydrolysis, which transform the drugs into metabolites. A family of enzymes called cytochrome P-450 (CYP) is responsible for these reactions. More than 50 CYP enzymes have been characterized but only six are responsible for most of the drug metabolism. These are CYP1A2, CYP2C9, CYP2D6, CYP2A6, CYP2E1, and CYP3A4; they are mainly found in the liver and convert lipid-soluble drugs to more water-soluble metabolites. Phase II reactions involve the addition or conjugation of subgroups, such as —OH, —NH, and —SH to the drug molecules. Enzymes other than P-450 are responsible for these reactions. These reactions give rise to more polar molecules, which are less lipid-soluble and are excreted from the body. Exhibit 5.5 describes some of the drug metabolism studies recommended by the Food and Drug Administration (FDA).

5.3.5 Excretion

Drugs are excreted from the body by the following routes:

- Kidneys
- Lungs
- Intestine and colon
- Skin

The kidneys are the most important organs for clearing drugs from the body. Water-soluble drugs are cleared more quickly than lipid-soluble drugs.

Exhibit 5.5 Metabolism Studies

The aim of metabolism studies is to (1) identify metabolic pathways and (2) investigate the possibility of drug–drug interactions.

Pharmacogenetics influences the therapeutic effects of drugs. A drug that is normally metabolized by the P-450 2D6 enzyme will not be metabolized in about 7% of the Caucasian population. Coadministration of drugs may have different effects, which are either (1) additive, or synergistic, or (2) antagonistic.

Some drugs are administered in a prodrug form. They are metabolized, and the metabolites elicit the interactions with receptors.

P-450 enzymes have been cloned and *in vitro* studies can be performed using these enzyme systems. Metabolic pathways can be studied by incubating the drug with the P-450 enzymes. Similarly, drug–drug interactions can be studied.

Source: Center for Drug Evaluation and Research. *Guidance for Industry, Drug Metabolism/Drug Interaction Studies in the Drug Development Process: Studies In Vitro*, FDA, Rockville, MD, 1997, http://www.fda.gov/cder/guidance/clin3.pdf [accessed September 20, 2007].

Volatile and gaseous by-products of drugs are exhaled by the lungs. Some drugs are reabsorbed into the intestine and colon and later passed out as solid waste. Another mechanism of clearance is for drugs to be excreted through the skin as perspiration. In terms of chemical reactions, drug elimination involves a number of processes such as conjugation, hydrolysis, oxidation, reduction, and proteolysis.

The clearance of a drug is given by the following expression:

$$CL = \frac{\text{Rate of drug elimination}}{\text{Drug concentration in blood}} \tag{5.7}$$

A typical drug clearance curve is shown in Fig. 5.9. The curve in Fig. 5.9 is a first-order curve; that is, the elimination rate is proportional to the amount of drug in the bloodstream. As the amount of drug in the blood is reduced, the elimination rate is also reduced. Another term often used is "half-life." This is the time taken to clear half (50%) of the remaining drug in the body. Mathematically, it is given by

$$t_{1/2} = \frac{0.693 \times V_d}{CL} = \frac{0.693}{k} \tag{5.8}$$

where k is the rate of drug elimination.

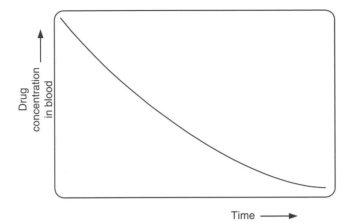

Figure 5.9   Clearance of drug from the bloodstream.

TABLE 5.1   Half-Life Calculations

	Amount of Drug in the Body	
Number of Half-Lives	Eliminated (%)	Remaining (%)
0	0.0	100.0
1	50.0	50.0
2	75.0	25.0
3	87.5	12.5
4	93.8	6.2
5	96.9	3.1
6	98.4	1.6

The half-life concept is further illustrated in Table 5.1.

In a first order simple approximation the concentration of drug in the body at time t is given by

$$C_{\mathrm{d}}(t) = C_{\mathrm{d}}(0)\,e^{-kt} \tag{5.9}$$

where $C_{\mathrm{d}}(t)$ is the concentration of drug in the body at time t, $C_{\mathrm{d}}(0)$ is the concentration of drug in the body at time 0, and k is rate of drug elimination.

The equation can be rearranged to give

$$k = \frac{\ln(C_{\mathrm{d}}1) - \ln(C_{\mathrm{d}}2)}{t_2 - t_1} \tag{5.10}$$

where $\ln(C_{\mathrm{d}}1)$ is the natural logarithm of C_{d} at time 1 and $\ln(C_{\mathrm{d}}2)$ is that at time 2, and t_1 and t_2 refer to time 1 and 2, respectively.

5.3.6 Application of Pharmacokinetics Results

By combining Figs 5.8 and 5.9, we obtain the situation depicted in Fig. 5.10. After a drug is absorbed, it enters the bloodstream and the concentration builds up until a steady state is reached. As time passes, the elimination process takes over and the concentration of the drug decreases.

Figure 5.10 would be the situation if a single dose of drug were given, but this is rarely the case. More than one dose is often administered to maintain the therapeutic level of the drug—the level that has been determined from pharmacodynamics studies of dose–response versus drug concentration. Before the drug is cleared by the excretion process, another dose is given to keep the drug concentration at a steady state and achieve maximal effects. This is illustrated in Fig. 5.11.

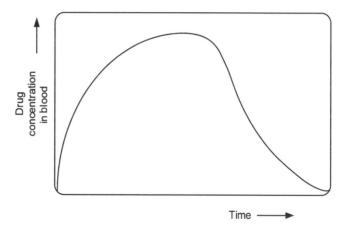

Figure 5.10 Plasma concentration of drug after a single dose.

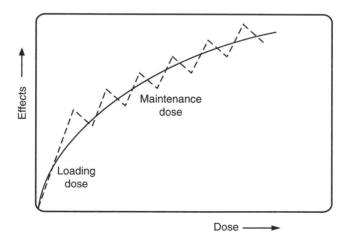

Figure 5.11 Multiple doses to maintain maximal effect.

Sometimes a larger dose is administered first; this is called a loading dose (see Section 5.3.3). It quickly builds up to the steady-state level. After that, smaller doses are given to maintain the steady state.

5.4 TOXICOLOGY

In addition to the preclinical research of pharmacodynamics and pharmacokinetics, study of the toxicology of a potential drug is critical to demonstrate that it is safe before it is given to humans in clinical trials. Toxicological studies show the functional and morphological effects of the drug. They are performed by determining the mode, site, and degree of action, dose relationship, sex differences, latency, and progression and reversibility of these effects.

We summarize in Exhibit 5.6 the International Conference on Harmonization (ICH) guidelines for toxicological and pharmacological studies. Appendix 6 shows the type of information on pharmacodynamics, pharmacokinetics, and toxicology that regulatory reviewers examine when a potential drug is filed for Investigational New Drug and New Drug Application approval. The information is extracted from the FDA *Guidance for Reviewers, Pharmacology/ Toxicity Review Format* (2001).

5.4.1 Toxicity

It is necessary to determine the toxicity of a drug. The maximum tolerable dose and area under the curve are established in rodents and nonrodents. There are two types of toxicity studies: single dose and repeated dose. Single

---

Exhibit 5.6   ICH Guidelines on Safety Studies

S1	***Carcinogenicity Studies***
S1A	Guideline on the need for carcinogenicity studies of pharmaceuticals
S1B	Testing for carcinogenicity in pharmaceuticals
S1C(R1)	Dose selection for carcinogenicity studies of pharmaceuticals & limit dose
S2	***Genotoxicity Studies***
S2A	Genotoxicity: specific aspects of regulatory genotoxicity tests for pharmaceuticals
S2B	Genotoxicity: a standard battery for genotoxicity testing of pharmaceuticals
S3	***Toxicokinetics and Pharmacokinetics***
S3A	Toxicokinetics: assessment of systemic exposure in toxicity studies
S3B	Pharmacokinetics: Guidance for repeated dose tissue distribution studies
S4	***Toxicity Testing***
S4	Single dose toxicity tests

S4A	Duration of chronic toxicity testing in animals (rodent and nonrodent)
S5	***Reproductive Toxicology***
S5(R2)	Detection of toxicity to reproduction for medicinal products and toxicity to male fertility
S6	***Biotechnological Products***
S6	Preclinical safety evaluation of biotechnology-derived pharmaceuticals
S7	***Pharmacology Studies***
S7A	Safety pharmacology studies for human pharmaceuticals
S7B	The nonclinical evaluation of the potential for delayed ventricular repolarization (QT interval prolongation) by human pharmaceuticals
S8	***Immunotoxicology Studies***
S8	Immunotoxicity studies for human pharmaceuticals Joint safety/efficacy (multidisciplinary) topic
M3(R1)	***Nonclinical Safety Studies for the Conduct of Human Clinical Trials for Pharmaceuticals***

Source: International Conference on Harmonization. *Safety Guidelines*. http://www.ich.org/cache/compo/502-272-1.html [accessed September 17, 2007].

dose acute toxicity testing is conducted for several purposes, including the determination of repeated doses, identification of organs subjected to toxicity, and provision of data for starting doses in human clinical trials.

The experiments are carried out on animals, usually on two mammalian species: a rodent (mouse or rat) and a nonrodent (rabbit). Two different routes of administration are studied: one is the intended route for human clinical trials, and the other is intravenous injection. Various characteristics of the animals are monitored, including weight, clinical signs, organ functions, biochemical parameters, and mortality. At the completion of the study, animals are killed and autopsies are performed to analyze the organs, especially the targeted organ for the drug.

Repeated dose chronic toxicity studies are performed on two species of animals: a rodent and nonrodent. The aim is to evaluate the longer-term effects of the drug in animals. Plasma drug concentrations are measured and pharmacokinetics analyses are performed. Vital functions are studied for cardiovascular, respiratory, and nervous systems. Animals are retained at the end of the study to check toxicity recovery. Table 5.2 shows the duration of the animal studies, which depends on the duration of the intended human clinical trial. Appendix 6 summarizes the information to be submitted to regulatory authorities.

TABLE 5.2   Duration of Repeated Dose Toxicity Studies to Support Phase I and II Trials in the European Union and Phase I, II, and III Trials in the United States and Japan[a]

	Minimum Duration of Repeated Dose Toxicity Studies	
Duration of Clinical Trials	Rodents	Nonrodents
Single dose	2–4 weeks	2 weeks
Up to 2 weeks	2–4 weeks (1 month)	2 weeks (1 month)
Up to 1 month	1 month (3 months)	1 month (3 months)
Up to 3 months	3 months (6 months)	3 months (3 months)
Up to 6 months	6 months	6 month (chronic)
>6 months	6 months	6–9 months

[a]There are slight differences in the requirements for the European Union, the United States, and Japan. Duration to support Phase III trials in the EU, when they differ from the other data, is given in parentheses. Readers are referred to *Guidance for Industry: M3 Nonclinical Safety Studies for the Conduct of Human Clinical Trials for Pharmaceuticals*, FDA, Rockville, MD, 1997. http://www.fda.gov/cder/guidance/1855fnl.pdf [accessed September 20, 2007].

5.4.2 Carcinogenicity

Carcinogenicity studies are carried out to identify the tumor-causing potential of a drug. Drugs are administered to animals continuously for at least 6 months. Rats are normally used, but another rodent study may be required. The studies are performed using the drug administration route intended for humans. Data for hormone levels, growth factors, and tissue enzymatic activities are gathered. At the end of the experiments, the animals are killed and the tissues examined. Appendix 6 summarizes the information to be submitted to regulatory authorities.

5.4.3 Genotoxicity

These studies are to determine if the drug compound can induce mutations to genes. A standard battery of tests includes the following:

- Assessment of genotoxicity in a bacterial reverse mutation test (Ames test, see Exhibit 5.7)
- Detection of chromosomal damage using *in vitro* method (mouse lymphoma tk test, a test to evaluate the potential of a drug to cause mutations to thymidine kinase [tk])
- Detection of chromosomal damage using rodent hematopoietic cells

5.4.4 Reproductive Toxicology

The aim of these studies is to assess the effect of the potential drug on mammalian reproduction. All the stages, from premating through conception, pregnancy and birth, to growth of the offspring, are studied. Rats are the

Exhibit 5.7   Ames Test

The Ames test is based on the reversion of mutations in the bacterium *Salmonella typhimurium*. Mutant strains of *S. typhimurium*, those with mutations in the *his* operon, are unable to grow without addition of the amino acid histidine. The drug to be tested is mixed with *S. typhimurium* and a small amount of histidine in a nutrient medium. After the histidine is consumed, the growth will stop if the drug is not a mutagen. However, if the drug is a mutagen, it will induce a reversion in the *his* operon and the bacterium will continue to grow.

predominant species used, and rabbit is the preferred nonrodent model. The route of administration is similar to the intended route for humans. At least three dosage levels and control groups are used (control groups are dosed with drug excipients or vehicles to provide a comparable basis for analysis). For the females, effects such as hormonal cycles, pregnancy, and embryo development are studied. For the males, effects on the reproductive organs are analyzed. Other parameters studied are detailed in Appendix 6.

5.5   ANIMAL TESTS, *IN VITRO* ASSAYS, AND *IN SILICO* METHODS

The use of animals for pharmacological and toxicological studies has yielded invaluable information for drug development. However, many drug candidates failed in Phase I and II clinical trials because the animal models were insufficient to represent human systems and functions for some drugs. Efficacy and acceptable toxicities derived from animal models were not replicated in humans (Exhibit 5.8). In recent years, the direction in development of drugs has shifted toward the use of *ex vivo, in vitro* assays and even *in silico* methods. Nevertheless, some tests must still be confirmed in animals.

Where animals are used, mice and rats are the preferred models. Other species used are hamsters, guinea pigs, and rabbits. These animals are bred in a specially controlled environment, under specific pathogen-free (SPF) conditions, to ensure that they do not carry infections or pathogens before being used in various tests. Different breeds or strains of animals are used, for example, BALB/c mice are used for immunity studies and Fischer 344 mice for carcinoma evaluation. Nude mice (in addition to the nude gene, which results in the absence of thymus and T-cell function) have two other mutations important in regulating the function of the immune system. More recent additions are transgenic animals with knockout genes. For example, mice with knockout *p53* genes have a high incidence of tumor growth.

Experimental use of animals is controlled under Good Laboratory Practice (GLP), and study protocols are submitted to the Animal Research Ethics Committee for approval. Studies using animals can only proceed with the approval of the Ethics Committee, which consists of technical personnel,

Exhibit 5.8   Clinical Trial Failures

Only one in 10 Investigational New Drugs (INDs) will become approved as drugs. Half the IND failures are due to unacceptable efficacy. One-third fail because of safety issues.

Toxicity failures occur for the following reasons:

- Toxicity in animals is not fully understood and potential toxicity in humans cannot be estimated.
- Toxicity in animals is understood and potential toxicity in humans is not acceptable.
- Acceptable therapeutic margins (efficacy versus toxicity) cannot be established.
- Toxicities in animals do not predict toxicity in human trials.

See also Exhibit 6.17.

Source: Johnson DE. Predicting human safety: screening and computational approaches, *Drug Discovery Today* 5:445–454 (2000).

Exhibit 5.9   Caco-2 Cell Assays

The Caco-2 cells are derived from human colorectal carcinoma. When these cells are cultured on semipermeable membranes, they grow into epithelial cells that are very similar to intestinal epithelial cells. The permeability of drugs across these Caco-2 cells provides model tools for the study of drug absorption.

including a veterinarian, as well as laypeople who evaluate the study from different perspectives.

In vitro assays are increasingly being used. Some of the reasons are cost, availability of more rapid results, and avoidance of negative publicity. Assays such as cytochrome P-450 enzymes, the Ames test, and the mouse lymphoma tk test are *in vitro* methods. For absorption studies, Caco-2 (Exhibit 5.9) and Madin-Darby canine kidney cell assays are now routinely used. Hepatocyte cell lines with metabolism capacity are being developed to test drug metabolism and toxicity. All these examples show that, where possible, pharmaceutical firms are gradually dispensing with animal studies.

With better understanding of drug functions and information from huge databases, predictive *in silico* ADME algorithms have been designed. These algorithms encompass information derived from *in vivo* and *in vitro* studies; they consider molecular interactions, biological data, pharmacological results, and toxicological endpoints. A description of some of these *in silico* methods is given in Exhibit 5.10.

The aim of all the laboratory and animal studies is to understand the effects of the potential drug in living systems. These studies cannot guarantee the safety and efficacy of the drug in humans, but they can enhance the reliability and predictive value. Results from these studies provide a basis for the starting dose for clinical trials in humans. *Guidance for Industry and Reviewers: Estimating the Safe Starting Dose in Clinical Trials for Therapeutics in Adult Healthy Volunteers* from the FDA (December 2002) outlines the derivation of the maximum recommended starting dose (MRSD) for a drug to be used in humans for the first time. This dose is based on the following derivation algorithm:

- Determine the no observed adverse effect level (NOAEL) in animals— the highest dose level that does not produce a significant increase in adverse effects.
- Convert the NOAEL to the human equivalent dose (HED) using the data from Table 5.3 (calculations are based on body surface areas).

Exhibit 5.10   *In Silico* Predictive Methods

Expert Systems

- *DEREK:* Expert system for the prediction of toxicity (genotoxicity, carcinogenicity, skin sensitization, etc.)
- *METAPC:* Windows based metabolism and biodegradation expert system
- *METEOR:* Expert system for the prediction of metabolic transformations
- *OncoLogic:* Rule-based expert system for the prediction of carcinogenicity

Data Driven Systems

- *lazar:* Open source inductive database for the prediction of chemical toxicity
- *MC4PC:* Windows based structure–activity relationship (SAR) automated expert system
- *PASS:* Predicts 900 pharmacological effects, mechanisms of action, mutagenicity, carcinogenicity, teratogenicity, and embryotoxicity
- *TOPKAT:* Quantitative structure toxicity relationship (QSTR) models for assessing various measures of toxicity

Source: Predictive Toxicology—Programs. http://predictive-toxicology.com/programs.html [accessed September 17, 2007].

TABLE 5.3 Conversion of Animal Dose to Human Equivalent Dose (HED) Based on Body Surface Area

Species	To Convert Animal Dose in mg/kg to Dose in mg/m², Multiply by kg/m² below:	To Convert Animal Dose in mg/kg to HED[a] in mg/kg, Either:	
		Divide Animal Dose by:	Multiply Animal Dose by:
Human	37	—	—
Child (20 kg)[b]	25	—	—
Mouse	3	12.3	0.08
Hamster	5	7.4	0.13
Rat	6	6.2	0.16
Ferret	7	5.3	0.19
Guinea pig	8	4.6	0.22
Rabbit	12	3.1	0.32
Dog	20	1.8	0.54
Primates			
Monkeys[c]	12	3.1	0.32
Marmoset	6	6.2	0.16
Squirrel monkey	7	5.3	0.19
Baboon	20	1.8	0.54
Micro-pig	27	1.4	0.73
Mini-pig	35	1.1	0.95

[a]Assumes 60 kg human. For species not listed or for weights outside the standard ranges, HED can be calculated from the formula:

$$HED = \text{Animal dose in mg/kg} \times (\text{Animal weight in kg/Human in kg})^{0.33}$$

[b]This is provided for reference only, as healthy children will rarely be volunteers for Phase I trials.
[c]For example, cynomolgus, rhesus, stumptail.
Note: Column 2 is for information only. For HED calculations, either column 3 or column 4 is used.

5.6 FORMULATIONS AND DELIVERY SYSTEMS

Development of manufacturing processes for the production of a drug (active pharmaceutical ingredients—APIs), initially to supply enough materials for laboratory testing, then for human clinical trials, and ultimately as production batches of drug products when approved by regulatory authorities, proceeds as soon as the clinical candidate is identified. There are two distinct manufacturing processes: synthetic chemistry for pharmaceuticals and recombinant DNA technology for biopharmaceuticals. Manufacturing processes are discussed in Chapter 10.

Apart from pharmacological and toxicological studies, the drug development process encompasses meticulous and methodical work in the following areas:

- Formulation of the drug product, which includes active pharmaceutical ingredients and excipients, into final form suitable to be administered to patients
- Study of drug delivery systems to improve effective presentation of the drug to patients for enhancing certain characteristics and improving patient compliance

The last two items are discussed next.

5.6.1 Formulations

Most drugs that are prescribed to us are formulated with active pharmaceutical ingredients and excipients. The formulations of selected drugs are presented in Exhibit 5.11. According to the *US Pharmacopoeia and National Formulary* definition, excipients are "any component, other than the active substance(s), intentionally added to the formulation of a dosage form." There are many reasons for the addition of excipients:

Exhibit 5.11   Selected Drug Formulations

Prilosec

An antiulcerant in 10, 20, and 40 mg doses

Active ingredient: Omeprazole

Excipients: Cellulose, disodium hydrogen phosphate, hydroxypropyl cellulose, hydroxypropyl methylcellulose, lactose, mannitol, sodium lauryl sulfate, etc.

Prozac

An antidepressant in 10, 20, and 40 mg doses

Active ingredient: Fluoxethine hydrochloride

Excipients: Starch, gelatin, silicone, titanium dioxide, iron oxide, etc.

Lipitor

Cholesterol reducer in 10, 20, 40, and 80 mg doses

Active ingredient: Atorvastatin calcium

Excipients: Calcium carbonate, candelilla wax, croscarmellose sodium, hydroxypropyl cellulose, lactose monohydrate, magnesium stearate, microcrystalline cellulose, polysorbate 80, simethicone emulsion

Celebrex

Anti-inflammatory in 100 and 200 mg doses

Active ingredient: Celecoxib

Excipients: Croscarmellose sodium, edible inks, gelatin, lactose monohydrate, magnesium stearate, povidone, sodium lauryl sulfate, and titanium dioxide

Source: Food and Drug Administration, Center for Drug Evaluation and Research, http://www.fda.gov/cber/ [accessed September 20, 2007].

- To control release of the drug substance in the body
- To improve the half-life of the drug substance (see Exhibit 4.8)
- To improve the assimilation process and bioavailability
- To enhance drug dissolution (e.g., with disintegration promoters)
- To extend the stability and shelf life of the drug (e.g., with antioxidants or preservatives)
- To aid in the manufacturing process (in the form of fillers, lubricants, wetting agents, and solubilizers)
- To mask the unpleasant taste of the active pharmaceutical ingredient
- To aid identification of the product

According to the International Pharmaceutical Excipients Council, the following are the most commonly used excipients in the United States:

- Simethicone emulsion—antifoam
- Selenium—antioxidant
- Vitamins A, C, and E—antioxidants
- Hydroxypropyl cellulose—binder
- Hydroxypropyl methylcellulose—binder
- Ethylcellulose—binder
- Lactose—binder
- Starch (corn)—binder
- Gelatin—capsule shell
- Silicon dioxide—colorant
- Titanium dioxide—colorant
- Microcrystalline cellulose—disintegrant

- Sodium starch glycolate—disintegrant
- Sodium carboxymethyl cellulose—disintegrant
- Polysorbate—emulsifier
- Calcium carbonate—filler
- Calcium phosphate—filler
- Talc—filler
- Calcium stearate—lubricant
- Magnesium stearate—lubricant
- Stearic acid—lubricant
- Sucrose—sweetener

The FDA maintains a database of approved excipients (*Drug Information: Electronic Orange Book*, http://www.fda.gov/cder/ob/default.htm). Standards and tests for regulatory acceptable excipients are included in the *US Pharmacopoeia and National Formulary*. Two such tests, dissolution and stability, are included in Exhibit 5.12 for reference. For new excipients to be included in a drug formulation, they have to satisfy one of the following criteria:

- Determination by the FDA that the substance is "generally recognized as safe" (GRAS) according to 21 CFR 182, 184, and 186

Exhibit 5.12   Dissolution and Stability Tests

FDA, Guidance for Industry (1997)—Dissolution Testing of Immediate Release Solid Dosage Forms: Dissolution tests using the basket method (50/100 rpm) or the paddle method (50/75 rpm) under mild test conditions are used to generate a dissolution profile at 15 minute intervals. The pH range is 1.2–6.8; pH up to 8.0 may be tested with justification. The temperature is $37 \pm 0.5\,°C$. Methods are described in the *US Pharmacopoeia*. Test requirements vary depending on solubility of drug products. *In vitro* test may need validation to confirm *in vivo* results.

Stability Tests on Active Ingredients and Finished Products EU Guidelines (1998)—Medicinal Products for Human Use, Vol. 3A, Quality and Biotechnology: Stability tests on drug products are performed to determine shelf life and storage conditions. Drug products are tested at various temperatures, for example, below $-15\,°C$, $2-8\,°C$, $25\,°C$, and $40\,°C$. High humidity testing at >75% is performed as well, and a combination of $40\,°C$ and 75% relative humidity. Photostability is tested by the exposure of drug products to visible and ultraviolet light sources. Properties and characteristics of the drugs are tested after temperature and humidity exposures to determine the storage conditions and shelf life.

- Approval by the FDA as a food additive under 21 CFR 171
- Excipients referenced in the New Drug Application, showing that they have been tested in laboratory and clinical trials

The foregoing applies mainly for small molecule drugs. In the case of large molecule drugs, formulations are undertaken first and foremost to improve stability as proteins are prone to undergo physical and chemical changes. These changes may involve aggregation due to dimerization, trimerization, and higher order associations, as well as crystallization and precipitation. Degradation processes such as deamidation, oxidation, hydrolysis, isomerization, proteolysis, and disulfide bond formation/dissociation may also occur. Large molecule drug formulations typically consist of buffers, surfactants, and stabilizers in liquid form or as lyophilized powder to be reconstituted with sterile water before administration via the parenteral route to patients (Exhibits 5.13 and 5.14).

For drugs to be administered by the parenteral route, there are some general requirements:

- Intravascular—clear solution, isoosmotic (serum and cellular fluid osmotic pressure is around 285–290 mOsm), pH ≈ 7.4
- Subcutaneous or intramuscular—suspensions allowable but avoid extreme pH; citrate as buffer should not be used as it causes pain

In order to study and derive an effective formulation, a suite of analytical methods must be developed to evaluate the formulation. Some suggested assays for biopharmaceutical drug formulation evaluations are the following:

- Bioassay—activity of formulation
- Immunoassay—purity assessment
- pH—chemical stability
- SDS-PAGE—protein characterization and purity
- HPLC—purity, identity, and stability
- IEF—modifications of protein
- N-terminal sequencing—identity, of protein
- UV—concentration and aggregation
- Circular dichroism—secondary and tertiary conformations

The use of antioxidants and preservatives and the types of container and closure are other aspects of formulations that are considered with regard to how the drug is to be delivered and the target patient group for the drug.

5.6.2 Drug Delivery Systems

Delivery systems have come a long way from pills, syrups, and injectables. As we have discussed earlier, the ADME process means that most drugs admin-

Exhibit 5.13   Selected Large Molecule Drug Formulations

Avastin: Avastin is used for the treatment of colorectal cancer. It is supplied in 100 and 400 mg dosages. The 100 mg formulation consists of 240 mg α,α-trehalose dihydrate, 23.2 mg sodium phosphate (monobasic, monohydrate), 4.8 mg sodium phosphate (dibasic, anhydrous), 1.6 mg polysorbate 20 and water-for-injection.

Source: Food and Drug Administration. *Avastin*. http://www.fda.gov/medwatch/SAFETY/ 2005/Jan_PI/Avastin_PI.pdf [accessed October 16, 2007].

Enbrel: Enbrel is used for the treatment of rheumatoid arthritis. It is supplied as a sterile, preservative-free, lyophilized powder. The powder is reconstituted with 1 mL sterile bacteriostatic water-for-injection (containing 0.9% benzyl alcohol) prior to parenteral injection.

Source: Food and Drug Administration. *Enbrel*. http://www.fda.gov/medwatch/SAFETY/ 2004/may_PI/Enbrel_PI.pdf [accessed October 16, 2007].

PEG-Intron: PEG-Intron is used for the treatment of hepatitis C. The product consists of a covalent conjugate of the recombinant interferon-α-2b with monomethoxy polyethylene glycol (PEG) supplied in vials with 74 μg, 118.4 μg, 177.6 μg, or 222 μg of the active ingredient and 1.11 mg sodium phosphate (dibasic, anhydrous), 1.11 mg sodium phosphate (monobasic, dihydrate), 59.2 mg sucrose, and 0.074 mg polysorbate 80. The powder is reconstituted with sterile water-for-injection.

Source: Food and Drug Administration. *PEG-Intron*. http://www.fda.gov/medwatch/ SAFETY/2005/Jan_PI/PEG-Intron_PI.pdf [accessed October 16, 2007].

Exubera (see also Exhibit 4.14): Exubera is an inhalable insulin for the treatment of type I and II diabetes. Each dose consists of 1 or 3 mg insulin in a powder formulation with sodium citrate (dehydrate), mannitol, glycine, and sodium hydroxide.

Source: Food and Drug Administration. *Exubera*. http://media.pfizer.com/files/products/ uspi_exubera.pdf [accessed October 16, 2007].

istered to us have tortuous paths to reach their targets, and in many instances the bioavailability is reduced. A traditional means to overcome the vagaries of ADME is to have larger doses or more frequent administrations. These types of treatment have complications: (1) potential for adverse events and (2) need to ensure patient compliance to take the medication regularly. New delivery systems are devised to overcome these problems.

Exhibit 5.14   Excipients for Lyophilized Formulations

Lyophilized excipients include the following:

- Bulking agent—mannitol
- Stabilizing agents—monosaccharides (glucose), disaccharides (sucrose, lactose, maltose, and trehalose)
- Surfactants (nonionic)—polyethylene sorbitan monolaurate (Tween 20, Tween 80), pluronic, Triton, sodium dodecyl sulfate (SDS)
- Buffering agents—phosphate, citric, glutaric, succinic, carbonic acid
- Chelating agents—to bind trace metals, EDTA 0.01–0.05%
- Antioxidants—to block specific chain reaction, 0.01–0.05%
- Preservatives—for multidose formulations, antimicrobial agents, phenol (0.3–0.5%), chlorobutanol (0.3–0.5%), benzyl alcohol (1.0–3.0%)
- Isotonic agents (osmolality 285–290 mOSm)—mannitol, sucrose, glycine, glycerol, and sodium chloride

TABLE 5.4   Bioavailability of Oral Drugs

Bioavailability	Dosage Form
Fastest	Solutions
	Suspensions
	Capsules
	Tablets
	Coated tablets
Slowest	Controlled release form

The oral route for drug administration is convenient and does not normally require a physician's intervention. Most protein-based drugs, however, are not administered via the oral route because they are destroyed by the low pH medium in the stomach. One means to overcome this is the use of an enteric coating for some drugs. Drugs are coated with cellulose acetate phthalate, which can withstand the acid environment in the stomach and yet readily dissolves in the slightly alkaline environment of the intestine. In this way, the protein-based drugs can have a safe passage to the intestine for absorption to take place. Table 5.4 shows the bioavailability of oral drugs in various dosage forms.

Another method is to prolong the release of the drug in the bloodstream. This will reduce the frequency for taking the drug, for example, from several times a day to once per day or even once per week. To achieve this, drug molecules are encapsulated within polymer matrices. These are known as

Exhibit 5.15   Polymeric Drug Delivery Systems

Two new developments are the dendrimers (highly branched, globular, synthetic macromolecules) and modified buckyballs. Together with hydrogels, they are tailored to provide targeted delivery.

The dendrimers form small micelles, which transport small molecules within their matrices or act as hubs for covalent bonding to drug molecules, extending like dendrites. In this way, they can shepherd high concentrations of drugs to targets.

Buckyballs are cage-like molecules of fullerenes. They are robust and can carry radioactive drugs to targets. Research is directed at using these buckyballs as delivery systems for the treatment of cancer.

Hydrogels are 3D cross-linked polymer networks. They can withstand acid conditions and release the entrapped drug molecules. Purdue University researchers have used a poly[methacrylic acid-g-poly(ethylene glycol)] hydrogel to encapsulate insulin, which could be released by pH trigger.

Source: Vogelson CT. Advances in drug delivery systems, *Modern Drug Discovery* 4(April):49–50, 52 (2001).

microspheres, polymer micelles, and hydrogels. The polymers are made with biodegradable materials and, through processes of hydrolysis, drug molecules are released at controlled rates as the polymer is degraded. The degradation process can be triggered by pH, temperature, electric field, or even ultrasound. Exhibit 5.15 provides further description on these polymeric delivery systems.

Other delivery systems are transdermal patches, metered dose inhalers, nasal sprays, implantable devices, and needle-free injections. A description of needleless injection is given in Exhibit 5.16.

5.7   NANOTECHNOLOGY

Nanotechnology is the science of matter with sizes in the range of 1–100 nm ($1-100 \times 10^{-9}$ m). These are scales of large molecules; for example, the sizes of some familiar matter are DNA, 1–2 nm; virus, 3–50 nm; and red blood cell, ~300 nm.

At these nano scales, matter behaves quite differently than it would at the macro level to which we are accustomed. Properties such as conductivity, magnetism, melting and boiling points, and reactivity may be dissimilar at the nano and macro scales due to the quantum mechanical behavior of small structures at molecular dimensions.

Exhibit 5.16   Needleless Injection

The sight of a hypodermic syringe is enough to send shivers down the spines of most patients, besides the agony of enduring the pain.

Needleless injections are new devices to bypass this problem. Drugs in powder or liquid form can be injected into the subcutaneous layer in the following ways:

- Propelled by a jet stream of compressed air
- Fired as pellets similar to that of bullets from rifles
- Electroporation (a temporary application of direct current, which disturbs the skin surface and allows penetration of the drug molecules)

Needleless injection is ideal for frequent injections, as in the cases of insulin and growth hormone, which are administered routinely.

Nanotechnology provides a means to manufacture particles with very high surface area to mass ratio and, together with their unique properties, may provide opportunities for more surface interactions and biochemical reactions to ensue. One use of nanotechnology is for drug delivery devices. Nano cages with embedded drugs can be delivered to their targets with high specificity and enable interactions to take place to alter the disease pathways.

Clinical trials are in progress where nano particle shells with chemotherapy drugs are tested on patients. It is believed that the nano shells can seek out cancerous cells and target them for destruction by the chemo drugs.

5.8   CASE STUDY #5

Zeprexa and Aranesp*

Examples of PD, PK, toxicology, and formulation for two selected drugs, Zyprexa (a small molecule drug) and Aranesp (a large molecule drug) are described next.

Source: (1) Food and Drug Administration, *Aranesp*. http://www.fda.gov/medwatch/ SAFETY/2005/Aranesp_PI_10-26-05.pdf [accessed August 22, 2007]. (2) Food and Drug Administration, *Zeprexa*. http://www.fda.gov/cder/foi/label/2001/darbamg091701lb.PDF [accessed August 22, 2007]. (3) Bunn HF. EPO binding to receptor: new agents that stimulate erythropoiesis, *Reviews in Translational Hematology*, 109:868–873 (2007). (4) Food and Drug Administration, *Zeprexa*. http://www.fda.gov/medwatch/SAFETY/2003/03Jul_PI/Zyprexa_PI.pdf [accessed August 22, 2007].

Zeprexa

Description: Zeprexa (olanzapine) is an antipsychotic drug. The chemical formula is 2-methyl-4(4-methyl-1-piperazinyl)10H-thieno[-2-3-b][1,5] benzodiazepine. The molecular weight is 312.44 Da.

PD: Zeprexa is a selective monoaminergic antagonist with high affinity for the following receptors: serotonin 5HT, dopamine, muscarinic, histamine, and adrenergic. Its action on schizophrenia is through the antagonism in serotonin and dopamine.

PK: Following an oral dose, plasma peak concentration is achieved in approximately 6 hours. About 40% is eliminated through first pass metabolism. The half-life is from 21 to 54 hours and plasma clearance is from 12 to 47 hours. Daily administration will lead to a steady-state plasma concentration in about a week with concentration twice that of the single dose. Metabolism of Zyprexa is by the cytochrome P-450 oxidation.

Toxicology: At 17 times the maximum human dose (on mg/kg basis), dogs developed reversible neutropenia and/or reversible hemolytic anemia between 1 and 10 months of treatment. Mice given doses twice the maximum human dose (on mg/kg basis) showed a decrease in lymphocytes and neutrophils in studies of 3 months' duration.

Formulation: The formulation consists of excipients such as carnauba wax, crospovidone, hydroxylpropyl cellulose, hydroxypropyl methylcellulose, lactose, magnesium stearate, microcrystalline cellulose, and other inactive ingredients.

Aranesp

Description: Aranesp is an erythropoiesis stimulating protein (EPO) produced using rDNA technology in CHO cells. It has 165 amino acids and the molecular weight is 30–37 kDa.

PD: Patients with chronic renal failure (CRF) and those receiving chemotherapy developed anemia due to deficiency in erythropoietin. Aranesp stimulates the production of red blood cells (RBCs). It mimics the natural erythropoietin and interacts with the progenitor stem cells to produce RBCs. The increased level of hemoglobin is observed after 2–6 weeks on Aranesp treatment.

PK: For therapeutic range of 0.45–4.5 μg/kg, maximum plasma concentration, half-life, and AUC are linear with respect to dose. Following subcutaneous injection, the absorption is slow and rate limiting. The half-life ranges from 27 to 89 hours. Peak plasma concentration is 34 hours after subcutaneous (SC) administration for CRF patients and 90 hours for cancer patients.

Toxicology: Animals treated with Aranesp showed no evidence of abnormal mitogenic and tumorigenic responses. In some studies, Aranesp appears to increase the beneficial effects of radiotherapy.

Formulation: Aranesp is formulated as a sterile, colorless, preservative-free protein solution for intravenous (IV) or subcutaneous (SC) administration. There are two formulations: the polysorbate solution includes excipients such as polysorbate 80, sodium phosphate monobasic monohydrate, sodium phosphate dibasic anhydrous, and sodium chloride in water-for-injection; while the albumin solution contains albumin, sodium phosphate monobasic monohydrate, sodium phosphate dibasic anhydrous, and sodium chloride in water-for-injection. The pH for both formulations is 6.2 ± 0.2.

5.9 SUMMARY OF IMPORTANT POINTS

1. Pharmacodynamics (PD) is the study of interactions between drugs and the body while pharmacokinetics (PK) describes the absorption, distribution, metabolism, and excretion (ADME) of drugs by the body.
2. PD studies allow us to understand the potency, effectiveness, therapeutic index, and safety margins of drugs. PK information on ADME provides us with an understanding of how drugs are transported, diffused into the bloodstream, and become available to the cells and act on the target sites.
3. Drugs are administered by various means: from oral to intravenous to topical. The oral route is a relatively slow process where a drug must be absorbed across the GI tract and then passed through the liver and metabolized before it becomes available to bind to receptors and perform its intended function. On the other hand, intravenous application is quick but has the potential of fast systemic reaction if adverse reactions occur. In the case of topical administration, the effects of the drug are localized.
4. Drug development has to evaluate the toxicity of drugs to the body. Animals, mainly rodents, are used to study toxicities. The evaluation should also consider the effect of drugs in causing cancers, tendency in inducing mutations, and the consequences on reproduction.

5. Over time more and more laboratory and cell-based assays are used to study the ADME properties and toxicities of drugs.

6. The active drug molecule is formulated with excipients to aid in the delivery process to receptors or enzymes. Excipients also help in maintaining the stability of the active component, in shipping, storage, and administration into the body. Other functions for the excipients are to modulate the bioavailability of the drug and prolong the half-life of the drug.

7. Nanotechnology is a new technique that may result in more targeted delivery of drug molecules to the active sites, thus improving bioavailability and reducing adverse events.

8. Most small molecule drugs are formulated for oral delivery while large molecule biopharmaceuticals are injected via parenteral means: intravenous, intramuscular, subcutaneous, and infusion.

5.10 REVIEW QUESTIONS

1. Distinguish between PD and PK. For PD, explain the term K_D and show how drugs interact with receptors. For PK, explain the mechanisms of ADME.

2. What do the terms potency and effectiveness of a drug mean? Explain the definitions of therapeutic index and safety margin.

3. Using acid–base theories for drugs, explain how ionization of a drug accounts for its solubility.

4. Using graphs, explain the absorption and clearance of a drug. What methods are used to prolong the availability of a drug in the body?

5. An IV dose of 500 mg is administered. The table below shows the drug concentration in blood, taken over a 10 hour period. Determine the rate of elimination and the half-life of the drug.

Time (h)	Concentration (μg/mL)
1	110
2	74
3	50
4	34
5	21
6	14
7	9
8	6
9	3
10	2

6. Explain the use of animals in drug testing. Provide examples to show alternative methods for testing the drugs. Discuss the pros and cons of each type of testing.
7. List the common excipients used and also the regulatory requirements for approving excipients.
8. Describe nanotechnology and clarify its role in drug application.

5.11   BRIEF ANSWERS AND EXPLANATIONS

1. Refer to Sections 5.2 and 5.3. Use Eqs. (5.1)–(5.3) to describe the term K_D. Refer to Sections 5.3.2– 5.3.5 to explain ADME.
2. Potency of a drug refers to the quantity of drug that generates a response, while effectiveness is the intensity of the response. Equations (5.4) and (5.5) show the therapeutic index and safety margin. In general, more specific drugs, such as those designed through the rational approach to bind to particular receptor(s), for example, Relenza and Tamiflu (Exhibit 3.7), are expected to have higher therapeutic index and safety margin.
3. Refer to Exhibit 5.2.
4. Use Figs. 5.8– 5.11 to explain the increase in drug concentration following administration and the subsequent clearance through the excretion route. The concept of drug half-life shows the elimination rate of a drug. One practice to prolong and sustain availability of drugs in the body is to increase the initial loading dose followed by reduced maintenance dosages. Another means is to prepare formulations for controlled release of drugs, for example, the pegylated interferon for hepatitis C treatment (Exhibit 4.8).
5. Using the natural logarithm for concentration, we obtain the following graph:

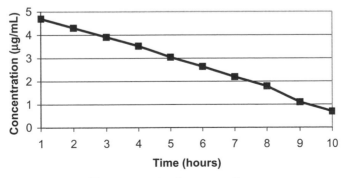

Drug concentration versus time.

This is a linear graph and the slope is the rate of elimination. Alternatively, we may substitute the values in Eq. (5.10):

$$k = \frac{\ln(C_d 1) - \ln(C_d 2)}{t_2 - t_1} = \frac{\ln(110) - \ln(2)}{10 - 1} = \frac{4.7 - 1.1}{9} = 0.4/h$$

$$t_{1/2} = \frac{0.693}{k} = 1.73\,h$$

6. Over the years the use of animal testing has yielded invaluable information on drug PD and PK before being administered to humans. For the study of many diseases, special breeds of animals are used as model systems to test the efficacy and safety of the drug candidates; refer to Section 5.5. Increasingly more and more *in vitro* assays are being used, partly to reduce pressure from animal rights groups and partly due to advances in assay development. It should also be noted that animal studies have to follow GLP, including the use of appropriate animal care facilities and protocols.

7. Refer to Section 5.6.1.

8. Refer to Section 5.7.

5.12   FURTHER READING

Appasani K. *Bioarrays: from Basics to Diagnostics*, Humana Press, Totowa, NJ, 2007.

Atkinson AJ Jr , Daniels CE, Dedrick RL, et al. *Principles of Clinical Pharmacology*, Academic Press, San Diego, CA, 2001.

Bauer LA. *Applied Clinical Pharmacokinetics*, McGraw-Hill, New York, 2001.

Butina D, Segall MD, Frankcombe K. Predicting ADME properties *in silico*: methods and models, *Drug Discovery Today* 7 (May):S83–S88 (2002).

Carstensen JT. *Advanced Pharmaceutical Solids, Drugs and Pharmaceutical Sciences*, Vol. 110, Marcel Dekker, New York, 2001.

Gad SC. *Drug Safety Evaluation*, Wiley-Interscience, Hoboken, NJ, 2002.

Goodsell DS. *Bionanotechnology: Lessons from Nature*, Wiley, Hoboken, NJ, 2004.

Gundertofte EK, Jorgensen, FS. *Pharmacokinetics Molecular Modeling and Prediction Of Bioavailability*, Kluwer Academic/Plenum Publishers, New York, 2000.

Julien RM. *A Primer of Drug Action*, 9th ed., Worth Publishers, New York, 2000.

Katzung BG, ed. *Basic and Clinical Pharmacology*, 10th ed., McGraw-Hill, New York, 2007.

Katzung BG, Trevor AJ. *Pharmacology, Examination & Board Review*, Prentice-Hill International, Englewood Cliffs, NJ, 1995.

Kramer JA, Sagartz JE, Morris DL. The application of discovery toxicology and pathology towards the design of safer pharmaceutical lead candidates, *Nature Reviews Drug Discovery* 6:636–649 (2007).

Langer R. Where a pill won't reach, *Scientific American* April:50–57 (2003).

Levine RR. Pharmacology—Drug *Actions and Reactions*, 6th ed., Parthenon Publishing Group, New York, 2000.

Marx U, Sandig V. *Drug Testing In Vitro, Breakthroughs and Trends In Cell Culture Technology*, Wiley-VCH, Weinheim, Germany, 2007.

Page CP, Curtis MJ, Sutter MC, et al, eds. *Integrated Pharmacology*, 2nd ed., Mosby, Edinburgh, UK, 2002.

Rang HP, Dale MM, Ritter JM. *Pharmacology*, 3rd ed., Churchill Livingstone, Edinburgh, UK, 1996.

Saunders LM, Hendren RW. *Protein Delivery Physical Systems, Pharmaceutical Biotechnology*, Vol. 10, Plenum Press, New York, 1997.

Van Dam D, De Devn PP. Model organisms: drug discovery in dementia: the role of rodent models, *Nature Reviews Drug Discovery* 5:956–970 (2006).

Vogelson CT. Advances in Drug Delivery Systems, *Modern Drug Discovery*, 4:49–50, 52 (2001).

CHAPTER 6

CLINICAL TRIALS

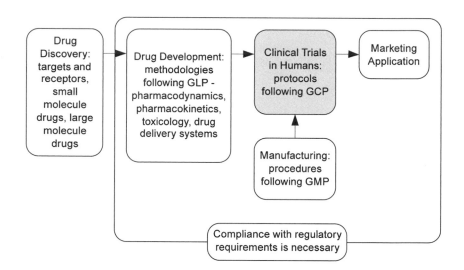

6.1	Definition of Clinical Trial	177
6.2	Ethical Considerations	177
6.3	Clinical Trials	181
6.4	Regulatory Requirements for Clinical Trials	186
6.5	Role of Regulatory Authorities	199

6.6   Gene Therapy Clinical Trial   199
6.7   Case Study #6   200
6.8   Summary of Important Points   204
6.9   Review Questions   205
6.10   Brief Answers and Explanations   205
6.11   Further Reading   206

6.1 DEFINITION OF CLINICAL TRIAL

After the lead compound has been optimized and tested in the laboratory, and pharmacological studies have been conducted to show that the lead compound has the potential to become a drug, it is ready for clinical trial in humans. Exhibit 6.1 presents some information about a typical clinical trial.

What is a clinical trial? According to the International Conference on Harmonization (ICH, see Section 7.11), the definition of a clinical trial or study is as follows:

Any investigation in human subjects intended to discover or verify the clinical, pharmacological and/or other pharmacodynamic effects of an investigational product, and/or to identify any adverse reactions to an investigational product, and/or to study absorption, distribution, metabolism, and excretion of an investigational product with the object of ascertaining its safety and/or efficacy.

6.2 ETHICAL CONSIDERATIONS

Before a drug is put forward for a clinical trial, there are ethical and regulatory constraints for the design and conduct of a clinical trial that have to be considered.

The United States National Institutes of Health (NIH) has stipulated seven ethical requirements to ensure that, before a trial begins, there is proper consideration of ethical issues and the trial subjects are protected. The essential tenet is that the potential exploitation of human subjects must be minimized and the risk–benefit ratio must be favorable. There are seven ethical requirements:

Exhibit 6.1   Typical Clinical Trial

In the past 20 years, the average number of trials per new drug has increased from 30 to more than 70. The number of patients recruited for testing a drug in a typical submission for marketing approval has increased from about 1500 to 5200.

Source: Clinical Trial, *Contract Pharma*, June 2001.

- Social value
- Scientific validity
- Fair subject selection
- Informed consent
- Favorable risk–benefit ratio
- Independent review
- Respect for human subjects

6.2.1   Social Value

This requirement is to ensure that the clinical trial is justified based on scientific research and will result in improvements in health or advancement of scientific knowledge. In this way, resources are not directed at nonmeaningful clinical research and human subjects are not being exploited.

6.2.2   Scientific Validity

The clinical trial should be conducted methodically with clear objectives and outcomes that are statistically verifiable. The preclinical and toxicological data should have been carefully analyzed and should confirm the scientific finding. The trial should not be biased and should be able to be executed without unreasonable caveats and conditions.

6.2.3   Fair Subject Selection

Selection of subjects is based on scientific objectives and not on whether the subject is privileged or vulnerable, or because of convenience (Exhibit 6.2). Inclusion and exclusion criteria are well thought out and designed solely to satisfy the scientific basis being put forward. There must be documented evidence to support the choice of selection criteria.

Exhibit 6.2   An Example of an Early Clinical Trial

In 1917, comparative studies were carried out in Georgia (USA) to evaluate the effects of diets on children with pellagra. Children were selected from orphanages. This practice would not be allowed today, as institutionalized children, who could not defend their rights, were taken advantage of.

Source: National Institutes of Health. *What Is a Clinical Trial?* http://www.cancer.gov/clinicaltrials/learning/what-is-a-clinical-trial [accessed September 18, 2007].

6.2.4 Informed Consent

Subjects are to be informed about the aims, methods, risks, and benefits of the trial. The availability of alternatives should be explained to the subjects. Subjects should not be pressured into enrolling in the trial, but rather should voluntarily join in and should be able to leave the trial at any time without duress or penalty. For young and incapacitated people who are not able to understand the requirements and implications of the trial, proxy decision from their representatives (parents or guardians) must be obtained.

6.2.5 Favorable Risk–Benefit Ratio

The risk–benefit ratio should be analyzed and, wherever possible, clinical trial subjects should be subjected to minimal risk and maximal benefit. The risk–benefit ratio should be based on proven scientific data gathered at the preclinical stage. A clinical trial should not be conducted if there is a doubt about the risk–benefit ratio.

6.2.6 Independent Review Board/Independent Ethics Committee (IRB/IEC)

An independent review is to ensure that an independent party assesses the clinical trial so the question of conflict of interest is addressed. The IRB/IEC acts as a third party to oversee the welfare of the trial subjects and to ensure that the trial is conducted in accordance with the study being put forward.

Members of the IRB/IEC may consist of clinicians, scientists, lawyers, religious leaders, and laypeople to represent different viewpoints and protect the rights of the subjects. The investigator is to inform the IRB/IEC if there are changes in the research activity. Such changes, if they present risks to the subjects, have to be approved before the trial continues. The IRB/IEC has the right to stop a trial or require that procedures and methods be changed.

6.2.7 Respect for Human Subjects

Subjects should be protected and their progress in the trial monitored closely, and appropriate treatments should be provided. New developments in the trial, either risks or benefits, must be relayed to the subjects without prejudice, and the subject's decisions should be honored.

Outcomes from the trial must be communicated to the subjects promptly and in an unbiased way. In addition to the ethical guidelines by the NIH, the World Medical Association has formalized a document called the *Declaration of Helsinki—Ethical Principles for Medical Research Involving Human Subjects* to describe the constraints on research involving human beings. Those countries that have signed this declaration are bound by the ethical principles. An extract of this document is given in Exhibit 6.3.

Exhibit 6.3   World Medical Association Declaration of Helsinki

Ethical Principles for Medical Research Involving Human Subjects: The World Medical Association has developed the Declaration of Helsinki as a statement of ethical principles to provide guidance to physicians and other participants in medical research involving human subjects. Medical research involving human subjects includes research on identifiable human material or identifiable data.

It is the duty of the physician to promote and safeguard the health of the people. The physician's knowledge and conscience are dedicated to the fulfillment of this duty.

The Declaration of Geneva of the World Medical Association binds the physician with the words "the health of my patient will be my first consideration," and the International Code of Medical Ethics declares that "medical progress is based on research, which ultimately must rest in part on experimentation involving human subjects. In medical research on human subjects, considerations related to the well being of the human subject should take precedence over the interests of science and society."

The primary purpose of medical research involving human subjects is to improve prophylactic, diagnostic, and therapeutic procedures and the understanding of the etiology and pathogenesis of disease. Even the best proven prophylactic, diagnostic, and therapeutic methods must continuously be challenged through research for their effectiveness, efficiency, accessibility, and quality.

In current medical practice and in medical research, most prophylactic, diagnostic, and therapeutic procedures involve risks and burdens.

Medical research is subject to ethical standards that promote respect for all human beings and protect their health and rights. Some research populations are vulnerable and need special protection. The particular needs of the economically and medically disadvantaged must be recognized. Special attention is also required for those who cannot give or refuse consent for themselves, for those who may be subject to giving consent under duress, for those who will not benefit personally from the research, and for those for whom the research is combined with care.

Research investigators should be aware of the ethical, legal, and regulatory requirements for research on human subjects in their own countries as well as applicable international requirements. No national ethical, legal, or regulatory requirement should be allowed to reduce or eliminate any of the protections for human subjects set forth in this Declaration.

6.3 CLINICAL TRIALS

Clinical trials are divided into four phases. These are Phase I to Phase IV (Fig. 6.1). These trials are conducted with specific purposes to evaluate the safety and effectiveness of the drug in defined population groups. A recent proposal is to conduct "Phase 0"—a microdosing trial on subjects. Exhibit 6.4 provides more details on this new topic.

6.3.1 Phase I

The Phase I clinical trial is the first experiment in which a drug is tested on the human body. The primary aim of the trial is to assess the safety of the new drug. Other areas of study include pharmacokinetics (absorption, distribution, metabolism, and excretion) and pharmacodynamics.

Normally, healthy volunteers are recruited for the Phase I trial. In many cases, volunteers are compensated financially for participation in the Phase I trial. However, in some situations, patients who are critically ill or have terminal disease are presented with the option to be included in the trial after due consideration of the risk–benefit ratio. Phase I trials are usually conducted

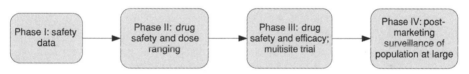

Figure 6.1   The four phases of clinical trials.

Exhibit 6.4   Phase 0, Microdosing

The term microdosing, or Phase 0, refers to the *in vivo* testing of drug candidates in humans at very low dosages. Typically, the dosages are 100 times less than the intended therapeutic dose. At this dosage whole-body reaction is unlikely to happen, yet with sensitive analytical techniques, cellular responses can be studied. Through this, pharmacokinetics are evaluated at very low risk. Such a study can be carried out before much time and expense are spent on preclinical animal studies and can enable viable drugs to be identified earlier and with less cost.

For this method, the drug candidate is labeled with a radioisotope, such as carbon-14. The ADME of the compound within the body can be monitored by analyzing samples using high sensitivity instrumentation, for example, accelerator mass spectroscopy.

with open label; that is, the subjects are aware of the drugs that they are being given.

The number of subjects is normally between 10 and 100 people. The starting doses are based on the results of preclinical work as described in Chapter 5. Doses are increased as the trial progresses for subjects recruited at later stages, as the effect of the experimental drug becomes apparent. Subjects are monitored closely to check their tolerance of the drug and incidents of side effects. Depending on the study, samples of blood, urine, or stool and other physiological information may be obtained for analysis to evaluate absorption, distribution, metabolism, and elimination of the drug in the body. Other observations about how the subject feels (e.g., pain, headache, fever, malaise, and irritability) and vital signs (blood pressure, heart rate) and behavioral matters are taken into account.

Depending on the complexity of the trial, the cost for Phase I is around US$10 million and the trial may last from several months to a year.

6.3.2   Phase II

The aim of the Phase II clinical trial is to examine the safety and effectiveness of the drug in the targeted disease group. A series of doses of varying strengths may be used.

It is now common to conduct Phase II trials with a control group in conjunction with the test group given the drug. The control group is given either the current standard treatment or placebo (an inert nondrug substance). Again, the risk–benefit profile has to be assessed as to whether the trial should use placebo or standard treatment to ensure the subjects' well-being is not compromised during the trial. Patients are randomized to either the control group or the drug group without bias. The randomization procedure is important, because the information will provide comparative data about the safety and effectiveness of the drug versus placebo or standard treatment. Phase II clinical trials can be divided into IIa and IIb, with IIb being an extension to the safety and efficacy studies assessed in IIa.

Another practice is to blind the trial, which means that the subjects are not privy to whether they receive the placebo or drug. In some trials, even the investigator is unaware of whether the subject is in the control or active group. This is called a double-blind trial. The rationale is to eliminate the possibility of bias affecting the trial results.

The result of the Phase II trial is information needed to determine the effective dose and the dosing regimen of frequency and duration. Specific clinical endpoints or markers are used to assess interaction of drug and disease. There are two types of markers: definitive and surrogate. For example, in the case of cancer or hypertension, the definitive markers are mortality and stroke, respectively, and the surrogate markers may be tumor size, or cancer-associated proteins p53, TGF-α in the case of cancer, and blood pressure or cholesterol level in hypertension. Statistical analysis is carried out to evaluate the influence of the drug on different patient groups, to determine the optimum conditions.

For Phase II, the number of patients is normally in the vicinity of 50–500. The trial may take 1–2 years or more to complete, depending on the study numbers and availability of patients. The cost for such a trial can be more than US$20 million.

The success rate of Phase I and II studies is estimated at around 30%. An example of a Phase II trial is presented in Exhibit 6.5.

6.3.3 Phase III

After the successful completion of the Phase II trial, the objective of Phase III is to confirm the efficacy of the drug in a large patient group. Phase III is an extension of Phase II, and the trial is normally conducted in several hospitals in different demographic locations, to determine the influence of ethnic responses, together with incorporation of new criteria for fine-tuning the trial. This trial is also known as a multisite trial.

Because the results are crucial to the determination of the drug's effectiveness, the Phase III trial is referred to as the pivotal trial, as it can make or break the success of a drug. The methodology of the trial has to be carefully prepared so that meaningful results can be gathered at the conclusion of the trial. Extensive statistical analyses are performed to evaluate the data. If for any reason the drug does not show significant advantage over current treatment, the result may be refined and certain subgroups are analyzed to determine if the effects are greater in one group than in the other. The study results provide comprehensive data for understanding the critical parameters of safety and effectiveness of the drug.

These results enable the pharmaceutical company to set the dosage, treatment frequency, duration, and target patient groups for the drug. The information and analyses gathered, together with chemistry, manufacturing, and control (CMC, see Chapter 8), are submitted to regulatory authorities to seek approval to market the drug. An example of a Phase III trial is presented in Exhibit 6.6.

Patient numbers for Phase III can vary from several hundreds to thousands. The larger number is normally for trials involving infectious diseases such as influenza or vaccines, as these can recruit up to tens of thousands of people to provide a larger sample size for detecting "rare" serious effects. Statistical proof to show the efficacy of the drug for the targeted patient group has to be established. At least two Phase III trials need to be conducted. Because of the magnitude of the trial, the duration may be 3–5 years and the cost is around US$50–100 million.

6.3.4 Phase IV

Phase IV clinical trials are postmarketing approval trials to monitor the efficacy and side effects of the drug in an uncontrolled real-life situation. This is

Exhibit 6.5   Phase II Study: Effect of EGb 761® on Patients with Mild to Moderate Alzheimer's Disease

The aim of this study is to measure the effect of EGb 761® versus placebo on the ratio of the isoform of the protein precursor of beta amyloid platelets, in patients with mild to moderate Alzheimer's disease.

Study Type: Interventional

Study Design: Treatment, randomized, double-blind, placebo control, parallel assignment, efficacy study

Official Title: Effect of EGb 761® on the Ratio of the Isoforms of the Protein Precursor of Beta Amyloid Platelets on Patients With Mild to Moderate Alzheimer's Disease. A Phase II, Randomized, Double-Blind Trial, on Parallel Groups Versus Placebo

Primary Outcome Measures: Effect of EGb 761® on the ratio of the isoform of the protein precursor of beta amyloid platelets

Secondary Outcome Measures: Efficacy of EGb 761® on the cognitive functions and safety of EGb 761® at a dosage of 240 mg per day

Total Enrollment: 40

Study Start: July 2005

Eligibility: Ages eligible for study, 50–85 years; genders eligible for study, both

Inclusion Criteria:

   Female or male 50–85 years old with a caregiver

   Mini Mental Status (MMS) test between 16 and 26 inclusive

   Clinical Dementia Rating (CDR) test inferior or equal to 1

   National Institute of Neurological and Communicative Disorders and Stroke/Alzheimer's Disease and Related Disorders Association (NINCDS/ADRDA) test positive for Alzheimer's disease

   Diagnostic and Statistical Manual of Mental Disorders, 4th edition (DSM IV) test positive for dementia

Exclusion Criteria:

   Patient already treated by medicines that could interfere with the study

   Low level of vitamin B_{12} and folate, which are considered as clinically relevant

   Clinically relevant pathologies (e.g., pulmonary illness, cardiovascular illness, evolutive cancer, neurological illness, blood illness)

Source: National Institutes of Health, ClinicalTrials.gov. *EGb 761®*. http://www.clinicaltrial.gov/ct/show/NCT00500500?order=3 [accessed Sepember 18, 2007].

Exhibit 6.6   Phase III Study: A Study of Lopinavir/Ritonavir Tablets Comparing Once-Daily Versus Twice-Daily Dosing in Antiretroviral-Experienced, HIV-1 Infected Subjects

The purpose of this study is to determine whether once-daily dosing of the lopinavir/ritonavir (Kaletra) tablet in combination with investigator-selected nucleoside/nucleotide reverse transcriptase inhibitors will reduce HIV viral load to very low levels in patients who have detectable viral loads with their current antiretroviral therapy.

Study Type: Interventional

Study Design: Treatment, randomized, open label, active control, parallel assignment, safety/efficacy study

Official Title: A Phase III, Randomized, Open-Label Study of Lopinavir/Ritonavir Tablets 800/200 mg Once-Daily Versus 400/100 mg Twice-Daily When Co-administered With Nucleoside/Nucleotide Reverse Transcriptase Inhibitors in Antiretroviral-Experienced, HIV-1 Infected Subjects

Primary Outcome Measures:

   Proportion of subjects responding (i.e., not demonstrating virologic failure) based on the FDA Time to Loss of Virologic Response Algorithm (time frame: 48 weeks)

   Frequency and percentage of treatment-emergent adverse events (time frame: 48 weeks)

   Frequency and percentage of very low and very high laboratory values compared between treatment groups (time frame: 48 weeks)

   Change from baseline to each visit for laboratory values will be summarized by treatment group (time frame: 48 weeks)

Secondary Outcome Measures:

   Proportion of subjects responding (i.e., not demonstrating virologic failure) at each visit based on the FDA Time to Loss of Virologic Response Algorithm (time frame: 48 weeks)

   Relationship between baseline genotypic resistance and virologic response (time frame: 48 weeks)

   Frequency and percentage of the emergence of mutations at virologic rebound that were not present at baseline (time frame: 48 weeks)

Total Enrollment: 600

Study Start: September 2006

Eligibility: Ages eligible for study, 18 years and above; genders eligible for study, Both

Inclusion Criteria:

Subject has provided written informed consent.

Subject is currently receiving an antiretroviral regimen that has not changed for at least 12 weeks.

Subject has never received lopinavir/ritonavir.

Subject is currently failing his/her antiretroviral regimen with the most recent two consecutive prestudy plasma HIV-1 RNA levels >400 copies/mL with the most recent being >1000 copies/mL, and in the investigator's opinion, should change therapy.

Screening plasma HIV-1 RNA > 1000 copies/mL.

There is no $CD4^+$ T-cell count restriction.

There is no significant history of cardiac, renal, neurological, psychiatric, oncologic, metabolic, or hepatic disease that would adversely affect his/her participating in the study.

Based on HIV-1 drug resistance genotypic test results at the Screening Visit and prior treatment history, the investigator considers lopinavir/ritonavir plus at least two NRTIs to be an appropriate treatment for the subject.

Subject does not require and agrees not to take any antiretroviral medication except lopinavir/ritonavir and NRTIs.

Source: National Institutes of Health, ClinicalTrials.gov. *Kaletra*. http://www.clinicaltrial.gov/ct/show/NCT00358917?order=20 [accessed September 18, 2007].

also known as a postmarket surveillance trial. Information about the effectiveness of the drug compared with established treatment, side effects, patient's quality of life, and cost effectiveness is collated.

Any adverse events are reported and acted on to ensure patients' welfare is not compromised by the drug. Serious events are reported to regulatory authorities within a specified time and, if deemed necessary, the drug is recalled or doctors and patients are notified.

6.4   REGULATORY REQUIREMENTS FOR CLINICAL TRIALS

Clinical trials are performed under Good Clinical Practice (GCP). Up to now, there has been no reference to the regulatory requirement. The reality is that every trial has to be approved and carried out under regulatory compliance to GCP requirements. Otherwise, the trials may be considered as noncompliant and become invalid.

A normal course of event in initiating a clinical trial is for the Sponsor (see below) to prepare an Investigator's Brochure and select an Investigator to conduct the trial. The Sponsor and Investigator then prepare the trial protocol, which is submitted to the Institutional Review Board or Independent Ethics Committee for approval. An application to the regulatory authority, such as the US Food and Drug Administration (FDA) or the Medicines and Healthcare Products Regulatory Agency (MHRA) of the United Kingdom, is then submitted (see Chapter 8).

Different countries have different requirements for clinical trials. However, the two main documents that most clinical trials are based on are the documents from the FDA and the ICH. The relevant documents are:

- FDA 21 CFR Parts 50, 56, 312
- ICH Harmonized Tripartite Guideline for Good Clinical Practice (Table 6.1 shows the ICH documents related to clinical trials)

In the United States, an IND (Investigational New Drug) application has to be filed with the FDA. For other countries, a notification has to be submitted to the respective regulatory authorities. For example, Clinical Trial Exemption (CTX) applications are required for the United Kingdom, Clinical Trial Notification (CTN) and CTX for Australia, and a Clinical Trial Certificate (CTC) for Singapore and the European Medicines Agency (EMEA). A more extensive discussion concerning regulatory authorities and the processes and procedures of applications is presented in Chapters 7 and 8. The relevant authority will review the application. A positive response from the authority is required before the trial can commence.

The scope of this book does not allow a discussion of all the requirements for GCP. Readers are referred to Exhibit 6.7 for the headings in the relevant regulatory documents to gain further understanding of the requirements. Some important issues, however, are discussed to clarify the important aspects and requirements for clinical trials in accordance with GCP. Some of these aspects are:

- Investigator
- Investigator's Brochure
- Informed consent
- Protocol
- Inclusion and exclusion criteria
- Case report form
- Randomization, placebo-controlled and double-blinded
- Monitoring
- Adverse events
- Statistics
- Sponsor
- Clinical research organization
- Surrogate markers

TABLE 6.1   ICH Clinical Study Efficacy Guidelines

Document	Title
E1	*The Extent of Population Exposure to Assess Clinical Safety for Drugs Intended for Long-Term Treatment of Non-Life Threatening Conditions*
E2A	*Clinical Safety Data Management: Definitions and Standards for Expedited Reporting*
E2B (R3)	*Clinical Safety Data Management: Data Elements for Transmission of Individual Case Safety Reports*
E2c (R1)	*Clinical Safety Data Management: Periodic Safety Update Reports for Marketed Drugs*
E2D	*Post-Approval Safety Data Management: Definitions and Standards for Expedited Reporting*
E2E	*Pharmacovigilance Planning*
	Clinical Study Reports
E3	*Structure and Content of Clinical Study Reports*
E4	*Dose–Response Information to Support Drug Registration*
	Ethnic Factors
E5 (R1)	*Ethnic Factors in the Acceptability of Foreign Clinical Data*
	Good Clinical Practice
E6 (R1)	*Good Clinical Practice*
	Clinical Trials
E7	*Studies in Support of Special Populations: Geriatrics*
E8	*General Considerations of Clinical Trials*
E9	*Statistical Principles for Clinical Trials*
E10	*Choice of Control Group and Related Issues in Clinical Trials*
E11	*Clinical Investigation of Medicinal Products in the Pediatric Population*
	Clinical Evaluation by Therapeutic Category
E12	*Principles for Clinical Evaluation of New Antihypertensive Drugs*
	Clinical Evaluation
E14	*The Clinical Evaluation of QT/QTc Interval Prolongation and Proarrhythmic Potential for Non-Antiarrhythmic Drugs*
	Pharmacogenomics
E15	*Terminology in Pharmacogenomics*

Source: International Conference on Harmonization. *Efficacy Guidelines.* http://www.ich.org/cache/compo/475-272-1.html [accessed September 4, 2007].

Exhibit 6.7   Examples of GCP Requirements

Main Heading from 21CFR Part 50—Protection of Human Subjects

Subpart A—General Provisions

Subpart B—Informed Consent of Human Subjects

Subpart C—Protection Pertaining to Clinical Investigations Involving Prisoners as Subjects

Main Heading from 21CFR Part 312—Investigational New Drug Application

Subpart A—General Provisions

Subpart B—IND

Subpart C—Administrative Actions

Subpart D—Responsibilities of Sponsors and Investigators

Subpart E—Drugs Intended to Treat Life-Threatening and Severely Debilitating Illnesses

Subpart F—Miscellaneous

Subpart G—Drugs for Investigational Use in Laboratory Research Animals or *In Vitro* Tests.

Main Heading from 21CFR Part 56—Institutional Review Board

Subpart A—General Provisions

Subpart B—Organization and Personnel

Subpart C—IRB Functions and Operations

Subpart D—Records and Reports

Subpart E—Administrative Action for Noncompliance

ICH Harmonized Tripartite Guideline for Good Clinical Practice: Section II

Introduction

Glossary

The Principle of ICH GCP

Institutional Review Board/Independent Ethics Committee (IRB/IEC)

Investigator

Sponsor

Clinical Trial Protocol and Protocol Amendments

Investigator's Brochure

Essential Documents for the Conduct of a Clinical Trial

6.4.1 Investigator

The Investigator is the person who conducts the trial. If there is a team in the investigation, then there is a Principal Investigator. This person is normally an expert in the field of the disease to be investigated. The Investigator's responsibility is to ensure that GCP is being implemented during the course of the trial and the subjects' rights and welfare are respected. Another important point is that the Investigator has to maintain impartiality. He/she is not an employee of the company (the Sponsor where the drug is developed), to show that there is transparency and no conflict of interest, nor is there financial gain if the drug is successful.

6.4.2 Investigator's Brochure

The Investigator's Brochure is a collection of information prepared and updated by the Sponsor for the Investigator. The information consists of all the data relevant to the drug under investigation, including properties of the drug, the PK and PD, and toxicity results on animals (Exhibit 6.8).

6.4.3 Informed Consent

This was described earlier and is a fundamental aspect that has to be included in a clinical trial.

6.4.4 Protocol

This document sets out how a trial is to be conducted. It contains the rationale for the clinical trial, the methodology on how the trial is designed, the number

---

Exhibit 6.8   Investigator's Brochure

Description of the Drug

Physical, chemical, and biological properties
Dosage form, storage conditions, stability

Pharmacology

Pharmacodynamics
Pharmacokinetics
Toxicology

Source: International Conference on Harmonization. *Guideline for Good Clinical Practice E6(R1)*. http://www.ich.org/LOB/media/MEDIA482.pdf [accessed September 17, 2007].

of subjects to be recruited, the biomarkers (refer to Exhibit 6.9) or endpoints to show effectiveness of the drug, the statistical methods to be used to analyze the data, how the subjects are protected in the trial, informed consent and confidentiality, as well as welfare and frequency of monitoring. Exhibit 6.10 summarizes the required information for a protocol.

6.4.5 Inclusion and Exclusion Criteria

These criteria set out the conditions under which a person may or may not be included in the trial. The criteria may include the disease type, medical history, age group, gender, and so on. It is necessary to set out the parameters for the criteria to enable meaningful analysis to be made for assessment of the safety and effectiveness of the experimental drug. Subjects are screened before commencement to ensure that they meet the recruitment criteria before being admitted to the trial.

6.4.6 Case Report Form

All the information relating to a subject is recorded in the Case Report Form. The commencement of the trial will include gathering baseline data from the

---

Exhibit 6.9   Biomarkers

According to the Biomarker Definitions Working Group, a biomarker is a characteristic that is objectively measured and evaluated as an indicator of normal biological processes, pathogenic processes, or pharmacologic responses to a therapeutic intervention. Biomarkers are used to measure a patient's pharmacological response to a drug to indicate the safety and effectiveness of the drug. They represent the endpoints of the patient's state of health.

In clinical trials, biomarkers are used to indicate a particular disease state and its progression. They may be used as surrogate markers in the evaluation of the effectiveness of a drug as representative of the natural endpoint such as survival rate or irreversible morbidity.

Biomarkers include, for example, cholesterol level, blood pressure, viral load, enzyme concentration, and tumor size.

Appendix 7 lists some of the biomarkers regularly tested in the laboratory.

Source: Biomarkers Definitions Working Group. Biomarkers and surrogate endpoints: preferred definitions and conceptual framework, *Clinical Pharmacology and Therapeutics* 69:89–95 (2001).

Exhibit 6.10   Clinical Trial Protocol

The following information is to be included (ICH GCP):

Protocol title

Name and address of Sponsor and Monitor

Name of authorized person

Name of Sponsor's medical expert

Name of Investigator responsible for the trial

Name of physician responsible for trial-related medical decisions

Name of clinical laboratory and other institutions involved in the trial

Name and description of the clinical trial protocol

Summary of results from nonclinical studies

Potential risks and benefits to human subjects

Description and justification for route of administration, dosage, and treatment plan

Compliance to GCP

Description of the population to be studied

Reference literature and related data

Standard operating procedures

Source: International Conference for Harmonization. *Good Clinical Practice*. http://www.ich.org/ [accessed September 8, 2007].

subjects. Then at each defined stage of a trial, the designed markers or endpoints are analyzed and recorded. These may include dosing information, observations, vital signs, blood analysis, targeted enzyme levels, hormonal changes, and so on. There are also records for patient's comments, adverse events, and the Investigator's spontaneous comments. The Case Report Forms are part of the regulatory document, and the data are statistically analyzed and submitted to regulatory authorities for marketing approval of the drug. An example of a hypothetical Case Report Form is presented in Exhibit 6.11.

6.4.7   Randomization, Placebo-Controlled and Double-Blinded

Some trials are conducted with open labels; that is, the subjects are aware of the type of drugs that they have been provided. However, in most trials, the subjects are divided into treatment and control groups using a statistical randomization process (Exhibit 6.12). The aim is to reduce bias in the studies. Subjects are divided into control and active groups.

In a double-blinded study, both the Investigator and the subjects are unaware of whether they receive the drug or the placebo. The randomization code is held in confidence and is opened at the end of the trial for data analysis or in cases where adverse events occurred.

Exhibit 6.11   An Example of a Case Report Form

CASE REPORT FORM

Personal Data
Patient's Last name: _____ First name: _____ Middle initial/s: _____
Age: _____ years   Sex: M/F   Race: _____ Ethnicity: _____
Address: _____
Telephone number: _____ Email: _____

Study Data
Study number: _____ IRB number: _____ Patient number: _____
Date of visit: _____ Details: _____

Clinical Data
Height: _____ Weight: _____
Symptoms, signs and adverse reactions: _____
Associated disease history: _____
Medication taken: _____

Laboratory Analysis
Hemoglobin: _____ Platelet count: _____
Bilirubin: _____ ALT: _____
Cholesterol/LDL: _____ Cholesterol/HDL: _____

Other Details:
Patient's comments: _____
Physician's name/address/telephone number: _____
Person completing this form: _____
Signature: _____ Date: _____

Exhibit 6.12   Randomization Techniques

Randomized Parallel Group Fixed Dose: Subjects are divided into several groups, such as placebo, 10 mg, 20 mg, and 40 mg. Subjects continue with this regimen for the duration of the trial.

Randomized Parallel Group Forced Titration: Subjects are divided into placebo and active groups. Active groups all start with the same dose, for example, 10 mg. One group continues with 10 mg, another group later increases to 20 mg and stays at this dose. A third active group then increases from 10 mg to 20 mg and finally to 40 mg progressively.

Randomized Parallel Group Optional Titration: Subjects are divided into placebo and active groups. Active groups all start with the same dose, say, 10 mg. Depending on response and safety assessment, dose can be increased to 20 mg and then 40 mg for selected subjects.

Randomized Crossover Design: Subjects are divided into placebo and active groups. After some time these two groups crossover: the initial placebo group now becoming the active group and vice versa. There may be a washout period before the crossover to enable the effect of the placebo and active to wash out. This method requires a smaller number of subjects and is useful in cases for studying rare or more stable illnesses.

Randomized Latin Square Design: This is a crossover design with dose ranging. For example, the regimens for six separate groups are: (1) placebo, 10 mg, 20 mg; (2) placebo, 20 mg, 10 mg; (3) 10 mg, 20 mg, placebo; (4) 10 mg, placebo, 20 mg; (5) 20 mg, placebo, 10 mg; and (6) 20 mg, 10 mg, placebo. This is a very powerful method to show the efficacy of the drug under trial.

Source: Monkhouse DC, Rhodes CT, eds. *Drug Products for Clinical Trials*, Marcel Dekker, New York, 1998.

6.4.8   Monitoring

An important aspect of the trial is the meticulous monitoring required. This is a process to interact with the subjects: monitoring their well-being, the effects of drug and placebo, adverse events, and so on. Information is recorded on the Case Report Forms. All the processes are recorded in accordance with Standard Operating Procedures, which describe how the trial is to be conducted, and GCP.

6.4.9 Adverse Events

These are the unintended reactions of the subjects as a consequence of taking the drug or placebo. Subjects are checked and, if the adverse events are serious, subjects may be temporarily removed from the trial. If there is a persistence of adverse event, the subject may be withdrawn from the trial. The randomization code may be broken (opened) to determine whether the subject has been given the drug or placebo. Refer to Exhibit 6.13 for a recent clinical trial that resulted in unexpected events. There are regulatory guidelines for accessing the adverse events to enable informed decisions to be made. Some examples of these "toxicity gradings" are presented in Appendix 8.

6.4.10 Statistics

Statistics plays a major role in the design of the clinical trial. The groups or subgroups to be studied, the frequencies, dosages, and the markers used to monitor drug efficacy are all important factors to consider. Statistical analysis provides the means to demonstrate, at a certain confidence level, whether the

---

Exhibit 6.13   A Monoclonal Antibody Trial—TGN1412

TGN1412 (CD28-SuperMAb) is a humanized monoclonal antibody that binds and acts as an agonist for the CD28 receptor of the immune system's T cell. It is intended to treat B cell chronic lymphocytic leukemia (B-CLL) and rheumatoid arthritis.

The first human clinical trial was conducted in March 2006 in the United Kingdom but unexpectedly caused systemic failures in the subjects, even though the dose administered was 500 times lower than the dose that is safe for animals, on a per kilogram basis. Of the eight subjects enrolled, two were on placebo. The six active subjects were hospitalized, and four showed signs of multiple organ dysfunction, with one having signs of developing cancer. The trial was stopped.

Later investigation confirmed the subjects experienced a "cytokine storm," and their white blood cells vanished almost completely upon administration of the MAb. Although earlier preclinical work using primates found no visible side effects at significantly higher dose, the case with humans appeared to be totally unexpected. This led the regulatory authority MHRA in April 2006 to suggest that the problem was most likely to be caused by "unforeseen biological actions in humans."

Source: Medicines and Healthcare Products Regulatory Agency. *Clinical Final Report on TGN1412*.http://www.mhra.gov.uk/home/idcplg?IdcService=SS_GET_PAGE&useSecondary=true&ssDocName=CON2023822&ssTargetNodeId=389 [accessed September 18, 2007].

drug is effective. This is normally reported in the form of a statistical power test, analyzing the Type I and Type II errors. A more detailed discussion about statistics in clinical trials is presented in Exhibit 6.14.

6.4.11   Sponsor

This is the organization or individual that initiates the clinical trial and finances the study. The organization may be a government department, pharmaceutical

---

Exhibit 6.14   Statistics for Clinical Trial

Based on the design of the trial protocol, statistics are used to calculate the number of people to be recruited for the trial, how the trial should be randomized (Exhibit 6.12), and finally analysis of the data. Statistics provide a nonbiased means to evaluate the trial results.

The objective of clinical trials is to demonstrate the safety and effectiveness of the drug compared to placebo or control. The statistical method normally used is known as hypothesis testing.

For example, we wish set up the null hypothesis (H_0) and claim that there is no difference (δ) between the control or placebo (μ_C) and the drug being trialed (μ_D). This is set against the alternative hypothesis (H_A), which states that indeed there is a difference (δ) between the control and drug under trial. Mathematically, the representation is given as follows:

$$H_0 : \mu_C - \mu_D = 0$$

$$H_A : \mu_C - \mu_D \neq 0$$

where μ_C and μ_D stand for the true mean of the population in the control group and the drug group, respectively.

In a trial, the parameters for comparison may be the mean level of cholesterol (in the case of a cholesterol-lowering drug), the amount of antibodies in the body (a vaccine trial), or the reduction in the size and severity of tumor (a cancer trial). A test statistic is normally used to compute and compare the means for the placebo and active groups.

One method is the Z distribution for testing the hypothesis with respect to differences in two means:

$$Z = \frac{\bar{x}_D - \bar{x}_C}{\sigma \sqrt{\dfrac{1}{N_D} + \dfrac{1}{N_C}}}$$

where $\bar{x}_D$ and $\bar{x}_C$ are the mean of their respective samples, N_D and N_C are the respective sample size for the drug and control groups, and σ is the standard error.

Another important aspect is to ensure that we limit the errors in drawing the wrong conclusion. These are described as Type I and Type II errors:

Type I Error (α, False Positive): The probability of wrongly concluding that a difference exists where in fact there is no real difference, thus putting a useless medicine onto the market. Normally, a 5% level of significance is chosen, which means there is a 95% confidence in the decision (i.e., $\alpha = 0.05$). The value of α may need to be even smaller for testing efficacy of a potentially dangerous medication. This value for α is customarily a condition required by the regulatory agency and is typically around 0.05–0.1.

Type II Error (β, False Negative): The probability of wrongly concluding that there is no difference when in fact there is a difference, which means keeping a good medicine away from patients, with the manufacturer missing an opportunity to market the drug. Type II error is normally limited to 5–20% (i.e., $\beta = 0.05$–0.2). The boundaries of Type II error are normally set by the company.

This leads to the term Power $(1 - \beta)$, which quantifies the ability of the study to find the true differences of various values of δ. It is the probability of rejecting the null hypothesis when it is false or determining that the alternative hypothesis is true when indeed it is true.

Clinical trials are carried out to show that the null hypothesis is false. The p value is the probability of having an effect by chance if the null hypothesis were actually true. The null hypothesis is rejected in favor of the alternative hypothesis when the p value is less than α.

Once the parameters for the hypothesis and Type I and Type II errors are set, the total number of subjects ($2N$) to be recruited to join the trial can be determined by the equation

$$2N = \frac{4(Z_\alpha + Z_\beta)^2 \sigma^2}{\delta^2}$$

where Z_α and Z_β are obtained from tables of the standardized normal distribution for given α and β; σ and δ are as defined previously.

The number of subjects may need to be more than the calculated figure as the trial has to account for dropouts and subject noncompliances over the duration of the trial.

company, university, or individual. Normally, however, the Sponsor is a pharmaceutical company.

6.4.12 Clinical Research Organization

This is the organization that is contracted by the Sponsor to conduct and monitor the trial. It also provides a certain measure of independence to the trial and enhances the validity of the trial results to be unencumbered by conflict of interest.

6.4.13 Surrogate Markers

Sometimes it is not possible to measure the direct effect of the drug. Endpoints or surrogate biomarkers are used to monitor the pharmacodynamics and pharmacokinetics of the drug. These markers may be changes in blood pressure, cholesterol level, concentrations of certain enzymes, proteins, blood glucose levels, and similar factors (see Table 6.2 for serum tumor markers and Appendix 7 for general biomarkers).

TABLE 6.2 Serum Tumor Markers

Cancer Antigen	Description	Indication
Cancer antigen (CA) 27.29	MAb to a glycoprotein present on apical surface of normal epithelial cells; CA 27.29 elevated in one-third of early-stage breast cancer and two-thirds of late-stage breast cancer	Breast cancer
Carcinoembryonic antigen (CEA)	Oncofetal glycoprotein expressed in normal mucosal cells and overexpressed in adenocarcinoma	Colorectal cancer
Cancer antigen (CA) 19.9	Intracellular adhesion molecule	Primarily pancreatic and biliary tract cancers
Alpha-fetoprotein (AFP)	Major protein of fetal serum	Hepatocellular carcinoma and nonseminomatous germ cell tumors
Beta subunit human chorionic gonodotropin (β-hCG)	Glycoprotein hormone	Germ cell tumors
Cancer antigen (CA) 125	Glycoprotein expressed in epithelium	Ovarian cancer
Prostate-specific antigen (PSA)	Glycoprotein produced by prostatic epithelium	Prostate cancer

Source: Perkins GL, Slater ED, Sanders GK, Prichard JG. Serum tumor markers, *American Family Physician* 68:1075–1082 (2003).

6.5   ROLE OF REGULATORY AUTHORITIES

Government bodies have on occasion accelerated clinical trials against advice from researchers, in response to public demands (Exhibit 6.15). The climate today is that due diligence regarding safety has to be performed before the drug is administered to human subjects and that clinical trial submissions have to be approved by the regulatory authorities before the trial commences.

Regulatory authorities play an important and active role to ensure regulatory compliance in the conduct of a clinical trial. Agencies such as the FDA inspect clinical studies. An inspection of a trial may reveal that the protocol is not being followed strictly, the Investigator may not be involved with the project as much as is expected, there may be a lack of patient care, changes to the protocol may not have been relayed to the IRB, and so on. In such cases, corrective actions have to be implemented immediately and the FDA must be satisfied before the trial can continue. Deficiencies found are reported on Form 583.

6.6   GENE THERAPY CLINICAL TRIAL

As genomic research progresses, the possibility of replacing a person's faulty genes with normal genes becomes a reality (Chapter 4). Currently, there are many ethical and scientific issues facing gene therapy.

Exhibit 6.15   Polio Vaccine Trial

In the 1950s, Dr. Jonas Salk and Dr. Albert Sabin from the University of Pittsburgh (USA) worked on polio vaccines. Salk used inactivated polio virus, whereas Sabin developed a live form of polio virus.

Scientists differed as to which method provided the better vaccine. Both Salk and Sabin agreed that more tests were needed before a mass vaccination program could begin.

The National Foundation, which funded the research, and the American public wanted a mass vaccination urgently. The average incidence of polio in the United States in 1949–1953 was 25.7 cases per 100,000 children. The National Foundation ordered 27 million doses of the Salk vaccine for a trial, and close to one million children were vaccinated (749,236 children from grades 1, 2, and 3 were offered vaccine, and 401,974 completed the trial).

The trial was one of the greatest triumphs in medical history. Church bells rang across the country when the trial results were announced. Within five years, polio was wiped out in the United States.

Source: Meier P. *The Salk Vaccine Trials*. http://www.math.uah.edu/siegrist/ma487/salk.html [accessed January 2, 2002].

Exhibit 6.16   Gene Therapy Trials

In 2007 there were 1309 gene therapy clinical trials. The breakdown is as follows:

Phase I	801
Phase I/II	258
Phase II	205
Phase II/III	13
Phase III	32
Total	1309

Source: Gene Therapy Clinical Trials. *The Journal of Gene Medicine*, Wiley, Hoboken, NJ, 2007.

For a gene therapy clinical trial, the FDA requires that the IND be filed as for normal drug trials. However, there are more stringent requirements on the source and tests being carried out on the gene to be inserted into the subject. There is also the need for closer monitoring, from both the Investigator and the FDA. In addition, the FDA has been conducting safety symposia to educate the Investigator on the safety issues of gene therapy. Exhibit 6.16 shows the number of gene therapy trials being conducted in 2007.

In situations where a drug is designed as a novel mechanism and the definitions of trial parameters are not well defined, the chances of failure are high. Exhibit 6.17 provides some insights into trial failures, showing that the road to clinical success is fraught with uncertainties.

6.7   CASE STUDY #6

Plavix and Gardasil*

This case study presents the clinical trials performed for Plavix (an antiplatelet) and Gardasil (a vaccine against cervical cancer).

Plavix: Two trials were carried out to determine efficacy: the CAPRIE and CURE studies.

Sources: (1) Food and Drug Administration, *Plavix NDA*. http://www.fda.gov/medwatch/safety/2006/Aug_PIs/Plavix_WARNINGS_PI.pdf [accessed August 23, 2007]. (2) Food and Drug Administration, *Gardasil*. http://www.fda.gov/cber/label/hpvmer040307LB.pdf [accessed August 23, 2007].

Exhibit 6.17   Reflections on Some Clinical Trials

An analysis of 656 Phase III clinical trials from 1990 to 2002 showed that 42% of them failed. Of the failed trials, 70 of them provided reasonably detailed data for further analysis. Out of these 70, 50% failed on efficacy and 31% had safety concerns, with the remaining 19% being neither safer nor more effective than the current drugs.

It was also revealed that when drugs were designed as novel mechanisms of action, they incurred twice the failure rate. In cases where the objective endpoints were less definitive, there were 10% more failures. When considered together—a novel mechanism and less objective endpoints—a 70% failure rate was experienced.

Source: The McKinsey Quarterly. *The Online Journal of McKinsey & Co.* http://www. mckinseyquarterly.com/article_page.aspx?ar=1879&L2=12&L3=62&srid=17&... [accessed May 1, 2007].

CAPRIE was a 19,185-patient, 304-center, international, randomized, double-blind, parallel-group trial comparing Plavix (75 mg daily) with aspirin (325 mg daily). The outcome was to compare the first occurrence of new ischemic stroke, new myocardial infarction, or other vascular death. The following are tabulated results:

Factor	Plavix	Aspirin
Patients	9599	9586
Ischemic stroke	438 (4.6%)	461 (4.8%)
Myocardial infarction	275 (2.9%)	333 (3.5%)
Other vascular death	226 (2.4%)	226 (2.4%)
Total	*939 (9.8%)*	*1020 (10.6%)*

Plavix resulted in overall reduction of outcome events.

The CURE study involved 12,562 patients randomized to receive Plavix (300 mg loading dose followed by 75 mg daily) or placebo and were treated for up to a year. Patients also received aspirin or other standard treatment such as heparin. The results showed that Plavix had a 20% relative risk reduction compared with placebo (582 cases of cardiovascular death, myocardial infarction, or stroke) versus 719 cases for placebo.

Gardasil: Four placebo-controlled, double-blind, randomized Phase II and Phase III trials were conducted to evaluate the efficacy. Cervical intraepithelial neoplasia (CIN) and adenocarcinoma *in situ* (AIS), vulvar intraepithelial neoplasia (VIN), and genital warts were used as the surrogate markers.

The two Phase II trials were Protocol 005 ($N = 2391$) and Protocol 007 ($N = 551$). The Phase III studies were FUTURE I (Protocol 013, $N = 5442$) and FUTURE II (Protocol 015, $N = 12{,}157$). Altogether, 20,541 women from 16 to 26 years of age were enrolled. Subjects were given Gardasil without prescreening for the presence of HPV infection.

Gardasil was efficacious in reducing the episodes of CIN and AIS, as shown in the table below:

Population	Gardasil		Placebo		Efficacy (%) (95% CI)
	N	Number of Cases	N	Number of Cases	
HPV 16- or 18-Related CIN or AIS					
Protocol 005	755	0	750	12	100.0
Protocol 007	231	0	230	1	100.0
FUTURE I	2200	0	2222	19	100.0
FUTURE II	5301	0	5258	21	100.0
Combined protocols	6467	0	8460	53	100.0
HPV 6-, 11-, 16-, 18-Related CIN or AIS					
Protocol 007	235	0	233	3	100.0
FUTURE I	2240	0	2258	37	100.0
FUTURE II	5383	4	5370	43	90.7
Combined protocols	7858	4	7861	83	95.2
HPB 6-, 11-, 16-, or 18-Related Genital Warts					
Protocol 007	235	0	233	3	100.0
FUTURE I	2261	0	2279	29	100.0
FUTURE II	5401	1	5387	59	98.3
Combined protocols	7897	1	7899	91	98.9

The analysis of results for prophylactic efficacy is shown in the table below, regardless of whether the women were HPV-naïve or not.

Endpoints	Analysis	Gardasil or HPV 16 L1 VLP Vaccine		Placebo		Reduction (%) (95% CI)
		N	Cases	N	Cases	
HPV 16- or 18-related CIN or AIS	Prophylactic efficacy	9342	1	9400	81	98.8
	HPV 16- and/or 18-positive on day 1	—	121	—	120	—
	General population impact	9831	122	9896	201	39.0

Endpoints	Analysis	Gardasil or HPV 16 L1 VLP Vaccine		Placebo		Reduction (%) (95% CI)
		N	Cases	N	Cases	
HPV 16- or 18-related VIN	Prophylactic efficacy	8641	0	8667	24	100.0
	HPV 16- and/or 18-positive on day 1	—	8	—	2	—
	General population impact	8954	8	8962	25	69.1
HPV 6-, 11-, 16-, 18-related CIN or AIS	Prophylactic efficacy	8625	9	8673	143	93.7
	HPV 16- and/or 18-positive on day 1	—	161	—	174	—
	General population impact	8814	170	8846	317	46.4
HPV 6-, 11-, 16- or 18-related genital warts	Prophylactic efficacy	8760	9	8786	136	93.4
	HPV 16- and/or 18-positive on day 1	—	49	—	48	—
	General population impact	8954	58	8962	184	68.5

The immune response of Gardasil was evaluated in 8915 women from 18 to 26 years of age (Gardasil $N = 4666$, placebo $N = 4249$) and 2054 adolescents from 9 to 17 years of age (Gardasil $N = 1471$, placebo $N = 583$). Overall, more than 99.5% of the subjects were seropositive with antibodies against HPV 6, HPV 11, HPV 16, and HPV 18. The table below shows the levels of antibodies with and without Gardasil vaccination.

Study Time	Gardasil		Aluminum-Containing Placebo	
	N	Antibody Titer (Geometric Mean) (mMU/mL)	N	Antibody Titer (Geometric Mean) (mMU/mL)
Anti-HPV 6				
Month 07	208	282.2	198	4.6
Month 24	192	93.7	188	4.6
Month 36	183	93.8	184	5.1
Anti-HPV 11				
Month 07	208	696.5	198	4.1
Month 24	190	97.1	188	4.2
Month 36	174	91.7	180	4.4
Anti-HPV 16				
Month 07	193	3889.0	185	6.5
Month 24	174	393.0	175	6.8
Month 36	176	507.3	170	7.7

| Study Time | Gardasil | | Aluminum-Containing Placebo | |
	N	Antibody Titer (Geometric Mean) (mMU/mL)	N	Antibody Titer (Geometric Mean) (mMU/mL)
		Anti-HPV 18		
Month 07	219	801.2	209	4.6
Month 24	204	59.9	199	4.6
Month 36	196	59.7	193	4.8

6.8 SUMMARY OF IMPORTANT POINTS

1. Clinical trials are conducted to test the effects of new drug candidates in humans. There are four phases to clinical trials:
   - Phase I—Safety study, 10–100 subjects, open label
   - Phase II—Safety and efficacy studies, 50–500 subjects, randomized, double-blinded
   - Phase III—Pivotal studies, multisite, 100s to 1000s of subjects, randomized, double-blinded
   - Phase IV—Postmarketing approval trial to monitor drug safety and efficacy at large
2. Regulatory authorities stipulate the need for ethical principles to be observed when conducting clinical trials. Clinical trials should never be conducted to gain knowledge *per se*. They should be based on risk–benefit considerations, informed consent, and respect for human individuals. Furthermore, subjects should be protected without being taken advantage of.
3. Clinical trials are conducted according to GCP. There should be a protocol that states the reason for the clinical trial, how it is to be conducted, the number of people to be included, eligibility criteria, medical tests and observations to be made, and information to be collected. Clinical trial protocol must be approved by the IRB/IEC before commencement.
4. Statistics analysis is an integral part of a clinical trial. A clinical trial protocol includes information on statistical parameters that the trial is expected to be based on and methods for the analysis of data.
5. An Investigator, not an employee of the Sponsor, is appointed to be responsible for the conduct of a trial. An appropriate quality system is followed and deviations from trial protocols are reported. Serious adverse events have to be reported to regulatory authorities within a specified time.

6.9 REVIEW QUESTIONS

1. Explain the reasons for ethical considerations before a clinical trial is conducted.
2. Discuss the use of biomarkers in clinical trials.
3. Describe the term "protocol" and list the parameters to be included in the document.
4. Explain randomization and justify the requirement for randomization and double-blinding in clinical trials.
5. An investigator is designing a clinical trial to test a cholesterol-lowering drug. She wants to compare the drug with placebo with a 95% confidence level; that is α is 0.05. She also limits the false-negative to 10%; that is, β is 0.10. From the literature, she knows the variability of cholesterol has a standard deviation of 50 mg/dL. How many people must she recruit in the study to demonstrate a 20 mg/dL difference between the drug and placebo?
6. Distinguish the various phases of clinical trials, I to IV. Provide a reason for conducting Phase IV trials.
7. Briefly explain GCP as applied to clinical trials.
8. Why is it necessary to regulate clinical trials?

6.10 BRIEF ANSWERS AND EXPLANATIONS

1. Due to problems in some early clinical trials, where subjects were taken advantage of, regulatory authorities require that clinical subjects be treated fairly. Ethical considerations should be undertaken to safeguard subjects' safety and well-being.
2. For some diseases it is not possible to measure directly the effect of the drug on trial, or the desired direct outcome may require a long time to eventuate. Biomarkers provide more convenient and timely signals in response to the trial drug and they can be measured at various time points to indicate the progressive treatment reaction.
3. Refer to Section 6.4.4 and Exhibit 6.10.
4. Refer to Section 6.4.7 and Exhibit 6.12 to explain randomization and the techniques used. Randomization and double-blinding are necessary to prevent bias in data collection so that statistical analysis based on normal distribution can be used to evaluate the trial results.
5. We use the following equation (from Exhibit 6.14) to calculate the total number of subjects to be recruited:

$$2N = \frac{4\left(Z_\alpha + Z_\beta\right)^2 \sigma^2}{\delta^2}$$

Using a standard normal distribution table, $Z_\alpha = Z_{0.05} = 1.96$; $Z_\beta = Z_{0.1} = 1.65$. Substituting into the equation, we have

$$2N = \frac{4(1.96 + 1.65)^2\, 50^2}{20^2}$$

$$2N = 326$$

Hence $N = 163$ is the number for each group. In reality, to account for dropouts and noncompliances, more subjects are normally recruited.

6. Refer to Section 6.3 to describe the phases of a clinical trial. Phase IV trials are necessary to maintain a close watch on the efficacy and adverse events of an approved drug when it is administered to the population at large. For example, even a small percentage of adverse events in Phase III trial for several thousand people may translate into a substantial number when a drug is made available to millions of people. A case in point is Vioxx and Bextra (see Section 2.9).

7. By following GCP, clinical trials are conducted by following the procedures in the protocol, data are collected and verified as intended, and deviations to procedures are addressed. The aim is to ensure data from clinical trials are valid and conclusions drawn are correct.

8. The regulation of clinical trials is to make certain that first and foremost subjects' welfare is not compromised. The need for review and approval by the IRB/IEC provides an independent party to decide the need and procedure contemplated for the clinical trials. The watchdog roles are steps taken to ensure trials are conducted ethically.

6.11   FURTHER READING

Bloom JC, Dean RA, eds. *Biomarkers in Clinical Drug Development*, Marcel Dekker, New York, 2003.

Cato A, Sutton L, Cato A III, eds. *Clinical Drug Trials and Tribulations*, 2nd ed., Marcel Dekker, New York, 2002.

Center for Drug Evaluation and Research. *Guideline for the Format and Content of the Clinical and Statistical Sections of an Application*, FDA, Rockville, MD, 1998.

Chow SC, Liu JP, *Design and Analysis of Clinical Trials: Concepts and Methodologies*, 2nd ed., Wiley, Hoboken, NJ, 2004.

Cohen A, Posner J, eds. *A Guide to Clinical Drug Research*, 2nd ed., Kluwer Academic Publishers, The Netherlands, 2000.

Dawson B, Trapp RG. *Basic and Clinical Biostatistics*, McGraw-Hill, Singapore, 2001.

DeMets DL, Furgerg CD, Friedman LM, eds. *Data Monitoring in Clinical Trials: A Case Studies Approach*, Springer, New York, 2006.

EMEA. *Ethical Considerations in Clinical Trials*, 2001.

Friedman LM, Furberg C. *Fundamentals of Clinical Trials*, 3rd ed., Mosby, St Louis, 1996.

Gad SC. *Drug Safety Evaluation*, Wiley, Hoboken, NJ, 2002.

Good PI. *A Manager's Guide to the Design and Conduct of Clinical Trials*, Wiley, Hoboken, NJ, 2002.

Hulley SB, et al. *Designing Clinical Research*, 3rd ed., Lippincott Williams & Wilkins, Philadelphia, 2007.

International Conference on Harmonization. *Guideline for Industry—Structure and Content of Clinical Study Reports*, ICH E3, 1995. http://www.ich.org/LOB/media/MEDIA479.pdf [accessed October 10, 2007].

Machin D, Day S, Green S, eds. *Textbook of Clinical Trials*, Wiley, West Sussex, England, 2004.

Monkhouse DC, Rhodes CT, eds. *Drug Products for Clinical Trials, An International Guide to Formulation, Production, Quality Control*, Marcel Dekker, New York, 1998.

Piantodosi S. *Clinical Trials—A Methodologic Perspective*, Wiley, Hoboken, NJ, 1997.

CHAPTER 7

REGULATORY AUTHORITIES

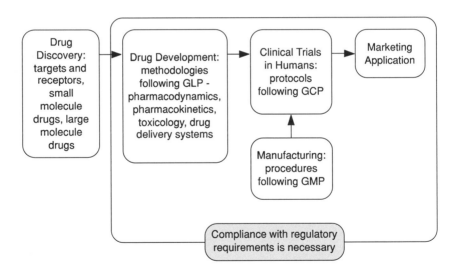

7.1   Role of Regulatory Authorities   209
7.2   US Food and Drug Administration   210
7.3   European Medicines Agency   214
7.4   Japan's Ministry of Health, Labor and Welfare   216
7.5   China's State Food and Drug Administration   217

Drugs: From Discovery to Approval, Second Edition, By Rick Ng
Copyright © 2009 John Wiley & Sons, Inc.

7.6   India's Central Drugs Standard Control Organization   219
7.7   Australia's Therapeutics Goods Administration   219
7.8   Canada's Health Canada   220
7.9   Other Regulatory Authorities   220
7.10   Authorities Other than Drug Regulatory Agencies   221
7.11   International Conference on Harmonization   222
7.12   World Health Organization   222
7.13   Pharmaceutical Inspection Cooperation Scheme   223
7.14   Case Study #7   225
7.15   Summary of Important Points   227
7.16   Review Questions   228
7.17   Brief Answers and Explanations   228
7.18   Further Reading   229

7.1   ROLE OF REGULATORY AUTHORITIES

All of us want the drugs that are prescribed for us to be safe and effective to treat our ailments. It is the role of public regulatory authorities to ensure that pharmaceutical companies comply with regulations. There are legislations that require drugs to be developed, tested, trialed, and manufactured in accordance to guidelines so that they are safe and patients' well-being is protected. There have been several occasions when drugs were not safe and people's health has been compromised; there were times when unscrupulous people or firms wrongly or carelessly manufactured drugs; children or vulnerable people have been recruited to clinical trials without consent, and insufficient tests were carried out on some drugs during development, leading to untold damage (see Exhibit 7.1 for an account of the thalidomide tragedy).

Regulatory authorities perform the watchdog role to ensure that animal studies comply with Good Laboratory Practice (GLP), clinical trials are

---

Exhibit 7.1   Thalidomide

Thalidomide was synthesized in Germany and became available in late 1957. It was prescribed for the treatment of insomnia and nausea in pregnant women.

However, it had not been discovered that the thalidomide drug molecule could cross the placental barrier and affect fetal development. As a result, thousands of babies were born with crippled extremities, disfigurement, and disabilities. Numerous fetuses were stillborn or died soon after birth.

The drug was banned in early 1962, but by then the lives of many people had been severely affected.

Refer to Exhibit 10.9 for the chemical structure of thalidomide.

performed in accordance with Good Clinical Practice (GCP), and drugs are manufactured under current Good Manufacturing Practice (cGMP) conditions. The regulatory authorities also carry out surveys to ensure that labels and advertising materials are accurate and in accordance with approved claims. Advertising materials should have clear explanations about the drug, indications and contraindications, dosage, and frequency of medication.

In this chapter, we explain the regulatory authorities in the major countries. The regulatory process is complicated and lengthy; this is especially the case where major industrialized nations have independently over the years set up their own systems of regulations and controls, which invariably have different requirements from those in other countries. However, processes are in place to harmonize the regulatory procedures in the major industrialized countries. In this way, regulatory requirements, technical documents, and review processes are consistent and can be mutually recognized by member countries. Eventually, harmonization will reduce duplicate requirements, reports, and the cost and time for regulatory reviews. This will translate to patients receiving access to new drugs more speedily and at less cost than now.

In Chapter 8 we examine more closely the regulatory processes for testing, trialing, and approving a drug for marketing.

7.2 US FOOD AND DRUG ADMINISTRATION

The US Food and Drug Administration (FDA) is required by the US Federal Food, Drug, and Cosmetic Act to regulate drug products in the United States. Its role is to ensure that drugs are developed, manufactured, and marketed in accordance with regulatory requirements so that they are safe and effective. The FDA has four centers and a regulatory office:

- Center for Drugs Evaluation and Research (CDER)
- Center for Biologics Evaluation and Research (CBER)
- Center for Devices and Radiological Health
- Center for Veterinary Medicine
- Office of Regulatory Affairs

Exhibit 7.2 presents a brief history of the FDA. For the purpose of regulation of drugs, the relevant centers are the CDER and CBER.

7.2.1 Center for Drug Evaluation and Research

The CDER oversees the research, development, manufacture, and marketing of synthetic small molecule drugs (drugs that are described in Chapter 3). As of June 30, 2003, the CDER is also responsible for the regulation of biologic therapeutic products. Most of these drugs are large protein-based molecules generated by hybridoma or recombinant DNA technology, such as monoclo-

Exhibit 7.2 A Brief History of the FDA

The FDA started from a single chemist in the US Department of Agriculture in 1862, with the appointment of Charles M. Wetherill by President Lincoln. By 2001, it had a staff of about 9100 and a budget of $1.294 billion. The FDA now has employees from diverse disciplines, including chemists, pharmacologists, physicians, microbiologists, veterinarians, pharmacists, and lawyers.

About a third of the agency's employees are stationed outside the Washington, DC area in over 150 field offices and laboratories, including five regional offices and 20 district offices.

The FDA regulates the following:

- Drugs (e.g., prescriptions, OTCs, generics)
- Biologics (e.g., vaccines, blood products)
- Medical devices (e.g., pacemakers, contact lenses)
- Food (e.g., nutrition, dietary supplements)
- Animal feed and drugs (e.g., livestock, pets)
- Cosmetics (e.g., safety, labeling)
- Radiation emitting products (e.g., cell phones, lasers)

nal antibodies, cytokines (interferon, interleukin), tissue growth factors, and other proteins described in Chapter 4. These products include the following:

- Monoclonal antibodies for *in vivo* use
- Cytokines, growth factors, enzymes, immunomodulators, and thrombolytics
- Proteins intended for therapeutic use that are extracted from animals or microorganisms, including recombinant versions of these products
- Other nonvaccine therapeutic immunotherapies

The CDER's involvement starts with the Phase I clinical study via the approval of an Investigational New Drug (IND) application. In its review process for the IND, the CDER checks that preclinical tests have been performed in compliance with GLP and that the toxicological studies are acceptable. When the clinical trials commence, the CDER monitors the conduct of the clinical trials through Phases I, II, and III, based on adherence to GCP. At the conclusion of Phase III trials, marketing applications from sponsor pharmaceutical organizations are evaluated by the CDER, relying on scientific data and clinical results. The marketing applications are:

- New Drug Applications (NDAs) for small molecule drugs
- Biologics License Applications (BLAs) for therapeutic biologic drugs

Risks (drugs have potential risks as they interfere with our bodily functions) and benefits evaluations are undertaken before drugs are approved for marketing. Expert reviews from external personnel are sought from time to time, to ensure that decisions are based on the latest scientific opinions. It also ensures that the advertising and marketing of drugs are in accordance with approved claims. Marketed drugs are monitored for unanticipated health risks. If unexpected health risks or adverse reactions are confirmed, the CDER informs the public or, in severe cases, directs the suppliers to remove drugs from the market. The manufacture of drugs is monitored to ensure compliance with cGMP.

The three categories of drugs regulated by the CDER are the following:

- Prescription drugs
- Generic drugs
- Over-the-counter (OTC) drugs

7.2.2 Center for Biologics Evaluation and Research

The CBER regulates nontherapeutic biologics—drugs that are described in Chapter 4—which are not regulated by the CDER. These include the following:

- Viral-vectored gene insertions (e.g., gene therapy)
- Drugs composed of human or animal cells or from physical parts of those cells
- Allergen patch tests
- Allergenics
- Antitoxins, antivenins, and venoms
- *In vitro* diagnostics
- Vaccines, including therapeutic vaccines
- Toxoids and toxins intended for immunization

In addition, the CBER controls the approval of human tissue for transplantation, blood and blood products, and devices related to blood products. These devices include automated cell separators, empty plastic containers, and blood storage refrigerators and freezers.

In contrast to the small molecule drugs, biologics are complex, large compounds with molecular weights >5 kDa and they are not easily characterized. They are dissimilar to small molecule drugs, which are chemically well-defined entities. Biologics are also labile (i.e., heat and shear sensitive) and are very dependent on the manufacturing process parameters and storage conditions.

The regulatory process is the filing of an IND for clinical trials. At the conclusion of clinical trials, the Sponsor files a Biological License Application (BLA) for marketing approval. The CBER evaluates a biologic in terms of risk versus benefits before approving it for marketing.

7.2.3 Pertinent FDA Processes and Controls

Drugs (small molecule drugs) are regulated in the United States as required by the Food, Drug and Cosmetic Act (FDCA) of 1938. Biologics, however, are regulated by the Public Health Service Act (PHSA) of 1944 and the FDCA. This is because the PHSA is concerned with medical products that are less well defined, necessitating more control in the handling and manufacturing processes.

The applicable regulations for drugs are codified in Title 21 of the US *Code of Federal Regulations* (CFR). These regulations promulgate the FDA's requirement in many aspects of drug clinical research, manufacturing, and marketing. Table 7.1 lists some of these applicable regulations. Readers should note that these regulations are updated from time to time by the FDA as a result of new requirements or information.

In addition, the FDA publishes Guidelines and Points to Consider (PTCs) documents to guide pharmaceutical organizations in many relevant areas, from testing methodologies, manufacturing requirements, and drug stability information, to filling in of forms and the requisite data.

The FDA also carries out inspections on establishments to ensure compliance with regulations. The establishments include laboratories, clinical trial centers, and manufacturing facilities. Further information on establishment inspection is discussed in Chapter 10.

TABLE 7.1 Selected Regulations from 21 CFR

Document Number	Description
21 CFR Part 11	Electronic Records, Electronic Signatures
21 CFR Part 50	Protection of Human Subjects
21 CFR Part 56	Institutional Review Board
21 CFR Part 58	Good Laboratory Practices for Non-clinical Laboratory Studies
21 CFR Part 202	Prescription Drug Advertising
21 CFR Part 203	Prescription Advertising
21 CFR Part 210	Current Good Manufacturing Practice in Manufacturing, Processing, Packaging or Holding of Drugs; General
21 CFR Part 211	Current Good Manufacturing Practice for Finished Pharmaceuticals
21 CFR Part 312	Investigational New Drug Applications
21 CFR Part 314	Applications for FDA Approval to Market a New Drug
21 CFR Part 600	Biological Products: General
21 CFR Part 610	General Biological Products Standards

Exhibit 7.3   Imatinib Mesylate (Gleevec)

Chronic myeloid leukemia (CML) occurs when there is a translocation of chromosomes 9 and 22 (also called Philadelphia translocation—a chromosomal abnormality). These two different chromosomes break off and reattach on the opposite chromosome. A consequence is that the activity of the *Bcr-Abl* gene, which encodes the enzyme tyrosine kinase, is turned on all the time. With this heightened activity, high levels of white blood cells are produced in the bone marrow.

Imatinib mesylate is a tyrosine kinase inhibitor (see Chapter 2 on receptors). It is used to block the growth of white blood cells.

Imatinib mesylate is manufactured by Novartis. Clinical trials showed that patients had their white blood cells reduced substantially after being treated with Gleevec.

Gleevec was approved by the FDA under the accelerated approval regulations for the treatment of CML.

Source: *FDA News: FDA Converts Gleevec in Second Line Setting to Regular Approval*, December 2003. http://www.fda.gov/bbs/topics/NEWS/2003/NEW00990.html [accessed October 5, 2007].

In some circumstances, the FDA processes drug reviews under the accelerated scheme. This mechanism is to review and approve drugs speedily for cases where effective therapies are lacking or in situations of rare diseases. One of the fastest approval times to date is the case of imatinib mesylate (Gleevec, Novartis—Exhibit 7.3) for the treatment of chronic myeloid leukemia (CML); it was approved in less than 3 months after the filing of an NDA with the FDA. Another example is the new AIDS drug indinavir (Crixivan, Merck), which was approved in a mere 42 days.

7.3   EUROPEAN MEDICINES AGENCY

There are several avenues for drug approval in Europe:

- *Centralized Procedure:* Under the European Community Regulation 726/2004 and Directive 2004/27/EC, the Centralized Procedure (also known as Community Authorization Procedure) is a single authorization procedure that is mandatory for medicinal products of the following categories:

  Derived from biotechnology processes, such as genetic engineering

  Intended for the treatment of HIV/AIDS, cancer, diabetes, or neuro-degenerative disorders

  Orphan medicines (medicines used for rare diseases)

- *Mutual Recognition Procedure:* A medicine is first authorized by one member state, according to the member state's own national procedure. The applicant can seek further authorizations through a mutual recognition procedure. When there is a dispute between member states on the issue of mutual recognition, the European Medicines Agency (EMEA) is called upon to arbitrate, and its decision is binding on the member states.
- *Decentralized Procedure:* This is applicable where authorization has not yet been approved in any member state. The applicant may apply for simultaneous authorization in more than one EU member state for medicines that do not fall within the mandatory scope of the centralized procedure.

The EMEA's key aims, according to the EU Enterprise Directorate-General publication, are the following:

- Protect and promote public health by providing safe and effective medicines for human and veterinary use
- Give patients quick access to innovative new therapy
- Facilitate the free movements of pharmaceutical products throughout the EU
- Improve information for patients and professionals on the correct use of medicinal products
- Harmonize scientific requirements to optimize pharmaceutical research worldwide

There are two committees within the EMEA:

- Committee for Medicinal Products for Human Use (CHMP)
- Committee for Medicinal Products for Veterinary Use (CVMP)

For our purposes, the committee for drug approval is the CHMP. Applications are submitted to the EMEA according to the centralized procedure. The review process is described in Section 8.3. In 2006 the CHMP provided 78 opinions (decisions) on medicinal products, of which 5 were negative (rejected).

Council Regulation EEC/2309/93 together with Directive 75/319/EEC require member states to establish a national pharmacovigilance system to collect and evaluate information on adverse reactions to medicinal products and to take appropriate action.

Clinical trial applications are not centralized. Submissions are made through individual member states. Refer to Section 8.3 for details of clinical trial application in Europe.

7.4 JAPAN'S MINISTRY OF HEALTH, LABOR AND WELFARE

The Japanese pharmaceutical market is the second largest in the world. It is larger than the combined markets of the United Kingdom, France, and Germany. Japan's Pharmaceutical Affairs Law aims to improve public health through regulations ensuring the quality, efficacy, and safety of drugs and medical devices. The Ministry of Health, Labor and Welfare (MHLW) is responsible for pharmaceutical affairs in Japan. There are three main parts of MHLW overseeing this charter: the Pharmaceutical and Food Safety Bureau (PFSB), the Health Policy Bureau, and the Pharmaceutical and Medical Device Agency (PMDA, KIKO).

PFSB oversees the policies to assure the safety and efficacy of drugs and medical devices. The Bureau's responsibility is to review and approve clinical trials and the importation and manufacture of drugs. The Health Policy Bureau handles production and distribution policies; that is, it deals with the manufacturers and distributors to ensure a high quality, efficient healthcare system. The PMDA reviews clinical protocols and details of drug submissions, including bioequivalence, as well as the testing and research on drugs. Within its organization, there are Offices for New Drugs, Biologics, OTC, and Generics. New drugs from vaccines or blood, their specifications, and test methods are examined by the National Institute of Health Sciences or the Infectious Disease Surveillance Center.

The Pharmaceutical Affairs and Food Sanitation Council (PAFSC) is an advisory body to MHLW on pharmaceutical and food matters. The First and Second Committees on New Drugs of the PAFSC meet about 8 times per year to review new drug applications as a consultative role to the PMDA. The First Committee is responsible for all therapeutics except those under the responsibility of the Second Committee, for example, antivirals, chemotherapy agents, and blood and biological products. New drugs are approved by the Pharmaceutical Affairs Department of the MHLW, based on the recommendations of the PAFSC. Marketing approval of drug products requires that the licensees of marketing businesses demonstrate compliance to Good Quality Practice (GQP) and Good Vigilance Practice (GVP). Refer to Section 8.4 for examples of the clinical trial and drug approval processes in Japan.

Foreign clinical results are acceptable except in areas where there are immunological and ethnic differences between Japanese and foreigners. The ethnic factors are divided into two components: intrinsic factors such as racial factors and physiological differences; and extrinsic factors, which include cultural and environmental issues. In these cases, the MHLW may require that some bridging comparative clinical trials be performed with dose ranging protocols. This will enable absorption, distribution, metabolism, and excretion studies to be carried out on Japanese individuals and provide better dosage and indication for the Japanese people. The MHLW also requires that application be accompanied by one year of real-time stability data and that sterility test results be included.

The standard processing period for drug approval according to the MHLW is as follows:

- One year for review
- One year for applicant response
- Total 2 years to approval

All new drug applications are expected to be in the CTD format according to ICH guidelines (see Section 7.11). Priority reviews are applicable for orphan drugs and those drugs for the treatment of serious illnesses. A restricted approval system has been implemented for emergency drugs to prevent spread of diseases. In this case, the standard review procedure is not applicable.

7.5 CHINA'S STATE FOOD AND DRUG ADMINISTRATION

China's pharmaceutical market is growing at a very fast pace. The current data show that the total market is around US$20 billion, and it is the ninth largest pharmaceutical country in the world. The Chinese government maintains price control on imported drugs. With China's entry into the World Trade Organization (WTO), tariffs have been reduced from 20% to 6.5%. The projection is that the market size will reach US$60 billion by 2010, and China will be the world's largest market by 2020.

The regulation of drugs in China is under the jurisdiction of the State Food and Drug Administration (SFDA). The SFDA is under the control of the State Council. Through the Drug Administration Law of the People's Republic of China, regulations are instituted for the control of clinical trials, registration, distribution, and marketing surveillance of new, generic, and OTC drugs (Exhibit 7.4). It also controls GMP manufacturing compliance, monitors adverse events, and prosecutes illicit, fraudulant, and unlicensed drug manufacturers through the Departments of Drug Safety and Inspection, and Drug Market Compliance. There are also strict controls on advertising of drugs; these prohibit the use of certain words, phrases, and unsubstantiated or unscientific claims.

Among the many departments of the SFDA, the relevant departments for drugs and medical devices are the Department of Drug Registration (DDR), the Department of Medical Devices, and the Department of Drug Safety and Inspection. The SFDA manages the regulation for "Western" drugs and Traditional Chinese Medicine (TCM) under the Division of Pharmaceuticals, Division of Biological Products, and Division of TCM of the DDR (Exhibit 7.5).

Drugs are classified into several categories. These are synthetic drugs, TCM, and biological products. The SFDA stipulates compliance to GMP for medical products, GCP for clinical trials, and GLP for nonclinical drug safety research.

Exhibit 7.4   Clinical Trials and Selected Drugs Approved in China

China offers a large pool of treatment naive patients for clinical trials. There are 1.3 billion people, of which 250 million are insured and another 250 million partially insured. Clinical trials are one-third the cost of that in the United States and recruitments are expected to be rapid. However, complicating factors are slower regulatory processes, limited qualified central laboratories for testing, and restriction of export of blood and serum samples outside China for testing.

Unlike in the United States and Europe, however, only certain research centers and hospitals are specially designated by the Chinese SFDA for the conduct of clinical trials.

In October 2003, the SFDA approved the world's first gene therapy— Gendicine (a recombinant human adenovirus type 5 mediated delivery of *p53* gene)—for the treatment of head and neck cancer. In 2005, another head and neck cancer drug, Oncorine (a recombinant oncolytic adenovirus type 5), was approved. In the same year, another recombinant human endostatin, Endostar, was approved for the treatment of small-cell lung cancer.

In July 2007, the SFDA strengthened its regulatory framework by introducing more stringent regulations in the approval process, requiring that (1) approval of drug licenses must be based on collective decisions, (2) drug evaluators will be made public and held accountable for their decisions, and (3) there should be no potential conflict of interest in the evaluators in reviewing the application.

Source: Jia H. China syndrome—a regulatory framework in meltdown, *Nature Biotechnology* 25:835–837 (2007).

Exhibit 7.5   Division of Pharmaceuticals, Division of Biological Products, and Office for Acceptance of Drug Registration

Division of Pharmaceuticals

- Draft and revise national standards and research guidelines of pharmaceuticals
- Evaluate and approve new drugs
- Approve and reregister controlled drugs
- Evaluate and approve clinical trials
- Approve and regulate pharmaceutical preparations dispensed by provincial medical institutions

Division of Biological Products

- Draft and revise national standards and research guidelines of biological products
- Evaluate and approve new biological products
- Evaluate and approve clinical trials of biological products
- Regulate and supervise national lot release of biological products

Office for Acceptance of Drug Registration

- Accept drug application and issue certificate for new drug, generic drug, imported drug, protected TCM products, and packaging material for drugs

Foreign drugs are required to have import registration. Foreign drug manufacturers and distributors file for examination and registration of their products with relevant data and documents. Clinical trials may need to be conducted based on evaluation by the Center for Drug Evaluation (CDE) (see Section 8.5).

7.6 INDIA'S CENTRAL DRUGS STANDARD CONTROL ORGANIZATION

Increasingly, India is becoming an important player in drug manufacture, in particular, the production of generics. Many of India's generics are now found in all parts of the world, challenging the dominance once held by the large pharmaceutical companies in Western countries.

Under India's Drug and Cosmetics Act, the central government of India, through the Central Drugs Standard Control Organization (CDSCO), is responsible for the approval of new drugs, clinical trials, maintenance of the standard of drugs, jurisdiction of importation of foreign drugs, approval of manufacturing licenses, and coordination of the activities of the State Drug Control Organizations. The central government is also responsible for the testing of drugs by the Central Drugs Labs, whereas the state authorities are responsible for the regulation of the manufacture, sale, and distribution of drugs.

Schedule Y of the Drugs and Cosmetics Rules sets up the requirements for clinical trials, and that of Schedule M for GMP compliance system.

7.7 AUSTRALIA'S THERAPEUTIC GOODS ADMINISTRATION

Australia's Therapeutic Goods Administration (TGA) has perhaps one of the most progressive and comprehensive regulatory systems in the world. Under the Therapeutic Goods Act, the TGA regulates prescription medicines, OTC medicines, complementary medicines, and medical devices. The roles of the TGA in medicines are as follows:

- Premarket evaluation and approval of registered products intended for supply in Australia
- Development, maintenance, and monitoring of the systems for listing of medicines
- Licensing of manufacturers in accordance with international standards of Good Manufacturing Practice
- Postmarket monitoring, through sampling, adverse event reporting, surveillance activities, and response to public inquiries
- Assessment of medicines for export

All medicines in Australia are listed or registered with the Australian Register of Therapeutic Goods (ARTG), except specifically exempted, and are provided with unique numbers in the database. Listed medicines are considered to be of lower risk and are self-selected by consumers. These medicines bear the "AUST L" numbers. In contrast, registered medicines include a "low risk" category, which are of the OTC type, or "high risk" medicines, which require prescription. Registered medicines have the "AUST R" numbers. Complementary medicines, such as alternative or traditional medicines, are either listed or registered depending on ingredients and claims.

Section 8.7 explains the clinical trial and drug approval processes in Australia.

7.8 CANADA'S HEALTH CANADA

All drugs sold in Canada must be authorized by Health Canada, which has several directorates: the Therapeutic Products Directorate (TPD) reviews and authorizes new pharmaceuticals and medical devices, the Biologics and Genetic Therapies Directorate (BGTD) evaluates biological and radiopharmaceutical drugs, and the Natural Health Products Directorate (NHPD) regulates natural health products such as vitamins and health supplements. For postmarket surveillance, the Marketed Health Products Directorate (MHPD) monitors adverse events and investigates complaints and problem reports.

7.9 OTHER REGULATORY AUTHORITIES

Table 7.2 shows the regulatory authorities in selected countries. A summary of the health system, both public and private in selected countries, is given in Appendix 9. It shows the %GDP each country spends on healthcare, per capita health expenditure, number of hospital beds, and doctors/1000 population.

TABLE 7.2   Selected International Regulatory Authorities

Country	Regulatory Authority
Argentina	National Administration of Drugs, Foods and Medical Technology
Brazil	Ministry of Health
Chile	Institute of Public Health
Denmark	Laegemiddelsturelsen
Egypt	Ministry of Health and Population
Finland	National Agency for Medicines
France	Agence du Medicament
Germany	Federal Institute of Drugs and Medical Devices
Greece	Ministry of Health and Welfare
Indonesia	Ministry of Health
Israel	Ministry of Health
Italy	Ministry of Health
Jamaica	Ministry of Health
Kenya	Ministry of Health
Korea	Food and Drug Administration
Malaysia	National Pharmaceutical Control Bureau
Mexico	Ministry of Health
Netherlands	Medicines Evaluation Board
New Zealand	Medicines and Medical Devices Safety Authority
Norway	Norwegian Board of Health
Philippines	Ministry of Health
Russia	Ministry of Health
Singapore	Health Sciences Authority
South Africa	Department of Health
Spain	Spanish Drug Agency
Sweden	National Board of Health and Welfare
Switzerland	International Office for Control of Medicaments
Taiwan	Department of Health
Thailand	Food and Drug Administration
United Kingdom	Medicines and Healthcare Products Regulatory Agency
Zimbabwe	Ministry of Health

7.10   AUTHORITIES OTHER THAN DRUG REGULATORY AGENCIES

Although pharmaceutical organizations have to comply with requirements of regulatory agencies, there are other authorities that control the manufacturing and marketing of drugs. For example, in the United States these include the following:

- State health authorities
- Occupational Safety and Health Administration (OSHA)
- Environmental Protection Agency (EPA)
- Local regulatory bodies

Compliance with all these authorities would assist in the approval of drugs for manufacturing and marketing.

7.11 INTERNATIONAL CONFERENCE ON HARMONIZATION

Specific plans for the formation of the International Conference on Harmonization (ICH) were conceived at the WHO International Conference of Drug Regulatory Authorities (ICDRA) in Paris in 1989. In April 1990, the ICH was formed in Brussels, with the aim of formulating a joint regulatory–industry initiative on international harmonization of drug regulations. The ICH is composed of representatives from the regulatory agencies and industry associations of the United States, Europe, and Japan. The ICH Steering Committee meets at least twice a year, with the location rotating among the three regions. It is charged with the responsibility to prepare harmonized guidelines that can be accepted by each region.

There are four major categories of guidelines: quality, safety, efficacy, and multidisciplinary. Details are provided in Exhibit 7.6.

The CTDs were implemented in July 2003. They are format-based documents for submission to the regulatory authorities; the country-specific process of review, for example, via the IND and NDA of the United States or the Centralized Procedure of the EMEA, is not affected. The harmonized CTDs help to reduce cost and accelerate approval time. Figure 7.1 shows the CTD structure: five modules with Module 1 for regional administrative information specific to each country, Module 2 on summary of quality, nonclinical and clinical, Module 3 on quality, Module 4 on nonclinical study reports, and Module 5 on clinical study reports.

7.12 WORLD HEALTH ORGANIZATION

The World Health Organization (WHO) is a specialized agency of the United Nations. There are 193 member states as of January 2008. The WHO is headquartered in Europe with four regional offices in Africa, the Americas, the Eastern Mediterranean, Southeast Asia, and the Western Pacific. The WHO is not a regulatory agency. Its functions are:

- To give worldwide guidance in the field of health
- To set global standards for health
- To cooperate with governments in strengthening national health programs
- To develop and transfer appropriate health technology, information, and standards

Exhibit 7.6   ICH Guidelines

ICH guidelines are divided into four major categories.

Current Status of Harmonization

- *Quality:* Ten topic headings—Stability, Analytical Validation, Impurities, Pharmacopoeias, Quality of Biotechnological Products, Specifications, GMP, Pharmaceutical Development, Quality Risk Management, Pharmaceutical Quality System; total of 24 guidelines
- *Safety:* Eight topic headings—Carcinogenicity Studies, Genotoxicity Studies, Toxicokinetics and Pharmacokinetics, Toxicity Testing, Reproductive Toxicology, Biotechnological Products, Pharmacology Studies, Immunotoxicology Studies; total of 13 guidelines
- *Efficacy:* Nine topic headings—Clinical Safety, Clinical Study Reports, Dose–Response Studies, Ethnic Factors, GCP, Clinical Trials, Clinical Evaluation by Therapeutic Category, Clinical Evaluation, Pharmacogenomics; total of 18 guidelines
- *Multidisciplinary:* Five cross-cutting topics that do not fit uniquely into one of the above categories

    M1—Medical Terminology (MedDRA)

    M2—Electronic Standards for Transmission of Regulatory Information (ESTRI)

    M3—Timing of Preclinical Studies in Relation to Clinical Studies

    M4—The Common Technical Document (CTD)

    M5—Data Elements and Standards for Drug Dictionaries

The WHO works with regulatory authorities in member states to set up policies and training programs to ensure drugs are safe, pure, and effective and are being distributed and administered as specified.

7.13   PHARMACEUTICAL INSPECTION COOPERATION SCHEME

The Pharmaceutical Inspection Cooperation Scheme (PIC/S) was formed in 1995 to enhance the work set up under the Pharmaceutical Inspection Convention (PIC) in 1970. The mission of the PIC/S is

"to lead the international development, implementation and maintenance of harmonized Good Manufacturing Practice (GMP) standards and quality systems of inspectorates in the field of medicinal products."

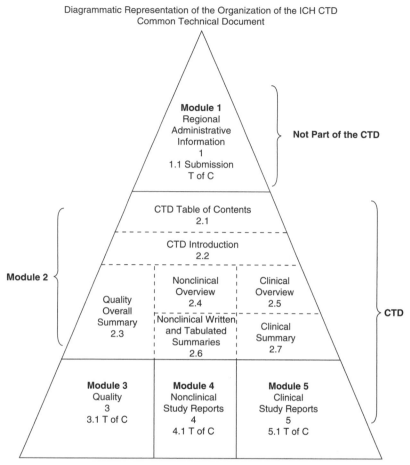

Diagrammatic Representation of the Organization of the ICH CTD
Common Technical Document

Figure 7.1 Common technical documentation. (*Source:* ICH Harmonised Tripartite Guideline. *Organisation of the Common Technical Document for the Registration of Pharmaceuticals for Human Use*, M4, Step 4 Version (2004).)

The member countries are Argentina, Australia, Austria, Belgium, Canada, Czech Republic, Denmark, Estonia, Finland, France, Germany, Greece, Hungary, Iceland, Ireland, Italy, Latvia, Liechtenstein, Malaysia, Malta, The Netherlands, Norway, Poland, Portugal, Romania, Singapore, Slovak Republic, South Africa, Spain, Sweden, Switzerland, and the United Kingdom. Inspection of pharmaceutical facilities by one member is mutually recognized by another member to streamline regulatory inspection processes. The partners/ observers to PIC/S are the EMEA, UNICEF, and WHO.

7.14 CASE STUDY #7

International Health Regulations (2005)*

In the early years of the 21st century, we have witnessed the spread of diseases quickly from one country to another, whether via human beings in the case of SARS, or through migrating livestock as evidenced by the proliferation of avian influenza across national borders; in addition to the cases of Ebola outbreak in Africa. These situations necessitated prompt and skillful control at early stages by more than one or a few countries alone, and in concerted effort to contain the spread.

On August 23, 2007, the WHO issued its World Health Report for 2007, which highlighted the international spread of disease. This report, entitled *A Safer Future: Global Public Health Security in the 21st Century*, tables six key recommendations for global public health security:

- Full implementation of the revised International Health Regulations (IHR 2005) by all countries
- Global cooperation in surveillance and outbreak alert and response
- Open sharing of knowledge, technologies, and materials, including viruses and other laboratory samples, necessary to optimize and secure global public health
- Global responsibility for capacity building within the public health infrastructure of all countries
- Cross-sector collaboration within governments
- Increased global and national resources for training, surveillance, laboratory capacity, response networks, and prevention campaigns

It is under these circumstances of SARS and avian flu threats, together with the emergence of Ebola and other viruses, that prompted the WHO and the member states to update the IHR in 2005. The realization is for member states to organize proactive measures to curb the spread of diseases.

An excerpt of the International Health Regulations (2005) is presented below:

The IHR (1969) addressed only four diseases: cholera, plague, yellow fever and smallpox by focusing on border controls and passive notification and control measures.

In contrast the IHR (2005), which have now been enforced since June 15, 2007, have an expanded scope that "covers existing, new and re-emerging diseases,

---

* *Source*: World Health Organization, *What Are the International Health Regulations?* http://www. who.int/features/qa/39/en/index.html [accessed July 26, 2007].

including emergencies caused by non-infectious disease agents." It is an international law which helps countries working together to save lives and livelihoods caused by the international spread of diseases and other health risks. The aim is to prevent, protect against, control and respond to the international spread of disease while avoiding unnecessary interference with international traffic and trade. The IHR (2005) are also designed to reduce the risk of disease spread at international airports, ports and ground crossings.

The IHR (2005) require Member States to notify WHO of all events that may constitute a public health emergency of international concern and to respond to requests for verification of information regarding such events. This will enable WHO to ensure appropriate technical collaboration for effective prevention of such emergencies or containment of outbreaks and, under certain defined circumstances, inform other States of the public health risks where action is necessary on their part.

The IHR (2005) have been agreed upon by consensus among WHO Member States as a balance between their sovereign rights and shared commitment to prevent the international spread of disease. Although the IHR (2005) do not include an enforcement mechanism *per se* for States which fail to comply with its provisions, the potential consequences of non-compliance are themselves a powerful compliance tool. Perhaps the best incentives for compliance are "peer pressure" and public knowledge. With today's electronic media, nothing can be hidden for very long. States do not want to be isolated. The consequences of non-compliance may include a tarnished international image, increased morbidity/mortality of affected populations, unilateral travel and trade restrictions, economic and social disruption and public outrage. Working together and with WHO to control a public health event and to accurately communicate how the problem is being addressed helps to protect against unjustified measures being adopted unilaterally by other States.

The key obligations for member states and the WHO are listed next.

Member States

- To designate a National IHR Focal Point
- To assess events occurring in their territory and to notify the WHO of all events that may constitute a public health emergency of international concern
- To respond to requests for verification of information regarding events that may constitute a public health emergency of international concern
- To respond to public health risks that may spread internationally
- To develop, strengthen, and maintain the capacity to detect, report, and respond to public health events
- To provide routine facilities, services, inspections, and control activities at designated international airports, ports, and ground crossings to prevent the international spread of disease

- To report to the WHO evidence of a public health risk identified outside their territory, which may cause international disease spread, manifested by exported/imported human cases, vectors carrying infection or contamination, or contaminated goods
- To respond appropriately to WHO-recommended measures
- To collaborate with other states/parties and with the WHO on IHR (2005) implementation

WHO

- Designating WHO IHR Contact Points at the headquarters or the regional level
- Conducting global public health surveillance and assessment of significant public health events, and disseminating public health information to states, as appropriate
- Offering technical assistance to states in their response to public health risks and emergencies of international concern
- Supporting states in their efforts to assess their existing national public health structures and resources, as well as to develop and strengthen the core public health capacities for surveillance and response, and at designated points of entry
- Determining whether or not a particular event reported by a state under the regulations constitutes a public health emergency of international concern, with advice from external experts if required
- Developing and recommending the critical health measures for implementation by states/parties
- Monitoring the implementation of IHR (2005) and updating guidelines so that they remain scientifically valid and consistent with changing requirements

7.15 SUMMARY OF IMPORTANT POINTS

1. The major regulatory agencies worldwide are:
   - Food and Drug Administration (FDA) of the United States
   - European Medicines Agency (EMEA) of the European Union
   - Ministry of Health, Labor and Welfare (MHLW) of Japan
   - State Food and Drug Administration (SFDA) of China
   - Central Drugs Standard Control Organization (CDSCO) of India
   - Therapeutic Goods Administration (TGA) of Australia
   - Health Canada of Canada

2. At the FDA, the Center for Drug Evaluation and Research (CDER) is responsible for the approval of small molecule drugs and therapeutic biologics while nontherapeutic biologics and blood products are under the jurisdiction of the Center for Biologics Evaluation and Research (CBER).
3. Drug approval at the EMEA is by the Committee for Medicinal Products for Human Use (CHMP).
4. The Pharmaceutical and Medical Device Agency (PMDA) of the MHLW is responsible for drug approval in Japan.
5. In China, the Department of Drug Registration (DDR) manages drug approval through the Division of Pharmaceuticals for "Western" drugs, the Division of Biological Products for biologics, and the Division of TCM for traditional Chinese medicine.
6. The other nonregulatory authorities related to drugs are the International Conference on Harmonization (ICH), which harmonizes regulations for the United States, the European Union, and Japan; the World Health Organization (WHO), which sets global guidance and standards on health matters and coordinates international health activities; and the Pharmaceutical Inspection Cooperation Scheme (PIC/S), which sets up mutual recognition on GMP inspections.

7.16 REVIEW QUESTIONS

1. Distinguish the different responsibilities of the CDER and CBER.
2. Explain the formation of the EMEA and its drug approval procedure.
3. How does the MHLW of Japan treat foreign clinical trial data in its approval process?
4. Describe the mechanism for the approval of drugs in China by the SFDA.
5. Outline the four categories of harmonized documents prepared by the ICH.
6. Describe the functions served by the PIC/S.
7. Provide a summary of the International Health Regulations (2005) and the roles for member states and the WHO in controlling the spread of diseases.

7.17 BRIEF ANSWERS AND EXPLANATIONS

1. Refer to Sections 7.2.1 and 7.2.2 about the CDER and CBER. It should be understood that small and large molecule drugs are legislated differently: the former under the Food, Drug and Cosmetic Act (FDCA) and the latter under the Public Health Service Act (PHSA).

2. The EMEA is an agency set up by the EU to evaluate and approve drugs for European countries under the centralized procedure for all biopharmaceuticals, new drugs for specific diseases, and orphan drugs (Section 7.3 and Exhibit 8.4). The other two routes to drug approval in Europe are the mutual recognition procedure and the decentralized procedure. The EMEA may arbitrate if disputes arise on drug approval decisions via the mutual recognition procedure.

3. Foreign data are acceptable to Japan's authority but there may be additional information with regard to immunological responses and ethnic factors required to gain approval (Section 7.4).

4. Drugs are separated into three categories in China: "Western" drugs are reviewed by the Division of Pharmaceuticals, biologics by the Division of Biological Products, and traditional Chinese medicine by the Division of TCM (Section 7.5). The approval process is described in Section 8.5.

5. Refer to Section 7.11 and Exhibit 7.6.

6. PIC/S is a cooperative entity for member countries to work on harmonization of documents and procedures and mutual recognition processes with respect to GMP inspections being performed.

7. Refer to Section 7.14.

7.18 FURTHER READING

Australia's Therapeutic Goods Administration website, http://www.tga.gov.au/ [accessed July 29, 2007].

Canada's Health Canada website, http://www.hc-sc.gc.ca/index_e.html.

Center for Biologics Evaluation and Research. *CBER's Report to the Biologics Community—2000*, FDA, Rockville, MD, 2000. http://ww.fda.gov/cber/inside/biolrpt.htm [accessed March 7, 2002].

Center for Drug Evaluation and Research. *Center for Drug Evaluation and Research Fact Book*, FDA, Rockville, MD, 1997. http://www.fda.gov/cder/reports/cderfact.pdf [accessed February 25, 2002].

China's State Food and Drug Administration website, http://www.sfda.gov.cn/eng/.

European Agency for the Evaluation of Medicinal Products website, http://www.emea.eu.int/.

European Commission Enterprise Directorate-General. *Pharmaceuticals in the European Union*, Office for Official Publications of the European Communities, Luxembourg, 2000.

EMEA, CHMP. *Guideline on therapeutic Areas Within the Mandatory Scope of the Centralised Procedure for Evaluation for Marketing Authorisation Application*, October 2005.

Food and Drug Administration website, http://www.fda.gov/.

Food and Drug Administration. *Activities of FDA's Medical Product Centers in 2001*, FDA, Rockville, MD, 2001. http://www.fda.gov/bbs/topics/ANSWERS/2002/ANS01132.html [accessed April 1, 2002].

Food and Drug Administration, Center for Biologics Evaluation and Research website, http://www.fda.gov/cber/.

Food and Drug Administration, Center for Drugs Evaluation and Research website, http://www.fda.gov/cder/.

India's Central Drugs Standard Control Organization website, http://www.cdsco.nic.in/html/central.htm.

International Conference on Harmonization website, http://www.ifpma.org/ich1.html.

Japan's Ministry of Health, Labor and Welfare website, http://www.mhlw.go.jp/english/index.html.

Pharmaceutical Administration and Regulations in Japan website (March 2007), http://www.nihs.go.jp/pmdec/youkoso.htm [accessed August 24, 2007].

Pharmaceutical Inspection Cooperation Scheme website, http://www.picscheme.org/index.php.

Pisano DJ, Mantus D. *FDA Regulatory Affairs: A Guide for Prescription Drugs, Medical Devices and Biologics*, CRC Press, Boca Raton, FL, 2004.

Trade Compliance Center. *Japan Report on Medical Equipment and Pharmaceuticals Market-Oriented, Sector Selective (MOSS) Discussions.* http://199.88.185.106/tcc/data/commerce_html/TCC_Documents/Japan_Moss/Japan_Moss.html [accessed June 13, 2001].

World Health Organization website, http://www.who.int/en/.

CHAPTER 8

REGULATORY APPLICATIONS

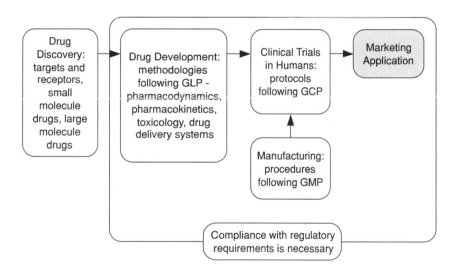

8.1	Introduction	232
8.2	Food and Drug Administration	233
8.3	European Union	250
8.4	Japan	263
8.5	China	264

8.6   India   266
8.7   Australia   269
8.8   Canada   269
8.9   Case Study #8   269
8.10   Summary of Important Points   273
8.11   Review Questions   274
8.12   Brief Answers and Explanations   274
8.13   Further Reading   275

8.1 INTRODUCTION

There are very few regulatory requirements that stipulate how an organization or institution should conduct drug discovery. In general, organizations and institutions are relatively unencumbered on the methods and techniques they adopt to discover new drugs, except in the case of gene therapy and stem cell research, where the regulations are specific and ethical limits are set. However, as drugs move along the pipeline from discovery to preclinical and clinical trials, there are strict procedures to follow.

In Chapter 5, we discussed the use of animals in preclinical studies. The applicable regulatory requirement is Good Laboratory Practice (GLP). In Chapter 6, we discussed clinical trials in humans. Here Good Clinical Practice (GCP) is required. Further along the pipeline, assuming that the drug shows efficacy with acceptable adverse events in the clinical trials, the drug will be registered and manufactured in compliance with Good Manufacturing Practice (GMP) for commercial sale. The processes for all these steps are governed by regulatory authorities.

The US Food and Drug Administration (FDA) has one of the most comprehensive and transparent regulatory systems in the world. In this chapter, we base most of our discussion on the FDA system. The emphasis in this chapter is on the processes for regulatory approvals. Before any new drug is trialed on human subjects, an Investigational New Drug (IND) application has to be filed. At the conclusion of clinical trials, the marketing approval for a drug is filed using a New Drug Application (NDA) for small molecule drugs or a Biologics License Application (BLA) for protein-based drugs. Other regulatory processes for Europe, Japan, China, India, Australia, and Canada are introduced in later sections of this chapter.

It should also be noted that US legislation excludes persons who have been debarred from being involved in drug product application as noted below:

Section 306(k) of the Federal Food, Drug, and Cosmetic Act (the Act) (21 U.S.C. 335a(k)), as amended by the Generic Drug Enforcement Act of 1992 (GDEA), requires that drug product applicants certify that they did not and will not use in any capacity the services of any debarred persons in connection with a drug product application. If the application is an abbreviated new drug application (ANDA), it must also include a list of all convictions described under section 306(a) and (b) of the Act (21 U.S.C. 335a(a) and (b)) that occurred within the previous 5 years and were committed by the applicant or affiliated persons responsible for the development or submission of the ANDA.

8.2 FOOD AND DRUG ADMINISTRATION

8.2.1 Drug Development Process

Figure 8.1 shows the drug development processes and the applicable regulatory steps. Before a drug is administered to humans, the FDA requires that preclinical research on animals be carried out. The information is necessary to assess the safety level of the drug. Based on this information, clinical trials on humans can be designed. The trial protocol will consider the safe dose, methods for dose ranging, route of drug administration, and toxicity effects.

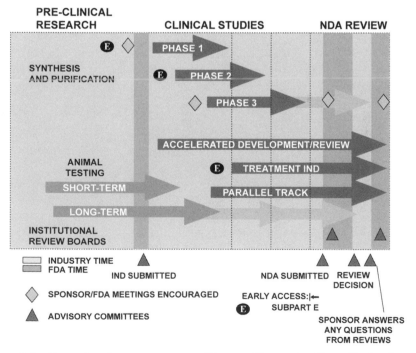

Figure 8.1 Drug development process. (*Source*: Center for Drug Evaluation and Research. The new drug development process, in *The CDER Handbook*, FDA, Rockville, MD. http://www.fda.gov/cder/handbook/develop.htm [accessed July 18, 2007].)

8.2.2 Investigational New Drug

An Investigational New Drug (IND) application to the FDA seeks permission for a human clinical trial to be conducted. An IND application is detailed under 21 CFR Part 312. The process for an IND is summarized in Fig. 8.2.

A firm or institution, called a Sponsor, is responsible for submitting the IND application. The relevant authorities are the Center for Drug Evaluation and

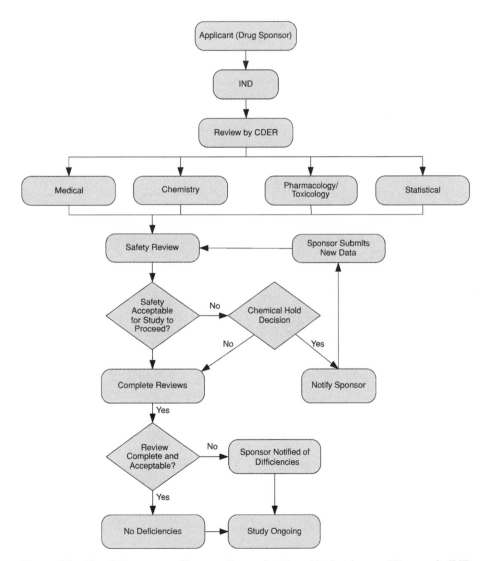

Figure 8.2 The IND process. (*Source*: Center for Drug Evaluation and Research. IND review process, in *The CDER Handbook*, FDA, Rockville, MD. http://www.fda.gov/cder/handbook/ind.htm [accessed July 18, 2007].)

Research (CDER) for small molecule synthetic drugs and therapeutic biologics, and the Center for Biologics Evaluation and Research (CBER) for non-therapeutic biologics (see Sections 7.2.1 and 7.2.2). A pre-IND meeting can be arranged with the FDA to discuss a number of issues:

- The design of animal research, which is required to lend support to the clinical studies
- The intended protocol for conducting the clinical trial
- The chemistry, manufacturing, and control of the investigational drug

Such a meeting will help the Sponsor to organize animal research, gather data, and design the clinical protocol based on suggestions by the FDA.

An IND is submitted on Form 1571. The materials to submit to the FDA are stated in Form 1571, Section 12 of Page 2 of this form:

1. Form 1571 [*21 CFR 312.23(a)(1)*]
2. Table of Contents [*21 CFR 312.23(a)(2)*]
3. Introductory Statement [*21 CFR 312.23(a)(3)*]
4. General Investigational Plan [*21 CFR 312.23(a)(3)*]
5. Investigator's Brochure [*21 CFR 312.23(a)(5)*]
6. Protocols [*21 CFR 312.23(a)(6)*]
   Study protocols [*21 CFR 312.23(a)(6)*]
   Investigator data [*21 CFR 312.23(a)(6)(iii)*]
   Facilities data [*21 CFR 312.23(a)(6)(iii)*]
   Institutional Review Board data [*21 CFR 312.23(a)(6)(iii)(b)*]
7. Chemistry, manufacturing, and control data [*21 CFR 312.23(a)(7)*]
8. Environmental assessment or claim for exclusion [*21 CFR 312.23(a)(7)(iv)(e)*]
9. Pharmacology and toxicology data [*21 CFR 312.23(a)(8)*]
10. Previous human experience [*21 CFR 312.23(a)(9)*]
11. Additional information [*21 CFR 312.23(a)(10)*]

Items 1–3 and 9 and 10 are self-explanatory and will not be discussed further. Items 4–6, and 9 on Investigational Plan, Investigator's Brochure, and Protocols and human experience are covered in Chapter 6, and Pharmacology and Toxicity data are discussed in Chapter 5. We will concentrate our discussion on Item 7.

Chemistry, Manufacturing, and Controls: As stated in 21 CFR Part 312, chemistry, manufacturing, and controls (CMC) information is to "describe the composition, manufacture, and controls of the drug substance and the drug product ... sufficient information is required to be submitted to assure the proper identification, quality, purity and strength of the investigational drug."

The FDA has various guidelines pertaining to the requirements of the data to be presented in the CMC for different drugs. In general, the CMC describes the drug, its chemistry, and characterization. Other requirements are the manufacturing processes, quality control testing and storage, stability, and labeling. We will highlight an example of a vaccine CMC according to the contents presented in *Guidance for Industry, Content and Format of Chemistry, Manufacturing and Controls Information and Establishment Description Information for a Vaccine or Related Product* (CBER, January 1999):

A. Description and Characterization
    1. Description
    2. Characterization (Physicochemical Characterization, Biological Activity)
B. Manufacturer
    1. Identification
    2. Floor Diagrams
    3. Manufacture of Other Products
C. Method of Manufacture
    1. Raw Materials
    2. Flow Charts
    3. Detailed Description (Sources, Cell Growth, Harvesting, Purification, etc.)
    4. Batch Records
D. Process Controls
    1. In-Process Controls
    2. Process Validation
    3. Control of Bioburden
E. Manufacturing Consistency
    1. Reference Standards
    2. Release Testing
F. Drug Substance Specification
    1. Specifications
    2. Impurities Profile
G. Reprocessing
H. Container and Closure System
I. Drug Substance Stability
    1. Contamination Precautions

The Sponsor has to explain how the drug is to be manufactured, tested, and stored. The important criterion is to ensure that it is safe for the subjects of the clinical trials. The CMC is a "living" document; it is updated as the clinical trials proceed from Phase I to Phases II and III and eventually to a licensed product. In essence, the CMC describes the adherence to Good Manufacturing Practice (GMP) for the manufacture of the trial drug. The subject of GMP is described in Chapters 9 and 10.

IND Review: Following submission of Form 1571, the FDA has 30 days to review the application. The topics reviewed include medical, chemistry, pharmacology and toxicology, and statistics. Medical review focuses on design of the clinical trial protocol, risk–benefit issues for the trial subjects, and supporting safety data from preclinical research. Chemistry review is based on the CMC to determine that appropriate controls are in place to manufacture, test, package, and label the drug for the trial. Pharmacological and toxicological review considers the mechanism of drug action, absorption, distribution, metabolism, and excretion (ADME), organs targeted or affected by the drug, and toxicological studies, including acute toxicity doses. Statistical review examines the design of the protocol with respect to subject numbers, doses, and biomarkers or indicators to demonstrate that sufficient data will be gathered for meaningful statistical analysis of the outcomes.

At the end of 30 days, the FDA informs the Sponsor of its review finding. There may be additional information that the FDA requires the Sponsor to submit, in which case the trial is put on clinical hold until all queries are satisfactorily answered. If the FDA considers the information provided does not support the conduct of a trial or subjects may be at risk in a trial, the clinical hold is not lifted and the IND is not approved.

Phase I, II, and III Trials: The clinical trials, from phases I to III, are conducted under one IND. At any stage of the trial, the FDA has the authority to put clinical hold on the trial until deficiencies or safety issues are resolved. The Sponsor can request meetings with the FDA at various stages:

- *End of Phase I Meeting:* After completing Phase I, the Sponsor meets with the FDA to discuss results of the trial and agree on a plan for Phase II studies.
- *End of Phase II/Pre-Phase III Meeting:* The meeting will evaluate the data obtained from Phase II studies. If the results are encouraging, Phase III is planned to gather further confirmation of the safety and efficacy of the drug. A more extensive protocol may need to be devised.
- *Pre-NDA/BLA Meeting:* This meeting is to prepare for the filing of the New Drug Application (NDA, for synthetic drug) or Biologics License Application (BLA, for protein-based drug). Results from Phase III are discussed. These data should support the safety and efficacy of the drug. A meeting at this stage can help to facilitate the FDA review process when the NDA or BLA is submitted.

Other Review Mechanisms: Although most drugs go through all the stages of Phases I, II, and III, there are special mechanisms in place to expedite development and approval of certain drugs. These mechanisms are divided into the following:

- *Accelerated Development/Review:* A drug for the treatment of serious or life-threatening diseases for which there are no alternative therapies may receive expedited review and approval. A condition for the approval is that the Sponsor undertakes to continue with further clinical trials after approval to confirm the efficacy of the drug.
- *Treatment Investigational New Drugs:* The FDA allows certain drugs to be administered to patients who have life-threatening illnesses that will lead to death without suitable treatment. An example is cancer patients receiving treatments with investigational new drugs (Exhibit 8.1).

Exhibit 8.1   Alimta and Zelnorm

Alimta: Recently, the FDA agreed to the use of a drug for treating a rare form of cancer under "compassionate" purposes before an NDA is submitted. The drug, pemetrexed (Alimta, Eli Lilly), has shown positive Phase III results in prolonging the lifespan of patients with pleural mesothelioma (a cancer linked to asbestos). Patients in trials, when treated with pemetrexed together with chemotherapy and vitamins, showed an average lifespan extension of 13 months after being diagnosed with pleural mesothelioma. This compares with 7 months for the current standard treatment of chemotherapy and vitamins.

Source: Food and Drug Administration. Drugs @ FDA, *Alimta*. http://www.accessdata.fda.gov/scripts/cder/drugsatfda/index.cfm?fuseaction=Search.DrugDetails [accessed July 25, 2007].

Zelnorm: On July 27, 2007, the FDA permitted the restricted use of tegaserod maleate (Zelnorm) to treat irritable bowel syndrome with constipation (IBS-C) and chronic idiopathic constipation (CIC) in women younger than 55 who meet specific guidelines. In reality, Zelnorm was approved in 2002 but the FDA asked the manufacturer, Novartis, to suspend marketing of the drug due to a higher incidence of serious symptoms for people treated with Zelnorm. But because Zelnorm's benefits for some cases may outweigh the risks, the FDA allows its restricted use.

Source: Food and Drug Administration. *Zelnorm Available for Restricted Use*. http://www.fda.gov/consumer/updates/zelnorm072707.html [accessed July 25, 2007].

• *Parallel Track:* Some patients do not fulfill the criteria to be enrolled in clinical trials, but their conditions qualify them to be treated in parallel with an ongoing clinical trial. AIDS patients are an example for this group.

8.2.3 New Drug Application/Biologics License Application

At the conclusion of the Phase III clinical trial, if the results demonstrate that the drug is safe and efficacious over existing treatment drugs, an application is made to the FDA to seek approval for marketing the drug. A New Drug Application (NDA) or a Biologics License Application (BLA) is filed. The NDA is covered under the Federal Food, Drug and Cosmetic Act, Section 505 (Exhibit 8.2) while the BLA is mandated by the Public Health Service Act, Section 351. The process for filing and reviewing of the NDA/BLA is presented in Fig. 8.3. The application is submitted using Form 356h (Fig. 8.4).

Form 356h is a harmonized form, and a Sponsor can use it for NDA, BLA, and Abbreviated New Drug Application (ANDA, see Section 8.2.5). Page 1 of the form requires Applicant Information, Product Description, Application Information, and Establishment Information. Page 2 requires the provision of a number of items to substantiate the application. The items to be submitted under Form 356h are as follows:

1. Index
2. Labeling
3. Summary
4. Chemistry section
   Chemistry, manufacturing, and controls information
   Samples
   Methods validation package
5. Nonclinical pharmacology and toxicology section
6. Human pharmacokinetics and bioavailability section
7. Clinical microbiology
8. Clinical data section
9. Safety update report
10. Statistical section
11. Case report tabulations
12. Case report forms
13. Patent information on any patent that claims the drug
14. A patent certification on any patent that claims the drug
15. Establishment description
16. Debarment certification
17. Field copy certification
18. User fee cover sheet
19. Other

Exhibit 8.2   NDA

There are three types of new drug applications under Section 505.

505(b)(1) Application: NDA that contains full reports of investigations of safety and effectiveness. The investigations the applicant relied on for approval were conducted by or for the applicant, or the applicant has obtained a right of reference or use for the investigations.

505(b)(2) Application: NDA for which some or all of the investigations the applicant relied on for approval were not conducted by or for the applicant, and the applicant has not obtained a right of reference or use for the investigations. Section 505(b)(2) expressly permits the FDA to rely, for approval of an NDA, on data not developed by the applicant, such as published literature or the FDA's finding of safety and/or effectiveness of a previously approved drug product.

505(j) Application: ANDA that contains information to show that the proposed product is identical in active ingredient, dosage form, strength, route of administration, labeling, quality, performance characteristics, and intended use, among other things, to a previously approved application (the reference listed drug—RLD). ANDAs do not contain clinical studies as required in NDAs but are required to contain information establishing bioequivalence to the RLD. In general, the bioequivalence determination allows the ANDA to rely on the FDA's finding of safety and efficacy for the RLD.

This is summarized in following table:

NDA—505(b)(1)	NDA—505(b)(2)	ANDA—505(j) (Refer to Case Study #10)
Preclinical	Preclinical	
Clinical	Clinical	
Pediatric use	Pediatric Use	
CMC	CMC	CMC
PK and bioavailability	PK Bioavailability	Bioavailability
Labeling	Labeling	Labeling
Patent information	Patent Info	
Exclusivity^a request	Patent Cert	Patent Cert
	Exclusivity Request and Statement	Exclusivity Statement

^{a}Exclusivity provides the holder of an approved new drug application limited protection from new competition in the marketplace for the innovation represented by its approved drug product: 5 years for NCE, 3 years for new indication of approved NCE.

Source: (1) Parise C. Office of Generic Drugs, FDA, *Regulatory Framework for the Submission of an Application.* http://www.fda.gov/cder/regulatory/pet/present/cecevan/sld001.htm [accessed August 3, 2007]. (2) Food and Drug Administration. *Frequently Asked Questions on the Pre-Investigational NDA Meeting.* http://www.fda.gov/cder/about/smallbiz/pre_IND_qa.htm [accessed August 3, 2007]. (3) Food and Drug Administration. *Frequently Asked Questions for New Drug Product Exclusivity.* http://www.fda.gov/cder/about/smallbiz/exclusivity.htm [accessed August 3, 2007].

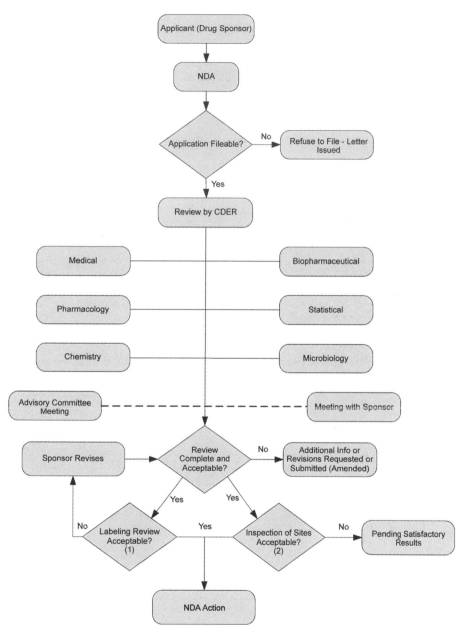

Figure 8.3 The NDA process. (*Source*: Center for Drug Evaluation and Research. NDA review process, in *The CDER Handbook*, FDA, Rockville, MD. http://www.fda. gov/cder/handbook/nda.htm [accessed July 18, 2007].)

DEPARTMENT OF HEALTH AND HUMAN SERVICES FOOD AND DRUG ADMINISTRATION **APPLICATION TO MARKET A NEW DRUG, BIOLOGIC, OR AN ANTIBIOTIC DRUG FOR HUMAN USE** *(Title 21, Code of Federal Regulation, Part 314 & 601)*	*Form Approved: OMB No. 0910-0338* *Expiration Date: September 30, 2008* *See OMB Statement on page 2.* **FOR FDA USE ONLY** **APPLICATION NUMBER**

APPLICANT INFORMATION

NAME OF APPLICANT	DATE OF SUBMISSION

TELEPHONE NO. *(Include Area Code)*	FACSIMILE *(FAX)* Number *(Include Area Code)*

APPLICANT ADDRESS *(Number, Street, City, State, Country, ZIP Code, Mail Code, and U.S. License number if previously issued)*:	AUTHORIZED US AGENT NAME & ADDRESS *(Number, Street, City, State, ZIP Code, telephone & FAX number)* IF APPLICABLE

PRODUCT DESCRIPTION

NEW DRUG OR ANTIBIOTIC APPLICATION NUMBER, OR BIOLOGICS LICENSE APPLICATION NUMBER *(If previously issued)*

ESTABLISHED NAME *(e.g. Proper name, USP/USAN name)*	PROPRIETARY NAME *(trade name)* IF ANY

CHEMICAL/BIOCHEMICAL/BLOOD PRODUCT NAME *(If any)*	CODE NAME *(If any)*

DOSAGE FORM:	STRENGTHS:	ROUTE OF ADMINISTRATION:

(PROPOSED INDICATIONS) FOR USE:

APPLICATION DESCRIPTION

APPLICATION TYPE *(check one)* ☐ NEW DRUG APPLICATION (CDA, 21 CFR 314.50) ☐ ABBREVIATED NEW DRUG APPLICATION (ANDA, 21 CFR 314.94)
☐ BIOLOGICS LICENSE APPLICATION TYPE (BLA, 21 CFR Part 601)

IF AN NDA, IDENTIFY THE APPROPRIATE TYPE ☐ 505 (b)(1) ☐ 505 (b)(2)

IF AN ANDA OR 505(b)(2), IDENTIFY THE REFERENCE LISTED DRUG PRODUCT THAT IS THE BASIS FOR THE SUBMISSION

Name of Drug _____   Holder of Approved Application _____

TYPE OF SUBMISSION *(check one)* ☐ ORIGINAL APPLICATION ☐ AMENDMENT TO A PENDING APPLICATION ☐ RESUBMISSION
☐ PRESUBMISSION ☐ ANNUAL REPORT ☐ ESTABLISHMENT DESCRIPTION SUPPLEMENT ☐ EFFICACY SUPPLEMENT
☐ LABELING SUPPLEMENT ☐ CHEMISTRY MANUFACTURING AND CONTROLS SUPPLEMENT ☐ OTHER _____

IF A SUBMISSION OF PARTIAL APPLICATION, PROVIDE LETTER DATE OF AGREEMENT TO PARTIAL SUBMISSION:

IF A SUPPLEMENT, IDENTIFY THE APPROPRIATE CATEGORY ☐ CBE ☐ CBE-30 ☐ Prior Approval (PA)

REASON FOR SUBMISSION

PROPOSED MARKETING STATUS *(check one)* ☐ PRESCRIPTION PRODUCT *(Rx)* ☐ OVER THE COUNTER PRODUCT *(OTC)*

NUMBER OF VOLUMES SUBMITTED _____   THIS APPLICATION IS ☐ PAPER ☐ PAPER AND ELECTRONIC ☐ ELECTRONIC

ESTABLISHMENT INFORMATION (Full establishment information should be provided in the body of the Application.)
Provide locations of all manufacturing, packaging and control sites for drug substance and drug product (combination sheets may be used if necessary). Include name, address, contact, telephone number, registration number (CFN), DMF number, and manufacturing steps and or type of testing (e.g. Final dosage from, Stability testing) conducted at the site. Please indicate whether the site is ready for inspection or, if not, when it will be ready.

Cross References (list related License Applications, INDs, NDAs, PMAs, 510(k)s, IDEs, BMFs, and DMFs referenced in the current application)

FORM FDA 356h (10/05)   PAGE 1 OF 4

Figure 8.4 Form 356h (Page 1). (*Source*: Food and Drug Administration, CDER. http://www.fda.gov/opacom/morechoices/fdaforms/356Hes.pdf [accessed September 26, 2007].)

The submission of Form 356h is the culmination of all the work and effort that has been put into discovering, developing, and trialing the drug. The information submitted is substantial, with many volumes prepared for separate sections; literally truckloads of documents are delivered to the FDA. However, the submission can now be streamlined through electronic means, for example, via eCTD. Instructions for electronic submission are detailed in the FDA document *Regulations and Instructions for Submitting Drug Applications Electronically*. It should be noted that every new drug in the United States has been approved via the NDA process since 1938, although there have been changes to the requirements for submission over the years. Before the introduction of BLA in 1998, biologics were approved under two separate submissions of Product License Application (PLA) and Establishment License Application (ELA).

Details for the required information to be submitted with Form 356h are stated in 21 CFR Part 314 for synthetic drugs and 21 CFR Part 601 for biopharmaceutical drugs. We will select a few key items for discussion.

Index: The index of Form 356h sets out how the extensive numbers of documents are to be referenced. A well-organized index system is important for the reviewers to search for the required information. This will expedite the review process, without the necessity for the FDA to stop the review time clock to seek clarification.

Labeling: Labeling is reviewed following requirements of 21 CFR Part 201. The requirements are as listed in Table 8.1.

Summary: The summary presents the case for the drug's approval. It includes discussion about the drug's mechanism of action, its effect on animals, results of clinical trials, manufacturing and tests methods, its stability, and proposed dosage and treatment protocol. The summary may run into hundreds of pages. It is one of the few documents being read by all the different reviewers; as such, a good summary will assist with the review process.

Chemistry Section: This is the CMC with updated information pertaining to the chemistry, manufacturing, and controls of the drug. The FDA recognizes that manufacturing processes and test methods go through various stages of optimization and refinement as the drugs are produced for Phase I and II clinical trials. By the Phase III stage, however, all the manufacturing processes are expected to be defined and test methods validated. A detailed explanation of the drug manufacturing processes is presented in Chapter 10. Some pertinent data are given next.

- *Drug Molecule:* Chemical composition, physical and chemical characteristics, and specifications

TABLE 8.1   Review of Labeling

Item	Explanatory Notes
Description	Proprietary and established name of drug; dosage form; ingredients; chemical name; and structural formula
Clinical pharmacology	Summary of the actions of the drug in humans; *in vitro* and *in vivo* actions in animals if pertinent to human therapeutics; pharmacokinetics
Indications and usage	Description of use of drug in the treatment, prevention, or diagnosis of a recognized disease or condition
Contraindications	Description of situations in which the drug should not be used because the risk of use clearly outweighs any possible benefit
Warnings	Description of serious adverse reactions and potential safety hazards, subsequent limitation in use, and steps that should be taken if they occur
Precautions	Information regarding any special care to be exercised for the safe and effective use of the drug; includes general precautions and information for patients on drug interactions, carcinogenesis/mutagenesis, pregnancy rating, labor and delivery, nursing mothers, and pediatric use
Adverse reactions	Description of undesirable effect(s) reasonably associated with the proper use of the drug
Drug abuse/dependence	Description of types of abuse that can occur with the drug and the adverse reactions pertinent to them
Overdosage	Description of the signs, symptoms, and laboratory findings of acute overdosage and the general principles of treatment
Dosage/administration	Recommendation for usage dose, usual dosage range, and, if appropriate, upper limit beyond which safety and effectiveness have not been established
How to be supplied	Information on the available dosage forms to which the labeling applies

Source: Adapted from Center for Drug Evaluation and Research. New Drug Application (NDA) Process, FDA, Rockville, MD. http://www.fda.gov/cder/regulatory/applications/nda.htm [accessed September 21, 2007].

- *Raw Materials:* List of all materials used, specifications and tests for these raw materials
- *Equipment:* List of equipment used, validation of the equipment, validated methods for cleaning, and procedures for contamination control
- *Analytical Methods:* Validation to assure that the analytical methods are appropriate for the tests

- *Manufacturing Processes:* Flow charts for production steps, controls for contamination, removal of impurities, purification steps, in-process tests, and batch records
- *Facility:* Controls on equipment, calibration policies, security of access, maintenance of clean environment, flow of materials, equipment, and products
- *Drug Stability:* Data to substantiate the stability of the drug for storage and transportation
- *Product Release Criteria:* Specifications, test methods, storage and shipping conditions

For biopharmaceuticals, further information is required. Listed below are some examples.

- *Cell Line:* Source, species, history, characteristics, cloning methods, vectors used, and genotype and phenotype of host cell system
- *Cell Bank:* Controls for working and master cell banks
- *Assays:* Validated methods of analysis (e.g., ELISA for MAb), QPCR for residual DNA, and potency assays for vaccines
- *Production:* Culture medium used, cell culture and fermentation techniques, in-process controls, purification steps, and cleaning of chromatographic columns and matrices

The CMC details all the manufacturing steps and controls being introduced, to ensure that the drug product is pure, consistent, safe, and effective. The Sponsor has to demonstrate that the manufacturing facility is set up and complies with cGMP regulations for the production of the drug when it is approved. The FDA has the right to obtain samples from the Sponsor for evaluation and test.

Nonclinical Pharmacology and Toxicology Section: This section is to present data in addition to that included in the IND. Long-term toxicology data are required. The Sponsor is also expected to provide study results of the drug on reproduction and effects on fetuses.

Clinical Results: Items 6–12 of Form 356h are all related to the clinical results. These are perhaps the most important sections of the submission to demonstrate the safety and efficacy of the drug for treating the target disease. Detailed analyses of clinical data are presented to support the application. Some of these analyses include the following:

- Kinetics studies to show the ADME mechanisms on target organs and tissues

- For anti-infective agents, reports on the *in vivo* and *in vitro* tests and the effects of the drug on the microorganisms
- Description of the statistical model adopted for analyses
- Statistical analyses of results from the clinical trials, showing statistical power of the test
- Comparison of the therapeutic index and safety data
- Report on adverse events, incapacity, and death, if any, and investigation of the cause

Drug Master File: As stated by the FDA, the Drug Master File (DMF) is submitted to the FDA to provide confidential information relating to the facilities and manufacturing processes and techniques for producing the drug material. However, it is not required by law or FDA regulations that a DMF be submitted accompanying the IND or NDA/BLA. In reality, however, most organizations prepare and submit the DMF with their applications.

The FDA *Guideline for Drug Master Files* (21 CFR Part 314.420) consists of the following sections:

- Contents
- Definitions
- Types of Drug Master Files
    Type I: Not applicable. Provision removed by FDA
    Type II: Drug substance, drug substance intermediate, and materials used in their preparation, or drug product
    Type III: Packaging materials
    Type IV: Excipient, colorant, flavor, essence, or materials used in their preparation
    Type V: Facilities for production, contract manufacturing facilities and testing facilities
- Authorization to Refer to a Drug Master File
- Processing and Reviewing Policies
- Holder Obligations
- Closure of a Drug Master File

The Type V DMF enables confidential information to be submitted to the FDA, for example, a contract manufacturing facility may provide proprietary information to the FDA without divulging it to the Sponsor client. The FDA reviews the DMF, but the DMF is never approved or disapproved. The holder of the DMF is notified of deficiencies for rectification. It is the holder's responsibility to update the DMF on an annual basis.

NDA/BLA Review: Review is undertaken by FDA staff from different offices within the CDER and CBER. These staff members are trained physicians, statisticians, chemists, biologists, pharmacologists, and other scientists. The FDA may consult with external review committees and experts but is not bound by their recommendations.

Since the introduction of the Prescription Drug User Fee Act (PDUFA, see Review Outcome below) in 1992, the FDA has set a target time for the review of the NDA/BLA. In general, the review time for the NDA/BLA is 12 months, including FDA time and Sponsor time to respond to deficiencies. The target for a priority NDA/BLA is 6 months. Figure 8.5 shows the review time of the NDA/BLA application. Drugs are eligible for priority review if they show significant improvement compared with marketed products in the treatment, diagnosis, or prevention of a disease.

For a new facility being set up to manufacture a drug under the NDA/BLA, the FDA is likely to perform a preapproval inspection (PAI) to ensure the facility has adequate procedures and controls to manufacture the drug under GMP according to *Compliance Program Guide 7346.832*. For an existing facility already manufacturing the drug, with the NDA/BLA for extension of treatment indications or other nonmanufacturing related matters, the FDA may waive the PAI.

When the NDA/BLA is approved, the Sponsor has the license to market the drug. It is the Sponsor's responsibility to inform the FDA of adverse events or any unexpected findings. The FDA has the responsibility to safeguard the public's health. It monitors adverse events, advertising, and manufacturing in accordance with GMP.

Review Outcome: The outcomes from the review can be classified into three categories:

- *Not Approvable Letter:* Application cannot be approved and deficiencies are detailed.

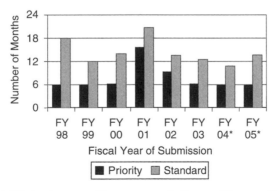

Figure 8.5 NDA/BLA drug median approval times. (*Source*: Food and Drug Administration. *FY 2006 Performance Report to the President and the Congress*, FDA, Rockville, MD, 2006. http://www.fda.gov/ope/pdufa/report2006/PDUFA2006perf.pdf [accessed September 20, 2007].)

- *Approvable Letter:* Deficiencies are minor and can be corrected or supplementary information has to be provided. Eventually the drug is approved.
- *Approval Letter:* The drug is approved. An example of the approval news for Tykerb, a kinase inhibitor in combination with capecitabine for the treatment of advanced or metastatic cancer, is presented in Exhibit 8.3.

Based on the Prescription Drug User Fee Act (PDUFA), the FDA collects fees from applicants to expedite the review and approval processes under strict guidelines. The PDUFA fees for fiscal year 2007 (October 1, 2006 to September 30, 2007) are shown in Table 8.2.

8.2.4   Orphan Drugs

Drugs are designated as orphan drugs for those diseases with a patient population of less than 200,000 in the United States. The FDA has a special provision

---

Exhibit 8.3   Tykerb Approval News

FDA Approves Tykerb for Advanced Breast Cancer Patients: The Food and Drug Administration (FDA) today approved Tykerb (lapatinib), a new targeted anticancer treatment, to be used in combination with capecitabine (Xeloda), another cancer drug, for patients with advanced, metastatic breast cancer that is HER2 positive (tumors that exhibit HER2 protein). The combination treatment is indicated for women who have received prior therapy with other cancer drugs, including an anthracycline, a taxane, and trastuzumab (Herceptin). According to the American Cancer Society, about 180,000 new cases of breast cancer are diagnosed each year. Approximately 8000–10,000 women die from metastatic HER2 positive breast cancer each year.

Tykerb, a new molecular entity (NME), is a kinase inhibitor working through multiple pathways (targets) to deprive tumor cells of signals needed to grow. Unlike, for example, trastuzumab—a monoclonal antibody, which is a large protein molecule that targets the part of the HER2 protein on the outside of the cell—Tykerb is a small molecule that enters the cell and blocks the function of this and other proteins. Because of this difference in mechanism of action, Tykerb works in some HER2 positive breast cancers that have been treated with trastuzumab and are no longer benefiting.

Source: Food and Drug Administration. FDA News, *FDA Approves Tykerb for Advanced Breast Cancer Patients*. http://www.fda.gov/bbs/topics/NEWS/2007/NEW01586. html [accessed August 16, 2007].

TABLE 8.2   PDUFA Fees for Fiscal Year 2007

Application	Fee (US$)
Applications requiring clinical data	896,200
Applications not requiring clinical data	448,100
Supplements requiring clinical data	448,100
Establishments	313,100
Products	49,750

Source: Food and Drug Administration. Prescription Drug User Fee Rates for Fiscal Year 2007. http://www.fda.gov/ohrms/dockets/98fr/E6-12397.pdf [accessed September 12, 2007].

for the development, marketing approval, and marketing of orphan drugs (refer to 21 CFR Part 316). The Orphan Drug Act provides incentives to organizations to research and test drugs that have limited commercial returns because of the small size of the patient group. In return for the commercial risks undertaken, there is assistance in the form of NDA fee waivers, tax credits for clinical research, and grants for the research. The FDA also provides market exclusivity (monopoly) to the organization to market the drug for 7 years.

8.2.5   Generics

A generic drug is defined as a drug that is equivalent to a prescription drug approved by the FDA, but for which the patent validity has expired. An ANDA approval is required (Fig. 8.6).

There is no requirement to provide preclinical or clinical data to demonstrate safety and efficacy of generic drugs. However, the review is based on bioequivalence and manufacturing control information. The Sponsor provides data to establish that the generic drug is equivalent to the off-patent prescription drug in terms of chemistry, dosage, bioavailability, absorption, distribution, metabolism, and excretion (ADME) characteristics, and toxicology. Information on manufacturing and control is submitted to demonstrate that production of generics complies with GMP.

8.2.6   Over-the-Counter Drugs

The approval process for over-the-counter (OTC) drugs is presented in Fig. 8.7. Some of the important points are the review of labeling to ensure it is clear and understandable by consumers, and public comment on the listing of the OTC drug. Monographs are prepared for OTC drugs; they list the raw materials used in the drug, dosage, indications of use, and labeling information.

Figure 8.6 Approval process for generics. (*Source*: Center for Drug Evaluation and Research. Generic drug (ANDA) approval process, in *The CDER Handbook*, FDA, Rockville, MD, 2007. http://www.fda.gov/cder/handbook/anda.htm [accessed July 20, 2007].)

8.3 EUROPEAN UNION

Similar to the US requirements, there are two regulatory steps to go through before a drug is approved to be marketed in the European Union. These two steps are clinical trial application and marketing authorization application. There are 27 member states in the European Union (as of August 2007); clinical trial applications are approved at the member state level, whereas marketing authorization applications are approved at both the member state or centralized levels.

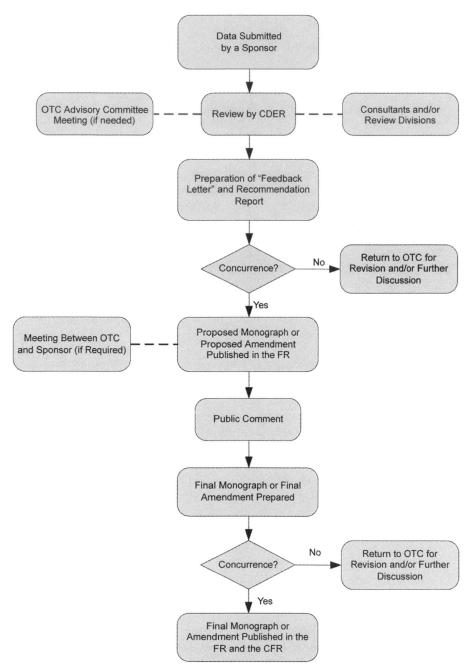

Figure 8.7 Approval process for OTC drugs. (*Source*: Center for Drug Evaluation and Research. OTC drug monograph review process, in *The CDER Handbook*, FDA, Rockville, MD, 2007. http://www.fda.gov/cder/handbook/otc.htm [accessed July 22, 2007].)

8.3.1   Clinical Trial Application

EU Directives 2001/20/EC and 2005/28/EC set out the new rules and regulations for the approval and conduct of clinical trials in Europe. Member states had to enact the Directives into national legislation and put them into effect by May 1, 2004.

A Sponsor submits a clinical trial application to the Competent Authority in each member state where the trials are to be conducted. The Competent Authority has 60 days to review and approve or reject the application. Application is in prescribed forms and covers the proposed clinical trial protocol, manufacturing, and quality controls on the drug, and supporting data, such as (1) chemical, pharmaceutical, and biological data, (2) nonclinical pharmacological and toxicological data, and (3) clinical data and previous human experience. The supporting data are submitted in the Common Technical Document (CTD) format (see Section 7.11).

Most of the information sought is similar to the FDA's IND requirements. One major difference is that a Qualified Person has to certify that the investigational medicinal product (IMP) is manufactured according to GMP. The Competent Authority has the right to inspect the manufacturing facility for GMP compliance, the preclinical facility for GLP compliance, and the clinical trial sites for GCP compliance.

In the United Kingdom, clinical trial applications are submitted to the Medicines and Healthcare Products Regulatory Agency (MHRA). There are several schemes for clinical trial application; the major two are the Clinical Trial Certificate (CTC) and Clinical Trial Exemption (CTX) schemes.

The CTC system was the scheme used for the control of clinical trials before the introduction of CTX in 1981. Most clinical trials are now conducted under the CTX scheme. The CTX scheme was devised to speed up review and approval of clinical trials to allow important drugs to enter trials with minimum delay. Identical data are submitted for the filing of CTC and CTX; the difference is that, for CTX, only a summary of raw data is required. Information for submission is in three parts:

- *Part I:* Application Form: Introduction, Background, and Rationale for Trial
- *Part II:* Composition, Method of Preparation, Controls of Starting Materials, Control Tests on Intermediate Products, Control Tests on the Finished Product, Stability and Other Information (Placebos, Comparator Products, Adventitious Agents, etc.)
- *Part III:* Experimental and Biological Studies: Nonclinical Pharmaceutical and Toxicological Studies

A CTC application is submitted on Form MLA 202; the approval is for 2 years. A CTX application is submitted on Form MLA 164; the approval is for 3 years. The conditions associated with the CTX scheme are as follows:

- A registered medical practitioner must certify the accuracy of the summary information.
- The Sponsor must inform the MHRA of any refusal by an ethics committee to permit the trial.
- The Sponsor must inform the MHRA of any adverse events and safety issues.

8.3.2 Marketing Authorization

Following successful clinical trials, the Sponsor has to apply for authorization to market the drug in Europe. Depending on the type of drug product and the intended market, there are four different types of marketing authorization applications (Fig. 8.8).

Centralized Procedure: This procedure, according to Regulation 726/2004 and Directive 2004/27/EC, is for drugs developed using biotechnology processes, novel drugs for specific treatments, and orphan drugs (see Section 7.3 and Exhibit 8.4). The marketing authorization is for the entire European community. The process for the Centralized Procedure is summarized in Fig. 8.9.

An application is submitted to the European Medicines Agency (EMEA). The EMEA evaluates the application and forwards its opinion (positive or negative for granting of a marketing authorization) to the European Commis-

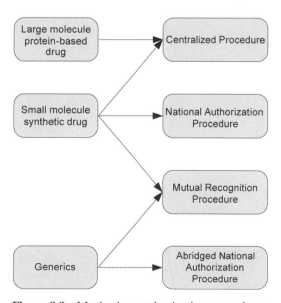

Figure 8.8 Marketing authorization procedures.

Exhibit 8.4 Drug Products According to EU Regulation 726/2004

Medicinal products developed by:

- Recombinant DNA technology
- Expression of proteins in prokaryotic and eukaryotic cells
- Hybridoma and monoclonal antibody methods

Medicinal products with a new active substance for treatment of:

- AIDS
- Cancer
- Neurodegenerative disorder
- Diabetes
- Autoimmune diseases and other immune dysfunctions (from May 20, 2008)
- Viral diseases (from May 20, 2008)

Orphan medicinal products pursuant to Regulation (EC) 141/2000

sion. The opinion is supported by the European Public Assessment Report (EPAR), which summarizes the scientific analyses and discussions during the evaluation process. The European Commission consults the relevant Standing Committees before granting the marketing authorization. The process takes up to 210 days.

Mutual Recognition Procedure: The Mutual Recognition Procedure is stated in Council Directive 93/39/EEC. In essence, once a drug is approved for marketing authorization by one member state, the company concerned can apply for marketing authorization in other member states through the mutual recognition procedure in place since 1998.

Identical applications are submitted to those member states where marketing authorizations are sought. The first member state that reviews the application is called the "Reference Member State." It notifies other states, called "Concerned Member States." Concerned Member States may suspend their own evaluations to await assessment by the Reference Member State. The decision of the Reference Member State is forwarded to the Concerned Member States. If the Concerned Member States reject mutual recognition, the matter is referred to the CHMP of the EMEA for arbitration. The EMEA forwards its opinion to the European Commission, which makes the final deci-

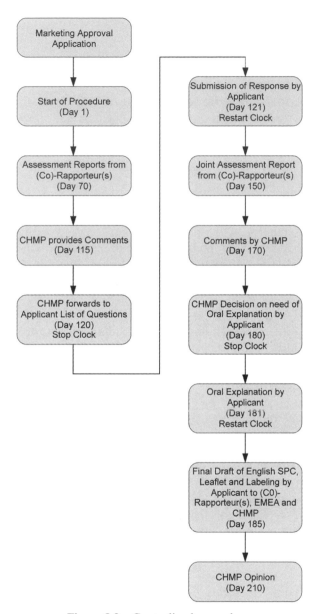

Figure 8.9   Centralized procedure.

sion. Altogether, the decision process may take up to 300 days if there is no objection, and 600 days when objections are raised.

National Authorization Procedure: To obtain marketing authorization in a country, the application must be submitted to the Competent Authority of that member state in its own language. For National Authorizations in more than one country, submissions have to be sent to each country in its own language. In many ways, the National Authorization system has been superseded by the Centralized and Mutual Recognition Procedures.

Abridged National Authorization Procedure: This procedure is for generics, and there is no necessity to provide preclinical or clinical results. However, evidence of bioavailability and bioequivalence and GMP manufacturing compliance has to be submitted. If the applicant has an abridged approval from a member state, the Mutual Recognition Procedure can be used.

Submission Details for Centralized Procedure: This procedure was effective from 1995 with further amendments effective from November 2005. Six months before submission, the pharmaceutical company wishing to apply for marketing authorization via the Centralized Procedure notifies the EMEA of its intention and expected submission date. This notification is required to be accompanied by a number of items, for example:

- Draft Summary of Product Characteristics (SPC)
- Justification for evaluation under the Centralized Procedure (refer to Parts A and B Products designation)
- Proposed DMF
- Manufacturing location

A Rapporteur and Co-rapporteur, whose roles are to coordinate the evaluation of the application, are appointed by the EMEA 3 months before the submission.

Submission for marketing authorization is in a prescribed format. It is accompanied by the payment of fees (see Table 8.3). Details for the submission dossier are described later.

There are two stages for the Centralized Procedure. The first phase is divided into the primary evaluation phase and secondary evaluation phase. In

TABLE 8.3 EMEA Application Fees, 2007

Application	Fee (Euro)
Full application	232,000
Marketing application (not full dossier, small molecule)	90,000
Marketing application (not full dossier, large molecule)	150,000
Scientific advice	69,600
Inspection	17,400

Source: EMEA. Guidelines on Fees Payable to EMEA. http://www.emea.europa.eu/htms/general/admin/fees/feesv.htm [accessed September 12, 2007].

the primary evaluation phase, the Rapporteur and Co-rapporteur coordinate the evaluation within the EMEA and communicate with the applicant with a list of questions. The secondary evaluation commences after the receipt of responses to the questions. The EMEA has the right to request drug samples for testing. The EMEA may also perform preauthorization inspection of the drug manufacturing facility to ensure compliance to GMP. At the end of stage 1, the EMEA sends its opinion, positive or negative, to the European Commission for decision-making, which is the second stage. Documents in eleven languages are sent by the EMEA to the European Commission. The European Commission checks to ensure the marketing authorization complies with European Community law and formalizes the EMEA decision into a decision for the entire European Community.

Preparation of Marketing Authorization Application Dossier: The application dossier is divided into four parts:

Part I: Summary of the dossier
Part II: Chemical/pharmaceutical/biological documentation
Part III: Toxicopharmaceutical documentation
Part IV: Clinical documentation

Part I: There are three subsections for Part I:

Part IA: Administrative data, packaging, samples
Part IB: SPC, package leaflets
Part IC: Expert reports

EU Directive 75/319/EEC requires that documents submitted for marketing authorization be drawn up and signed by experts with technical and professional qualifications.
Part IA is self-explanatory and will not be discussed further.
The major headings for Part IB are:

- Trade name of the medicinal product
- Qualitative and quantitative composition
- Pharmaceutical form
- Clinical particulars
- Pharmacological properties
- Pharmaceutical particulars
- Marketing authorization holder
- Marketing authorization number
- Date of first authorization/renewal of authorization
- Date of revision of text

Part IC is divided into three subparts: Parts IC1, IC2, and IC3, containing expert reports in prescribed format.

Part IC1 contains an expert report on the chemical, pharmaceutical, and biological documentation. Topics presented include composition, method of preparation, control of starting materials, control tests on intermediate products, control tests on finished product, stability, and information on the pharmaceutical expert. An example is given in Exhibit 8.5.

Part IC2 contains an expert report on the toxicopharmacological (preclinical) documentation. Topics presented include pharmacodynamics, pharmacokinetics, toxicity, and information on the preclinical expert. An example is given in Exhibit 8.6.

Part IC3 contains an expert report on the clinical documentation. Topics presented include clinical pharmacology, clinical trials, postmarketing experience, and information on the clinical expert (see Exhibit 8.7).

Part II: Part II is the report concerning chemical, pharmaceutical, and biological documentation. The report details the composition, method of development of formulation, manufacturing processes under GMP, analytical test procedures, bioavailability, and bioequivalence. It should be noted that all analytical test procedures need to be validated, and the validation studies must be provided.

There are four different drug products under Part II: chemical active substance(s), radiopharmaceutical products, biological medicinal products, and vegetable medicinal products. For example, the GMP production report for biological medicinal products includes description of the genes used, strain of cell line, cell bank system, fermentation and harvesting, purification, characterization, analytical method development, process validation, impurities, and batch analysis (GMP production of biopharmaceuticals is described in Chapter 10). A DMF (Exhibit 8.8) is submitted.

Part III: This part is to ensure that safety tests have been carried out according to GLP. The data to be submitted are toxicity (single dose and repeated dose), reproduction function, embryo–fetal and perinatal toxicity, mutagenic potential, carcinogenic potential, pharmacodynamics, pharmacokinetics, and local tolerance.

Part IV: This is the clinical documentation. All phases of clinical trials must be carried out in accordance with GCP. The clinical data are pharmacodynamics, pharmacokinetics, clinical trials (including all individual data), and postmarketing experience.

Approval of Marketing Authorization: Assessment of the application by the Committee for Medicinal Products for Human Use (CHMP) is published initially as a Summary of Opinion—positive or negative. After the granting of a Marketing Authorization by the European Commission, a more detailed report is published as the European Public Assessment Report (EPAR).

Exhibit 8.5   Expert Report on the Chemical, Pharmaceutical, and Biological Documentation

Part II Concerning Chemical, Pharmaceutical, and Biological Documentation for Chemical Active Substances

Part II A: Composition
1. Composition of the Medicinal Product
2. Container (Brief Description)
3. Clinical Trial Formula(e)
4. Development Pharmaceutics

Part II B: Method of Preparation
1. Manufacturing Formula
2. Manufacturing Process
3. Validation of the Process

Part II C: Control of Starting Materials
1. Active Substance(s)
   a. Specifications and Routine Tests
   b. Scientific Data
2. Excipient(s)
   a. Specifications and Routine Tests
   b. Scientific Data
3. Packaging Material
   a. Specifications and Routine Tests
   b. Scientific Data

Part II D: Control Tests on Intermediate Products (if necessary)

Part II E: Control Tests on the Finished Product
1. Specifications and Routine Tests
   a. Product Specifications and Tests for Release
   b. Control Methods
2. Scientific Data
   a. Analytical Validation of Methods
   b. Batch Analysis

Part II F: Stability
1. Stability Tests on Active Substance(s)
2. Stability Tests on the Finished Product

Part II G: Bioavailability/Bioequivalence

Part II H: Data Related to the Environmental Risk Assessment for Products Containing, or Consisting of Genetically Modified Organisms

Part II Q: Other Information

Exhibit 8.6   Expert Report on the Toxicopharmacological (Preclinical) Documentation

Part III Toxicopharmacological Documentation

Part III A: Toxicity
  1. Single Dose Toxicity Studies
  2. Repeated Dose Toxicity Studies

Part III B: Reproductive Function

Part III C: Embryo–Fetal and Perinatal Toxicity

Part III D: Mutagenic Potential
  1. *In Vitro*
  2. *In Vivo*

Part III E: Carcinogenic Potential

Part III F: Pharmacodynamics
  1. Pharmacodynamic Effects Relating to the Proposed Indications
  2. General Pharmacodynamics
  3. Drug Interactions

Part III G: Pharmacokinetics
  1. Pharmacokinetics After a Single Dose
  2. Pharmacokinetics After Repeated Dose
  3. Distribution in Normal and Pregnant Animals
  4. Biotransformation

Part III H: Local Tolerance

Part III Q: Other Information

Part III R: Environmental Risk Assessment/Ecotoxicity

Exhibit 8.7   Expert Report on the Clinical Documentation

Part IV Clinical Documentation

Part IV A: Clinical Pharmacology
  1. Pharmacodynamics
    a. A Summary
    b. Detailed Research Design (Protocol)
    c. Results Including:

        i. Characteristics of the Population Studied
       ii. Results in Terms of Efficacy
      iii. Clinical and Biological Results Relevant to Safety
      iv. Analysis of Results
  d. Conclusions
  e. Bibliography
2. Pharmacokinetics
  a. A Summary
  b. Detailed Research Design
  c. Results
  d. Conclusions
  e. Bibliography

Part IV B: Clinical Experience
1. Clinical Trials
  a. A Summary
  b. Detailed Description of the Research Design
  c. Final Results Including:
        i. Characteristics of the Population Studied
       ii. Results in Terms of Efficacy
      iii. Clinical and Biological Results Concerning Safety
      iv. Statistical Evaluation of the Results
       v. Tabulated Patient Data, Including Clinical and Laboratory Monitoring Results
  d. Possible Discussion
  e. Conclusion
2. Postmarketing Experience (if available)
  a. Adverse Reaction and Monitoring Event and Reports
  b. Number of Patients Exposed
3. Published and Unpublished Experience

Part IV Q: Other Information

The EPAR shows the scientific conclusion reached by CHMP at the end of the centralized evaluation process. It is available to the public, with commercial confidential information deleted. The EPAR gives a summary of the reasons for the CHMP opinion in favor of granting a marketing authorization for a specific medicinal product. It results from the Committee's review of the documentation submitted by the applicant and from subsequent discussions held during CHMP meetings. The EPAR is updated throughout the authorization period as changes to the original terms and conditions of the authorization (i.e., variations, pharmacovigilance issues, specific obligations) are made. The

Exhibit 8.8 European DMF

The DMF is used for the following active substances:

- New active substances
- Existing active substances not described in the *European Pharmacopoeia*, but described in the pharmacopoeia of a member state

The DMF consists of a confidential part and a nonconfidential part. The confidential part is to protect valuable intellectual property or "knowhow" of the active substance manufacturer.

Content of the DMF	Restricted Part (Expert Report), Confidential	Applicants Part (Expert Report), Nonconfidential
Names and sites of active substance manufacturer	+	+
Specification and routine test		+
Nomenclature		+
Description		+
Previous use in medicinal products	+	
Manufacturing method		
Brief outline (flow chart)		+
Detailed description	+	
Quality control during manufacture	+	
Process validation and evaluation of data	+	
Development chemistry		
Evidence of structure		+
Potential isomerism		+
Physicochemical characterization		+
Analytical validation		+
Impurities		+
Batch analysis		+
Stability		+

EPAR also contains a summary written in a manner that is understandable by the public.

Accelerated Assessment: The EU introduced the accelerated assessment in November 2005. The aim is to speed up the regulatory procedure to enable patients access to new medicines. The EMEA review time for accelerated assessment is 150 days. See Exhibit 8.9 for the approval of Soliris (eculizumab),

Exhibit 8.9   Soliris

Soliris is for the treatment of paroxysmal nocturnal hemoglobinuria (PNH). PNH is a chronic disease where a patient's oxygen-carrying red blood cells are missing the normally present complement inhibitors. The cells are therefore abnormally fragile and inadvertently destroyed by normal complement activation.

PNH is caused by a mutation in certain types of adult blood cells. Because of this mutation, certain types of proteins, including complement inhibitors, are unable to attach to the surface of the cell, as is normally the case. More specifically, the PNH mutation prevents the assembly of a fatty tail, known as a glycosyl-phosphatidylinositol (GPI) anchor, a necessary step in surface attachment of some proteins.

Consequently, proteins with this GPI anchor are diminished or absent, two of which are crucial in protecting blood cells from inappropriate complement destruction. Without these two protective proteins, PNH red blood cells, in particular, are easily burst by complement, resulting in low red blood cell count (anemia), fatigue, bouts of dark colored urine, and various other complications.

Soliris is a protein-based drug that specifically blocks cleavage of the C5 component of the complement system, thereby preventing the final stages of complement activation.

Source: EMEA. *Doc. Ref. EMEA/184876/2007*, London, 2007.

an antibody for the reduction of hemolysis—destruction of red blood cells in patients with rare blood disorder. Under the accelerated scheme, Soliris was granted a positive opinion in 147 days.

8.4   JAPAN

Drug approval processes go through IND and NDA procedures in Japan. The MHLW of Japan has set up the Pharmaceutical and Medical Device Agency (PMDA), which provides technical consultation services for clinical trials. There are four types of consultations: before IND, at the end of Phase II studies, before NDA, and consultation on individual protocols.

Japan has adopted the ICH GCP guidelines for clinical trials since 1997. It upholds the Helsinki Declaration to ensure the rights, welfare, and privacy of subjects are protected in clinical trials. Japan accepts foreign clinical trials, but bridging trials may need to be performed to take into consideration effects of ethnic factors.

An NDA submitted to the MHLW is reviewed by the PMDA. PMDA personnel have the authority to inspect the drug manufacturing facility and clinical trial sites to assess compliance. In the process, the PMDA consults the Pharmaceutical Affairs and Food Sanitation Council (PAFSC). Results of the review are forwarded to the Pharmaceutical and Food Safety Bureau (PFSB), which prepares the final approval through the Minister of the MHLW. Figure 8.10 shows the drug approval process in Japan. The procedure for manufacturing and distribution of drugs for overseas manufacturers is presented in Fig. 8.11.

8.5 CHINA

Two regulatory processes exist: one for imported products and the other for locally manufactured products. For imported drugs, the registration package includes the following:

- Application form
- Technical data
- CMC
- Nonclinical pharmacology and toxicology
- PK/PE, local clinical data
- Labeling
- Samples from three different batches

The regulations regarding the registration of Western drugs, both synthetic and protein based, are complex, with several different levels of review. We now discuss (Fig. 8.12) the application for registration to import a "Western" drug into China.

An application on the prescribed form is submitted to the Department of Drug Registration (DDR) of the State Food and Drug Administration (SFDA). The DDR evaluates the completeness of the document and then forwards it to the Center for Drug Evaluation (CDE) for technical review. External experts may be consulted, and the CDE compiles a technical report for the DDR.

The National Institute for the Control of Pharmaceutical & Biological Products (NICPBP) performs tests on the drug samples submitted. Based on the test results and the report from the CDE, the DDR approves the conduct of clinical trials at designated hospitals in China (Fig. 8.13).

At the conclusion of clinical trials, the results are evaluated by the CDE, which submits a report to the DDR. Based on the report, the DDR makes a final recommendation to the Director of the SFDA for approval to import the drug into China. The overall process may take 1–2 years.

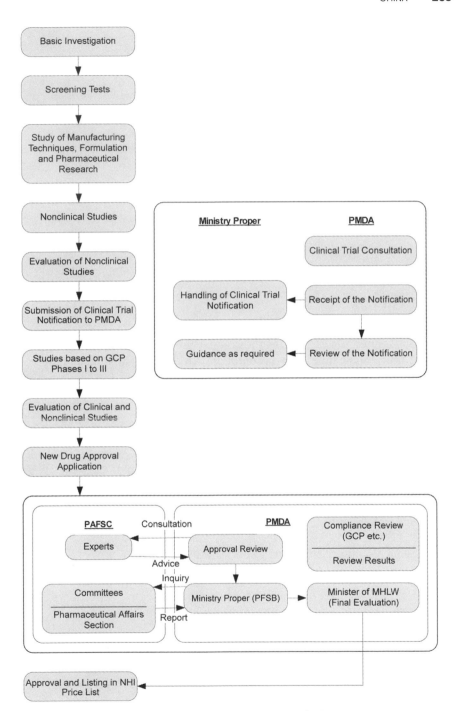

Figure 8.10 Drug approval process in Japan.

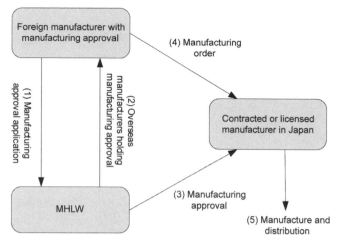

Figure 8.11 Manufacturing and distribution of drugs in Japan.

Almost all new drugs entering the country must go through domestic testing in some form. The Chinese government approved 50–60 hospitals and medical centers where trials can be performed; and study must be conducted at a minimum of three sites.

8.6 INDIA

Application for the importation and manufacture of new drugs or to undertake clinical trial is made on Form 44 to the Central Drugs Standard Control Organization (CDSCO) together with 50,000 rupees. Data to be submitted according to Schedule Y include the following:

- Chemical and pharmaceutical information
- Animal pharmacology
- Animal toxicology
- Human/clinical pharmacology (Phase I)
- Exploratory clinical trials (Phase II)
- Confirmatory clinical trials (Phase III, including published review articles)
- Bioavailability, dissolution and stability study data
- Regulatory status in other countries
- Marketing information
- Proposed product monograph

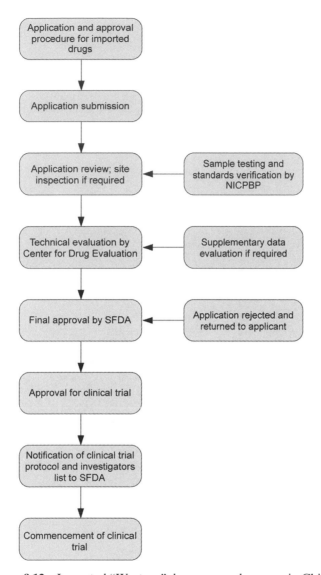

Figure 8.12 Imported "Western" drug approval process in China.

• Drafts of labels and cartons
• Application for test license

Local clinical trial may be waived by the licensing authority in the interest of the public good, in which case data from preclinical studies are to be evaluated. The approval for import permission is given on Form 45 or 45A, clinical trial on Form 46 and/or 46A, and new bulk drug substance on Form 54a.

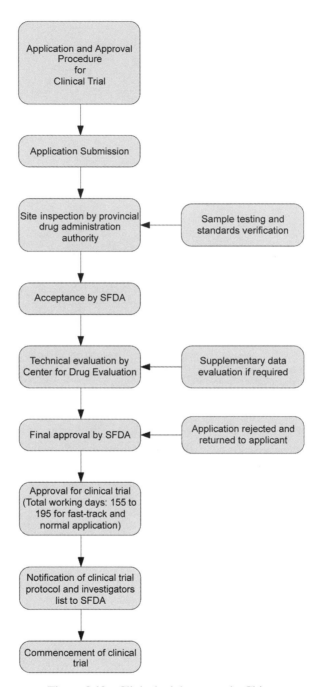

Figure 8.13   Clinical trial process in China.

8.7 AUSTRALIA

All drugs to be imported into, supplied in, or exported from Australia must be included in the Australian Register of Therapeutic Goods (ARTG). The sponsoring company for the drug must apply to the TGA to show the safety and efficacy of the drug before it can be accepted by the ARTG. Submissions of data are based on the CTD format. The review process is depicted in Fig. 8.14.

For clinical trials, two systems exist: the Clinical Trial Notification (CTN) scheme and the Clinical Trial Exemption (CTX) scheme. The CTN relies on approval by a local research institution whereas the CTX is approved by the TGA. Figure 8.15 shows these two schemes.

The 2007 TGA fees payable for various applications are tabulated in Table 8.4.

8.8 CANADA

Clinical Trial Application (CTA) has to be submitted to Health Canada seeking permission to conduct clinical trials. The submission should include information regarding drug characteristics, test data, animal studies, and clinical protocol. A clinical trial may be stopped when either it is shown to be unsafe or dramatic benefits are obtained. The approval process may be fast-tracked if a drug is shown to have substantial benefits, such as for treatment of life-threatening or severely debilitating conditions.

After a drug has demonstrated its efficacy in Phase III, a New Drug Submission (NDS) may be submitted. The information submitted includes preclinical, clinical, chemistry, and manufacturing data for Health Canada to evaluate. Samples may be required for testing and assessment. Generic manufacturers are required to submit an abbreviated NDS, showing bioequivalence to established drugs. There is a Priority Review, which allows for expedited review of drugs for life-threatening or severely debilitating conditions.

Approval for a drug to be marketed in Canada is in the form of a Notice of Compliance (NOC); and a Drug Identification Number (DIN) is issued. Drugs deemed not to have shown sufficient safety and efficacy are not approved and a Notice of Non-Compliance is issued.

8.9 CASE STUDY #8

Counterfeit Drugs*

One of the many challenges facing regulatory authorities worldwide is that of counterfeit drugs. The World Health Organization (WHO) defined

---

* *Sources*: (1) World Health Organization, *Counterfeit medicines*, http://www.who.int/mediacentre/factsheets/fs275/en/ [accessed August 28, 2007]. (2) Sheridan C. Bad medicine, *Nature Biotechnology* 25:707–709 (2007).

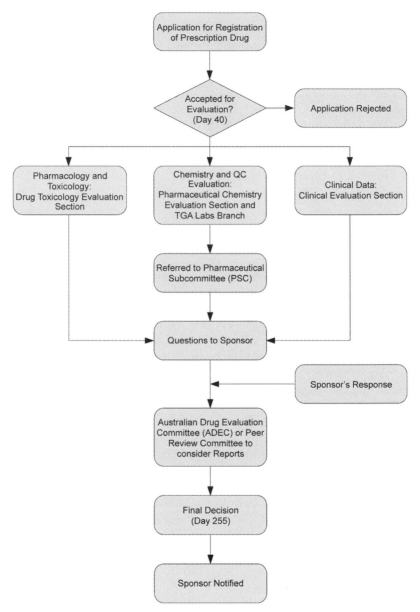

Figure 8.14 Drug approval process in Australia.

counterfeit drugs as those "manufactured below established standards of quality and therefore dangerous to patients' health and ineffective for the treatment of diseases. They are deliberately and fraudulently mislabeled and may contain the correct ingredients but fake packaging, or with the wrong ingredients, or without active ingredients or with insufficient active ingredients."

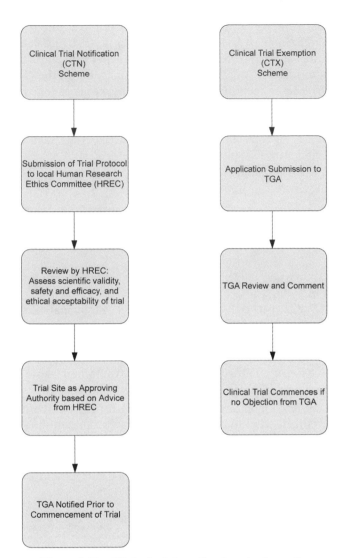

Figure 8.15 Clinical trial applications in Australia.

TABLE 8.4   TGA Fees, 2007

Application	Fee (A$)
NCE	170,200
Extensions of indications	101,200
New generics	65,000
Clinical trial CTX 30 days	1,240
Clinical trial CTX 50 days	15,300
CTN	250

Source: Therapeutics Goods Administration. Fees and Payments. http://www.tga.gov.au/fees/fees06.htm [accessed September 12, 2007].

The current counterfeit drugs market is estimated to be about US$40 billion and is expected to increase to almost US$80 billion by 2010. Most of the counterfeit drugs are found in developing countries, where the enforcement of the regulatory authorities is weak or ineffective. In general, the distribution of counterfeit drugs as a proportion of the total market is as follows:

- Industrialized and developed countries with effective regulatory systems—less than 1% of total market; mainly lifestyle drugs such as hormones, steroids, and antihistamines
- Some developing countries in Asia, Latin America, and Africa, ranging from 10% to 30% of total market
- Former Soviet countries—estimated to be about 20% of total market
- Online purchase from the Internet with dubious source with as high as 50% of purchases being counterfeit

The top 10 countries in terms of counterfeit seizures or counterfeit discoveries are given in Table 8.5.

The WHO has set up the International Medicinal Products Anti-Counterfeiting Taskforce (IMPACT) with all 193 member states to combat counterfeit drugs. Five key areas were identified:

- Legislative and regulatory infrastructure—strengthen legislation, increase penal sanctions, and empower law enforcement agencies
- Regulatory implementation—improve control on safety and efficacy of drugs and the distribution channels; develop better coordination between local, regional, and central authorities
- Enforcement—monitor and track borders for counterfeit activities by working with World Customs Agency, INTERPOL, and other enforcement networks

TABLE 8.5 Top 10 Countries with Counterfeit Drugs

Ranking	Country	Number of Seizures
1	Russia	93
2	China	87
3	South Korea	66
4	Peru	54
5	Colombia	50
6	United States	42
7	United Kingdom	39
8	Ukraine	28
9	Germany	25
10	Israel	25

Source: Pharmaceutical Security Institute, Vienna, Virginia, 2006.

- Technology—involve pharmaceutical companies and distributors to develop innovative solutions, such as radiofrequency identification (RFID) to track movements of drugs, tamper-proof packaging to deter tampering, and printing technologies aimed at end user compliance
- Risk communication—IMPACT to develop and coordinate effective mechanisms to alert and respond to counterfeit drug activities; inform and educate users and healthcare professionals to be alert and report suspicious cases

8.10 SUMMARY OF IMPORTANT POINTS

1. In the United States the drug approval processes are as follows:
   - *Clinical Trial*
      (i)   Investigational New Drug Application (IND): submit information on the drug, preclinical data (PD, PK, and toxicology), manufacturing procedures, test methods, specifications, contamination controls, and stability data.

   - *Marketing Approval*
      (i)   New Drug Application (NDA) for new small molecule drug: submit information on drug, preclinical data (PD, PK, and toxicology), chemistry, manufacturing, and control (CMC), clinical data, statistical analysis, safety information, and validation of processes and test methods.
      (ii) Biologics License Application (BLA): submit information as for NDA.
      (iii) Abbreviated NDA (ANDA) for generics: submit information on drug, comparability studies, and CMC.
      (iv) Orphan drug: submit either NDA or BLA with assistance and provision of exclusivity.
      (v)   OTC drug: submit raw material list and samples showing labeling, dosage, and indication.

2. In Europe the drug approval processes are as follows:
   - *Clinical Trial*
      (i)   Clinical Trial Application (CTA): submit information similar to that for IND, but requirement of Qualified Person (QP) to certify drug manufacture complies to GMP.

   - *Marketing Approval*
      (i)   Centralized Procedure: required for all biopharmaceuticals, specific novel drugs, and orphan drugs; approval is for the entire EU;

submit information similar to NDA/BLA in CTD format and inclusion of Expert Reports.

(ii) Mutual Recognition Procedure: application for marketing approval in other member states through a mutual recognition process after approval by one or more member state(s). Dispute is arbitrated by EMEA, which makes final determination.

(iii) National Authorization Procedure: submission is to one member state in its own language.

(iv) Abridged National Authorization Procedure: applicable for generics and submission in one member state; mutual recognition process required for extension to other states.

3. Essentially similar procedures with some specific country-based requirements for Japan, China, India, Australia, and Canada.

4. There are increasing numbers of counterfeit drugs on the market. Although these are mostly in the developing countries with lax and inadequate regulatory controls, counterfeit drugs are also sold in developed countries through online systems—particularly lifestyle drugs. The WHO, together with all 193 member states, are working toward containing and combating the counterfeit drug trade.

8.11 REVIEW QUESTIONS

1. Compare and contrast NDA and BLA. Why are there different applications for small and large molecule drugs?
2. Explain the meaning of an orphan drug. Why is it necessary to have a separate approval route for orphan drugs?
3. What is meant by bioequivalence studies and why they are needed for generics applications?
4. Describe the Centralized Procedure in Europe for drug approval. Compare and contrast the Centralized Procedure with NDA/BLA.
5. Describe the Mutual Recognition Procedure and the resolution of disputes between member states.
6. Explain the clinical trial procedures in Australia.
7. How do the WHO and member countries tackle the problem of counterfeit drugs?

8.12 BRIEF ANSWERS AND EXPLANATIONS

1. Refer to Section 8.2.3. The NDA is for small molecule drugs and the BLA is for large molecule drugs. They are legislated under different acts.

More information with respect to cell line/bank, test methods, and production processes are required for the BLA.

2. Refer to Section 8.2.4. The orphan drug approval mechanism is implemented to provide incentives to pharmaceutical companies to research and develop drugs for diseases with small patient groups where otherwise commercial returns are considered to be lower than other diseases with large patient pools.

3. Bioequivalence studies are designed to evaluate the PD and PK of drugs against reference off-patent drugs. They have to demonstrate that the generics behave similarly to the original drugs in terms of active component, formulation, mechanism of actions, bioavailability, and ADME. Since the generics are based on off-patent drugs, regulatory authorities have waived the need for preclinical and clinical trials. However, bioequivalence studies are conducted to establish that there are no unintended reactions from the generics.

4. Refer to Section 8.3.2. Note the requirement for the Qualified Person and Expert Reports in Europe. There are no separate applications in the Centralized Procedure for small and large molecule drugs.

5. Refer to Section 8.3.2.

6. The clinical trial application in Australia follows the CTN and CTX schemes. Refer to Section 8.7.

7. Refer to Section 8.9. The five strategic areas identified to combat counterfeit drugs are:
   - Legislative and regulatory infrastructure
   - Regulatory implementation
   - Enforcement
   - Technology
   - Risk communication

8.13 FURTHER READING

Cameron AM. The European Clinical Trials Directive, *Global Outsourcing Review* 4:50–52 (2002).

EMEA. How Are Medicines Authorized In Europe? http://www.emea.europa.eu/Patients/routes.htm [accessed June 14, 2007].

European Commission. *Notice to Applicants, Medicinal Products for Human Use, Presentation and Content of Dossier*, Volume 2B of *The Rules Governing Medicinal Products in the European Union*, EC, 1998.

European Commission. *Notice to Applicants, Medicinal Products for Human Use, Procedures for Marketing Authorization*, Volume 2A of *The Rules Governing Medicinal Products in the European Union*, EC, 1998.

Food and Drug Administration. *Accelerated Approvals*, FDA, 21 CFR Parts 314.500 and 601.4.

Food and Drug Administration. *Application for FDA Approval to Market a New Drug*, FDA, 21 CFR Part 314.

Food and Drug Administration, Center for Biologics Evaluation and Research website, http://www.fda.gov/cber/index.html.

Food and Drug Administration, Center for Drug Evaluation and Research website, http://www.fda.gov/cder/.

Food and Drug Administration. *Cover Form for the Technical Review of Drug Master Files*, FDA, Rockville, MD, 1998.

Food and Drug Administration. *Guidance for Industry, Changes to an Approved Application: Biological Products*, FDA, Rockville, MD, 1997.

Food and Drug Administration. *Guidance for Industry, Content and Format of Chemistry, Manufacturing and Controls Information and Establishment Description Information for a Vaccine or Related Product*, FDA, Rockville, MD, 1999.

Food and Drug Administration. *Guidance for Industry, Cooperative Manufacturing Arrangements For Licensed Biologics*, FDA, Rockville, MD, 1999.

Food and Drug Administration. *Guidance for Industry, Forms For Registration of Producers Of Drugs and ListIng Of Drugs In Commercial Distribution*, FDA, Rockville, MD, 2001.

Food and Drug Administration. *Guidance for Industry, IND Meetings For Human Drugs and Biologics*, FDA, Rockville, MD, 2001.

Food and Drug Administration. *Guidance for Industry, ProvidIng Regulatory Submissions In Electronic Format—General*, FDA, Rockville, MD, 1999.

Food and Drug Administration. *Guidance for Industry, Submitting Type V Drug Master Files to the Center for Biologics Evaluation and Research*, FDA, Rockville, MD, 2001.

Food and Drug Administration. *Guidance to Industry, IND Meetings for Human Drugs and Biologics—Chemistry, Manufacturing, and Controls Information*, FDA, Rockville, MD, 2001.

Food and Drug Administration. *Guideline for Drug Master Files*, FDA, Rockville, MD, 1989.

Food and Drug Administration. *Implementation of Biologics License; Elimination Of Establishment License and Product License Public Workshop*, FDA, Rockville, MD, 1998.

Food and Drug Administration. *Investigational New Drug Application*, FDA, 21 CFR Part 312.

International Conference on Harmonization. Organization of the common technical document for the registration of pharmaceuticals for human use, in *Harmonized Tripartite Guideline*, ICH, 2002.

International Conference on Harmonization. The common technical document for the registration of pharmaceuticals for human use: quality, quality overall summary of module 2, module 3: quality, in *Harmonized Tripartite Guideline*, ICH, 2002.

International Conference on Harmonization. The common technical document for the registration of pharmaceuticals for human use: safety, nonclinical overview and nonclinical summaries of module 2, organization of module 4, in *Harmonized Tri-partite Guideline*, ICH, 2002.

Japan Pharmaceutical Manufacturers Association. *New Drug Development and Approval Process*. http://www.jpma.or.jp/12english/guide_industry/new_drug/new_drug.html [accessed August 8, 2002].

Lehman, Lee, Xu. *Food & Drug FAQ*. http://www.lehmanlaw.com/FAQ/faq/FD.htm [accessed December 22, 2002].

Maeder T. The orphan drug backlash, *Scientific American*, May:80–87 (2003).

Medicines and Healthcare products Regulatory Agency website, http://www.mca.gov.uk/.

Sietsema WK. *Preparing the New Drug Application—Managing Submissions Amid Changing Global Requirements*, FDA News, Falls Church, VA, 2006.

CHAPTER 9

GOOD MANUFACTURING PRACTICE: REGULATORY REQUIREMENTS

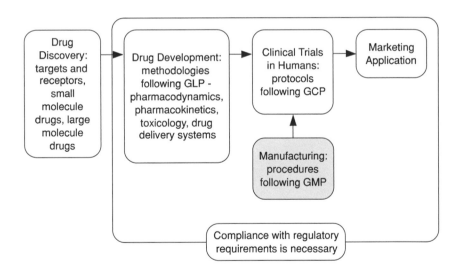

9.1   Introduction   279
9.2   United States   279
9.3   Europe   283
9.4   International Conference on Harmonization   283
9.5   Core Elements of GMP   287
9.6   Selected GMP Systems   297
9.7   The FDA's New cGMP Initiative   310

Drugs: From Discovery to Approval, Second Edition, By Rick Ng
Copyright © 2009 John Wiley & Sons, Inc.

9.8   Case Study #9   313
9.9   Summary of Important Points   315
9.10 Review Questions   316
9.11 Brief Answers and Explanations   316
9.12 Further Reading   317

9.1 INTRODUCTION

In the earlier chapters, we discussed how a drug is discovered, followed by research on pharmacodynamics and pharmacokinetics, then through clinical trials on humans, leading finally to filing the application and approval being given for the drug to be marketed. This is a long journey of some 10–12 years, with many risks of failure along the way. Now, after having been granted the marketing approval, a pharmaceutical firm is ready to manufacture the drug for sale, but it must do so in accordance with Good Manufacturing Practice (GMP).

GMP is a quality concept and consists of a set of policies and procedures for manufacturers of drug products. These policies and procedures describe the facilities, equipment, methods, and controls for producing drugs with the intended quality. The guiding principle for GMP is that quality cannot be tested into a product, but must be designed and built into each batch of the drug product throughout all aspects of its manufacturing processes. Manufacturers are required to abide by the GMP regulatory guidelines to ensure drugs are pure, consistent, safe, and effective. Regulatory guidelines are dynamic; they are revised and updated from time to time to implement new research, data, or information. Therefore, manufacturers have to keep abreast with regulatory developments by following current Good Manufacturing Practice (cGMP).

On a global level, GMP regulations are very similar for various countries. There are, however, differences in emphasis and implementation in specific areas. We will explain the GMP regulations from the United States, Europe, and the International Conference on Harmonization (ICH) in this chapter. In Chapter 10, we will discuss the manufacturing processes for small molecule synthetic and large molecule protein-based drugs.

9.2 UNITED STATES

GMP regulations came into effect in the United States in 1963. They have since undergone several major revisions. The implementation of GMP is the result of a number of tragedies to ensure that drugs are safe for the patients and effective for treatment. Some of these tragedies are described in Exhibit 9.1.

The Food and Drug Administration (FDA) is charged with the responsibility for ensuring drug manufacturers comply with GMP regulations in the

Exhibit 9.1   Some Drug Tragedies

In 1902, several children died after being administered contaminated diphtheria antitoxin.

In 1937, 107 people died when the drug sulfanilamide was wrongly formulated.

In 1955, 10 children died after being given improperly inactivated polio vaccine.

In the 1960s, untold physical damage was caused by thalidomide (see Chapter 6).

United States. GMP is defined by the FDA as "a federal regulation setting minimum quality requirements that drug, biologics and device manufacturers must meet. It describes in general terms known and accepted quality assurance principles for producing these products. Its components are scientific understanding, documentation, analysis and measurements and personnel matters. Its intended result is total quality assurance and product control."

The US FDA GMP is codified in the following regulations:

- 21 CFR Part 210: Current Good Manufacturing Practice in Manufacturing, Processing, Packing, or Holding of Drugs; General
- 21 CFR Part 211: Current Good Manufacturing Practice for Finished Pharmaceuticals
- 21 CFR Part 600: Biological Products: General
- 21 CFR Part 610: General Biological Products Standards

Further details for each of these sets of regulations are presented in Exhibit 9.2. Selected items of these regulations (as part of ICH Q7) are discussed in later sections. It should be noted that the applicable regulations for small molecule drugs are 21 CFR Parts 210 and 211, and for biopharmaceuticals the regulations are 21 CFR Parts 210, 211, 600, and 610. The reason is that protein-based biopharmaceuticals are less well-defined chemically and they are sensitive to the storage and manufacturing environment as well as the manufacturing processes. Biopharmaceuticals are normally prepared under aseptic conditions, as they are sensitive to degradation under normal sterilization processes. Special techniques and analytical methods are required for the production and testing of biopharmaceuticals.

According to 21 CFR 210.1(a), the regulations "contain the minimum current good manufacturing practice for methods to be used in, and the facilities or controls to be used for, the manufacture, processing, packing, or holding of a drug to assure that such drug meets the requirements."

Exhibit 9.2   FDA Current Good Manufacturing Practice

21 CFR Part 210: Current Good Manufacturing Practice in Manufacturing,
Processing, Packing, or Holding of Drugs; General

210.1   Status of current good manufacturing practice regulations
210.2   Applicability of current good manufacturing practice regulations
210.3   Definitions

21 CFR Part 211: Current Good Manufacturing Practice for
Finished Pharmaceuticals

SUBPART A: GENERAL PROVISIONS

211.1   Scope
211.3   Definitions

SUBPART B: ORGANIZATION AND PERSONNEL

211.22   Responsibilities of quality control unit
211.25   Personnel qualifications
211.28   Personnel responsibilities
211.34   Consultants

SUBPART C: BUILDING AND FACILITIES

211.42   Design and construction features
211.44   Lighting
211.45   Ventilation, air filtration, air heating and cooling
211.48   Plumbing
211.50   Sewerage and refuse
211.52   Washing and toilet facilities
211.56   Sanitation
211.58   Maintenance

SUBPART D: EQUIPMENT

211.63   Equipment design, size, and location
211.65   Equipment construction
211.67   Equipment cleaning and maintenance
211.68   Automatic, mechanical, and electronic equipment
211.72   Filters

SUBPART E: CONTROL OF COMPONENTS AND DRUG PRODUCT CONTAINERS AND CLOSURES

211.80   General requirements
211.82   Receipt and storage of untested components, drug product containers,
            and closures
211.84   Testing and approval or rejection of components, drug product
            containers, and closures
211.86   Use of approved components, drug product containers, and closures
211.89.1 Rejected components, drug product containers, and closures
211.94   Drug product containers and closures

SUBPART F: PRODUCTION AND PROCESS CONTROLS

211.100   Written procedures; deviations
211.101   Charge-in of components
211.103   Calculation of yield
211.105   Equipment identification
211.110   Sampling and testing of in-process materials and drug products
211.111   Time limitation on production

211.113 Control of microbiological contamination
211.115 Reprocessing

SUBPART G: PACKAGING AND LABELING CONTROL

211.122 Materials examination and usage criteria
211.125 Labeling issuance
211.130 Packaging and labeling operations
211.132 Tamper-resistant packaging requirement for over-the counter human
        drug products
211.134 Drug product inspection
211.137 Expiration dating

SUBPART H: HOLDING AND DISTRIBUTION

211.142 Warehousing procedures
211.150 Distribution procedures

SUBPART I: LABORATORY CONTROLS

211.160 General requirements
211.165 Testing and release for distribution
211.166 Stability testing
211.72   Filters

21 CFR Part 610: General Biological Products Standards

610.1   Test prior to release required for each lot
610.2   Requests for samples and protocols; official release
610.9   Equivalent methods and processes
610.10   Potency
610.11   General safety
610.11a Inactivated influenza vaccine, general safety test
610.12   Sterility
610.13   Purity
610.14   Identity
610.15   Constituent materials
610.16   Total solids in serums
610.17   Permissible combinations
610.18   Cultures
610.19   Status of specific products; Group A streptococcus
610.20   Standard preparations
610.21   Limits of potency
610.30   Test for mycoplasma
610.40   Test for hepatitis B surface antigen
610.41   History of hepatitis B surface antigen
610.45   Human immunodeficiency virus (HIV) requirements
610.46   "Lookback" requirements
610.47   "Lookback" notification requirements for transfusion services
610.50   Date of manufacture
610.53   Dating periods for licensed biological products
610.60   Container label
610.61   Package label
610.62   Proper name; package label; legible type
610.63   Divided manufacturing responsibility to be shown
610.64   Name and address of distributor
610.65   Product for export

In addition to the regulations under 21 CFR, the FDA publishes Guidance for Industry and documents called Points to Consider (PTCs) as guidelines and recommendations to industry to adopt as part of the compliance program.

9.3   EUROPE

The principles and guidelines for GMP for human medicinal products were laid down in EU Directive 91/356/EEC on June 13, 1991. The following are basic requirements:

- Quality management
- Personnel
- Premises and equipment
- Documentation
- Production
- Quality control
- Contract manufacture and analysis
- Complaint and product recall
- Self-inspection
- Inclusion of annexes
  Manufacture of sterile medicinal products
  Manufacture of biological medicinal products for human use
  Manufacture of radiopharmaceuticals
  Manufacture of veterinary medicinal products other than immunologicals

9.4   INTERNATIONAL CONFERENCE ON HARMONIZATION

We discussed in Section 7.11 the tripartite harmonization of guidelines by the United States, Europe, and Japan. The GMP Guidance is one of these guidelines; it is described in the ICH Q7 document called *GMP Guidance for Active Pharmaceutical Ingredients*. Because of the wide implications of this guidance, the Steering Committee of the ICH invited experts from Australia, India, and China and industrial representatives from the generics industry, self-medication industry, and PIC/S (Pharmaceutical Inspection Cooperation Scheme, Section 7.13) to participate in the preparation of this document. Hence, the Q7 document has been endorsed as a truly international document for GMP.

The United States, the European Union, and Japan have implemented this GMP Guidance, and the details are presented in Exhibit 9.3. The Q7 GMP Guidance sets out the requirements for GMP manufacturing. Details are summarized in Exhibit 9.4. Most of the requirements of ICH Q7 are derived from

Exhibit 9.3   Implementation of ICH Q7 GMP Guide

European Union: Adopted by CPMP (a former committee, now changed to CHMP), November 2000; issued as CPMP/ICH/1935/00 http://dg3.eudra.org/.

Ministry of Health, Labor and Welfare, Japan: Adopted November 2, 2001, PMSB Notification No. 1200. http://www.nihs.go.jp/dig/ich/ichindex.htm.

Food and Drug Administration (USA): Published in the *Federal Register*, Vol. 66, No. 186, September 25, 2001, pp. 49028–49029.

CDER: http://www.fda.gov/cder/guidance/index.htm.
CBER: http://www.fda.gov/cber/guidelines.htm.

Exhibit 9.4   ICH Q7—Guidance for Active Pharmaceutical Ingredients

Introduction
    Objective
    Regulatory Applicability
    Scope
Quality Management
    Principles
    Responsibilities of the Quality Unit(s)
    Responsibilities for Production Activities
    Internal Audits (Self Inspection)
    Product Quality Review
Personnel
    Personnel Qualifications
    Personnel Hygiene
    Consultants
Buildings and Facilities
    Design and Construction
    Utilities
    Water
    Containment

Lighting

Sewerage and Refuse

Sanitation and Maintenance

Process Equipment

Design and Construction

Equipment Maintenance and Cleaning

Calibration

Computerized System

Documentation and Records

Documentation System and Specifications

Equipment Cleaning and Use Record

Records of Raw Materials, Intermediates, API Labeling and Packaging
Materials

Master Production Instructions

Batch Production Records

Laboratory Control Records

Batch Production Record Review

Materials Management

General Controls

Receipt and Quarantine

Sampling and Testing of Incoming Production Materials

Storage

Reevaluation

Production and In-Process Controls

Production Operations

Time Limits

In-Process Sampling and Controls

Blending Batches of Intermediates or APIs

Contamination Control

Packaging and Identification Labeling of APIs and Intermediates

General

Packaging Materials

Label Issuance and Control

Packaging and Labeling Operations

Storage and Distribution

Warehousing Procedures

Distribution Procedures

Laboratory Controls
   General Controls
   Testing of Intermediates and APIs
   Validation of Analytical Procedures
   Certificate of Analysis
   Stability Monitoring of APIs
   Expiry and Retest Dating
   Reserve/Retention Samples
Validation
   Validation Policy
   Validation Documentation
   Qualification
   Approaches to Process Validation
   Process Validation Program
   Periodic Review of Validated Systems
   Cleaning Validation
   Validation of Analytical methods
Change Control, Rejection, and Reuse of Materials
   Rejection
   Reprocessing
   Reworking
   Recovery of Materials and Solvents
   Returns
Complaints and Recalls
Contract Manufacturers (Including Laboratories)
Agents, Brokers, Traders, Distributors, Repackers, and Relabelers
   Applicability
   Traceability and Distributed APIs and Intermediates
   Quality Management
Repackaging, Relabeling, and Holding of APIs and Intermediates
   Stability
   Transfer of Information
   Handling of Complaints and Recalls
   Handling of Returns
Specific Guidance for APIs Manufactured by Cell Culture/ Fermentation
   General

Cell Bank Maintenance and Record Keeping
Cell Culture/Fermentation
Harvesting, Isolation, and Purification
Viral Removal/Inactivation Steps
APIs for Use in Clinical Trials
General
Quality
Equipment and Facility
Control of Raw materials
Production
Validation
Changes
Laboratory Controls
Documentation
Glossary

the 21 CFR and EU GMP Directive. The important additional sections in ICH Q7 are Internal Audits (Self-Inspection), Contract Manufacturers, and Agents, Brokers, Traders, Distributors, Repackers, and Relabelers. The section on APIs for Use in Clinical Trials clarifies the regulatory authorities' expectations for drugs designated for clinical trials, as opposed to approved drugs manufactured on a routine production basis. With the increasing importance of biopharmaceuticals, a section on the production via cell culture methods, harvesting, and viral removal has been added as Specific Guidance for APIs Manufactured by Cell Culture/Fermentation.

Since late 2005, the ICH has added three more quality guidelines: Q8— *Pharmaceutical Development*, Q9—*Quality Risk Management*, and Q10— *Pharmaceutical Quality System*. All these guidelines are intended to bolster the quality of drugs to be manufactured, steering manufacturers in the direction to improve compliance, safety, and consistency of the drugs.

9.5 CORE ELEMENTS OF GMP

The core elements of the ICH Q7 GMP Guidance are discussed next.

9.5.1 Introduction: Scope

The Guidance applies to the manufacture of active pharmaceutical ingredients (APIs) for use in human drug products. It is detailed in Table 9.1.

TABLE 9.1   Application of ICH Q7 to API Manufacturing

Type of Manufacturing	Application of the Guidance to Steps (shown in gray) Used in This Type of Manufacturing			
Chemical manufacturing	Introduction of the API starting material into process	Production of intermediates	Isolation and purification	Physical processing and packaging
API derived from animal sources	Cutting, mixing, and/ or initial processing	Introduction of the API starting material into process	Isolation and purification	Physical processing and packaging
API extracted from plant sources	Cutting and initial extractions	Introduction of the API starting material into process	Isolation and purification	Physical processing and packaging
Biotechnology: fermentation/ cell culture	Maintenance of working cell bank	Cell culture and/or fermentation	Isolation and purification	Physical processing and packaging
"Classical" fermentation to produce an API	Maintenance of the cell bank	Introduction of the cells into fermentation	Isolation and purification	Physical processing and packaging

Source: Adapted from International Conference for Harmonization. *Good Manufacturing Practice Guide for Active Pharmaceutical Ingredients*, ICH Q7, 2000.

9.5.2   Quality Management

The first and foremost element for GMP is the quality system. This can be divided into Quality Assurance (QA) and Quality Control (QC). QA is a total system approach. It sets out the compliance policies and procedures for all facets of drug manufacturing. QC is the practical extension of QA. The role of QC is concerned with inspection and testing of the manufacturing environment, raw materials, in-process intermediates, and finished products.

All personnel involved in GMP production of drugs have to take ownership of quality. It is a requirement that processes and equipment for drug manufacturing must be approved and operated by trained, qualified personnel. Quality-related activities have to be recorded to enable traceability of data and information. Deviations and excursions of processes and results from specified conditions or criteria have to be reported, investigated, and resolved. Drug products have to be tested and must meet specifications before being released by an authorized person, normally from the QA department. Responsibilities for the QA and QC departments and production activities need to be defined. Approved procedures are to be followed and processing conditions and data recorded.

Two important aspects are internal audits and product quality review. Internal audits are implemented to regularly monitor the compliance activities in drug manufacture and to ensure rectification to these activities if deviations occur. Trending and statistical analysis of data provide early warning of impending problems. Product quality review checks the relevance and adequacy of the manufacturing activities. It provides input to update and improve the quality system.

9.5.3 Personnel

Personnel engaged in GMP manufacturing of drug products are required to be formally trained in quality practices. They are only assigned to tasks for which they have been trained. This is to guarantee that drugs are manufactured by qualified personnel and quality is built into each step of the manufacturing process.

Personnel are the main source of contaminants to drug products, and hence personnel cleanliness is an important factor (see Exhibit 9.5). Any personnel suffering from infectious diseases or having open wounds are assigned to non-GMP production activities to reduce the possibility of contamination.

9.5.4 Buildings and Facilities

Buildings must be designed with regard to the needs for manufacturing with minimum risk of contamination. There must be demarcation of areas for different activities. Such segregation reduces the possibility of contamination and materials mix-ups.

For the manufacture of drug products, certain processes have to be performed in clean areas. Specifications for environmental airborne par-

---

Exhibit 9.5 Human-Caused Particles

About 10^7 dead cells shed each day

About 2000 microorganisms per square centimeter

Number of $0.3\,\mu m$ particles shed during specific activities:

Motionless	100,000
Getting up	1,000,000
Walking	5,000,000

Source: Adapted from Hofmann FK. *GMP Compliance*, Centre for Continuous Education, Vista, CA, 2001.

TABLE 9.2   Airborne Environmental Cleanliness Requirements

EU 91/356/EEC Annex 1	Maximum Permissible Number of Particles Or Microorganisms			
	Aseptic Core (Grade A)	Aseptic Process Area (Grade B)	Clean Preparation Area (Grade C)	Support Area (Grade D)
At Rest				
0.5 μm particles/m^3	3,500	3,500	350,000	3,500,000
5 μm particles/m^3	None	None	2,000	20,000
In Operation				
0.5 μm particles/m^3	3,500	350,000	3,500,000	Unclassified
5 μm particles/m^3	None	2,000	20,000	Unclassified
Viable organisms cfu/m^3	<1	<10	<100	<200
ISO 14644-1: 1999	ISO 5	ISO 7	ISO 8	—
In Operation				
0.5 μm particles/m^3	3,520	352,000	3,520,000	Unclassified

ticulates and viable microorganisms in cleanrooms are provided in EU Directive 91/356/EEC, Annex 1 and ISO 14644-1:1999 (which replaced FS 209E in January 2002). Details for these specifications are summarized in Table 9.2.

The cleanliness is graded in accordance with the nature of operations: for example, Grade A for aseptic preparation and filling, Grade B for background environment to aseptic preparation and filling, Grade C for preparation of solutions to be filtered, and Grade D for handling of components after washing. The "at rest" condition is defined as the condition where installation is complete with production equipment installed and operating, but with no operating personnel present. The "in operation" condition is when equipment is functioning and a specified number of personnel are present. The viable microorganisms are the permissible number of colony forming units (cfu) on a culture plate for a cubic meter of air sample. There are other limits for microorganisms present on surfaces and personnel that need to be monitored. Specifications for these microorganism limits for clearances in operation are recommended in EU Directive 91/356/EEC Annex 1 (Table 9.3).

To control the cleanliness levels, the heating, ventilation, and air-conditioning (HVAC) system circulates air that is filtered through high efficiency particulate air (HEPA) filters, which remove up to 99.99% of particles 0.3 μm and larger. The number of air exchanges is also controlled, at a minimum of 20 air changes per hour, depending on room classifications.

TABLE 9.3   Recommended Limits for Microbial Contamination

Grade	Air Sample (cfu/m³)	Settle Plate, Diameter 90 mm (cfu/4 h)	Contact Plate, Diameter 55 mm (cfu/plate)	Personnel Glove Print, 5 fingers (cfu/glove)
A	<1	<1	<1	<1
B	<10	<5	<5	<5
C	<100	<50	<25	N/A[a]
D	<200	<100	<50	N/A

[a]N/A = Not applicable.

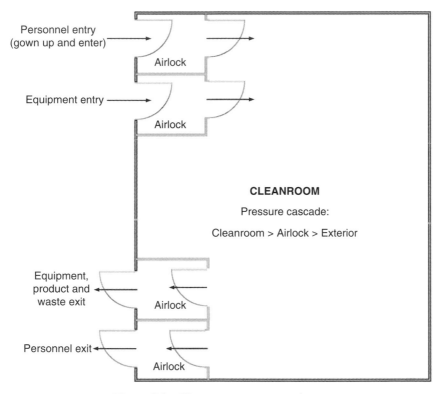

Figure 9.1   Cleanroom pressure scheme.

Cleanrooms are pressurized to prevent contaminant from entering. The FDA specifies a minimum of 0.05 inch water (12.5 Pa) difference in pressure between cleanrooms of different classifications, with the more critical, cleaner rooms having higher pressures. A schematic diagram showing the pressure gradient through air locks is shown in Fig. 9.1.

Both the temperature and relative humidity are normally controlled by the HVAC system, for example, to $21 \pm 2\,°C$ and 30–50%, respectively, for operator comfort and to reduce growth of microorganisms at drier conditions. The facility is also designed to prevent product from escaping into the environment. Wastes (both solids and liquids) are decontaminated and exhaust air is filtered before discharge. In some facilities, the direction of flow of personnel, materials, products, and equipment is controlled to prevent cross-contamination.

For a facility manufacturing biopharmaceuticals, appropriate designs according to biosafety level (BSL1 to BSL4; refer to Exhibit 9.6 for biosafety definitions) have to be implemented.

Utilities such as gases and air piped to reaction vessels are filtered ($\leq 0.2\,\mu$m filters) to control the risk of microbial contamination. Water is considered a

---

Exhibit 9.6   Biosafety Levels

BSL 1: Biosafety Level 1 is suitable for work involving well-characterized microorganisms not known to consistently cause disease in healthy adults, and of minimal potential hazard to laboratory personnel and the environment. Safety equipment: none required. Microorganisms include *Bacillus subtilis, Naegleria gruberi*, and infectious canine hepatitis virus.

BSL 2: Biosafety Level 2 is suitable for work involving microorganisms of moderate potential hazard to personnel and the environment. Safety equipment: Class I or II biosafety cabinets or other physical containment devices; laboratory coats, gloves, face protection as needed. Microorganisms include hepatitis B virus, HIV, salmonellae, and mycoplasma.

BSL 3: Biosafety Level 3 is for work with indigenous or exotic microorganisms, which may cause serious or potentially lethal disease if inhaled. Safety equipment: Class I or II biosafety cabinets or other physical containment devices; protective laboratory clothing, gloves, respiratory protection as needed. Microorganisms include *Mycobacterium tuberculosis, Bacillus anthracis*, and *Coxiella burnetii*.

BSL 4: Biosafety Level 4 is for work with dangerous and exotic microorganisms that pose a high individual risk of aerosol-transmitted laboratory infections and life-threatening disease. Safety equipment: Class III biosafety cabinet or Class I or II biosafety cabinets with full-body, air-supplied, positive pressure personnel suit. Microorganisms include Marburg virus, Ebola virus, Congo-Crimean hemorrhagic fever virus, and Nipah virus.

Source: National Institutes of Health. *Biosafety in Biological and Microbiological Laboratories*. http://bmbl.od.nih.gov/sect2.htm [accessed September 21, 2007].

raw material in the manufacture of drug products. It requires a more detailed discussion and is presented in Section 9.6.1.

9.5.5 Process Equipment

Appropriate equipment is used for manufacturing drugs. The equipment must be maintained and calibrated at defined periods to ensure it functions as intended. After each production batch, equipment is cleaned to prevent cross-contamination from residues. Ideally, this entails that equipment should be designed without difficult-to-clean areas. A discussion of cleaning is given in Section 9.6.2.

For computerized process equipment, the regulatory requirements are very specific, and these are detailed in Section 9.6.3.

9.5.6 Documentation and Records

Documentation comprises procedures, instructions, test methods, batch records, and so on that are documented and controlled. Documentation is prepared, reviewed, and approved by qualified personnel. Approved copies of documents are distributed to relevant departments and superseded copies are retrieved and archived. The retention period for each type of document is specified. Documents are issued with document and version numbers for ease of identification and reference. Master copies of documents are filed at secured locations with authorized access. Master copies stored in electronic media require validation in accordance with FDA regulation 21 CFR Part 11 (see Section 9.6.3) to assess the security of access and data integrity. Operators are trained and retrained to only apply the latest approved documents.

Records include materials transfer records, batch records, materials/intermediates/finished product test records, shipping records, water test records, and environmental test records. They provide an audit trail for reviewing all the information related to the production of any batch of drug product. The data are required to be reviewed for product release.

9.5.7 Materials Management

Materials are managed to assure the following:

- Materials received match those that were ordered.
- Identification labels are attached.
- Where required, certificates of approval are provided.
- Materials are quarantined and stored under specified conditions prior to QC inspection and test.
- Approved materials are segregated from rejects.
- Materials are transferred to relevant departments for use.

- Receipt and quarantine of finished products is documented.
- Storage of approved products pending shipment follows specific guidelines.
- Shipment of products to designated receivers is documented.

It is necessary to ensure suppliers of materials have in place appropriate quality systems and that they are reliable. External audits may be required to inspect and confirm the supplier's facility and quality system.

9.5.8   Production and In-Process Controls

Materials, processes, and control parameters for drug production are stated in written documents. Production personnel follow procedures and record materials used, amounts weighed, and date of operation. Equipment, reaction vessels, and the production area are cleaned and their status recorded in logbooks. Throughout the production stages, equipment conditions (e.g., pH, pressure, stirring speed, and temperature) are also recorded. Adjustments to in-process control parameters, if permitted, are entered onto batch records.

Samples of intermediates and finished products taken for analysis are recorded, stating the time, date, and conditions for these samples. Deviations in operating conditions and out of specification (OOS) conditions in samples are reported and investigated. Figure 9.2 shows a mechanism for production and in-process controls.

Equipment, raw materials, intermediates, finished products, and packaging materials may require sterilization. A discussion of the sterilization process is presented in Section 9.6.4.

9.5.9   Packaging, Identification, and Labeling of APIs, Intermediates, and Finished Products

Proper identification of raw materials, intermediates, and finished products is necessary to prevent misuse and mix-ups. Labels are controlled and accounted for to prevent mislabeling. If required, packages may be sealed to provide an alert of mishandling or unauthorized tempering.

9.5.10   Laboratory Controls

The aim of laboratory controls is to ensure that only approved materials are used and only intermediates and drug products that meet specifications are released.

Laboratory controls commence with sampling and testing of incoming materials according to written procedures. Some aspects of the tests include identity, quantity, purity, activity, heterogeneity, stability, sterility, and safety. Intermediates and finished products are tested in accordance with preset

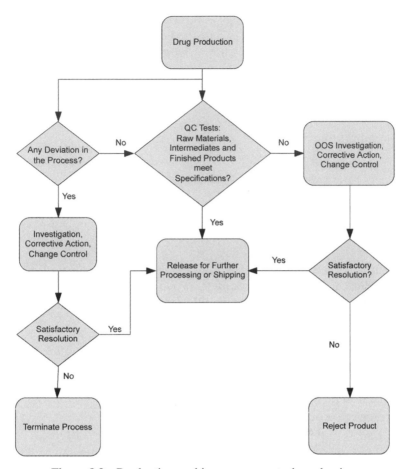

Figure 9.2   Production and in-process control mechanism.

specifications. Out of specification (OOS) results are investigated. OOS conditions may be due to analyst or operator error, inappropriate method, production problems, or an inherent problem with the samples. It may also be necessary to trace the origins of materials: for example, in the case of raw materials for use in medium formulation for growing cells in protein drug production—if animal or plant components are used—the locations and herd or plantation where these components are obtained have to be recorded.

Test methods used in the laboratory are generally derived from pharmacopoeias such as the *US Pharmacopoeia, British Pharmacopoeia*, or *European Pharmacopoeia*. For test methods that are not from recognized pharmacopoeias, validation of the analytical methods is required. The validation includes testing for accuracy, specificity, ruggedness, robustness, precision, detection limit, quantitation limit, and range. A discussion of analytical methods validation is presented in Section 9.6.5.

Samples are retained for possible future evaluation; normally they are retained for 1 year after the expiry date of the batch or 3 years after distribution of the production batch, whichever is longer. Drug stability governs the effective date and storage conditions of the drug. Programs to evaluate stability of drug are an integral part of tests. Details for stability programs are discussed in Section 9.6.7.

9.5.11 Validation

The following is the FDA definition for Process Validation (FDA *Guidelines on General Principles of Process Validation*, May 1987):

> Process validation is establishing documented evidence which provides a high degree of assurance that a specific process will consistently produce a product meeting its predetermined specifications and quality attributes.

A pharmaceutical company has to adopt a proactive policy of validation for its facilities, production processes, production equipment and support systems, analytical methods, and computerized systems. A properly validated approach will help to assure drug product quality, optimize the processes, and reduce manufacturing cost.

The approach to validation commences with the Validation Master Plan (VMP), which details the following:

- Validation policy
- Organization of validation activities
- Personnel responsibilities
- Facilities, systems, equipment, and processes to be validated
- Documentation structure and formats
- Change control processes
- Planning and scheduling

There are various phases of validation:

- *Design Qualification (DQ):* This provides documented verification that the design of the facilities, equipment, or systems meets the requirements of the user specifications and GMP.
- *Installation Qualification (IQ):* This provides documented verification that the equipment or systems, as installed or modified, comply with the approved design and that all the manufacturer's recommendations have been duly considered.
- *Operational Qualification (OQ):* This provides documented verification that the equipment or systems perform as intended throughout the anticipated operating ranges.

- *Performance Qualification (PQ):* This provides documented verification that the equipment and ancillary systems, when connected together, can perform effectively and reproducibly based on the approved process method and specifications.

DQ is performed by the supplier of the equipment or system at the supplier's factory as part of the factory acceptance test (FAT). IQ (based on site acceptance test—SAT), OQ, and PQ are performed on-site at the GMP facility. For a GMP manufacturing facility, the validation activities include the facility design, HVAC system, environment control, laboratory and production equipment, water system, gases and utilities, cleaning, and analytical methods. Validation protocols (IQ, OQ, and PQ) are prepared for each item, listing all critical steps and acceptance criteria. Deviations are reviewed and resolved before the validation activity proceeds to the next phase.

9.5.12 Change Controls

Regulatory authorities recognize that, in spite of all the control systems put in place, deviations and changes are sometimes inevitable. A robust GMP system includes procedures to handle, review, and approve changes in raw materials, specifications, analytical methods, facilities, equipment, processes, computer software, and labeling and packaging. All the changes have to be documented with references for traceability.

Proposed changes have to be reviewed with reference to risk assessment and approved before being implemented. There may be justification to retest or revalidate the affected system, equipment, or process to ensure that quality of the drug product is not compromised. It is necessary to perform ongoing monitoring of changes for a period and assess the long-term impact of the changes to ensure the control system is put in place.

9.6 SELECTED GMP SYSTEMS

In this section, we describe selected systems to illustrate the implementation of GMP concepts for these systems.

9.6.1 Water System

Two grades of water are used in drug manufacture: purified water (PW) and water-for-injection (WFI). In general, oral dosage drugs are prepared using PW, and parenteral injection drugs using WFI. Figure 9.3 illustrates a typical water system for generating pharmaceutical PW and WFI.

Incoming potable water (drinkable water) normally contains undissolved particulate matter and dissolved organic and inorganic compounds, as well as

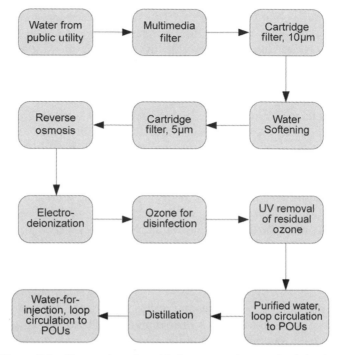

Figure 9.3 Generation of purified water and water for injection.

microorganisms. Several stages of treatment and purification are needed to produce PW and WFI.

Multimedia filters, which consist of a top layer of coarse and low density anthracite, layers of silica, and then dense finest medium vitreous silicate, remove about 98% of particulates >20 μm. These filters are regularly backwashed to avoid buildup of particulates. Finer filters (5–10 μm) are used to remove suspended matter and colloidal materials. To prevent scaling due to water hardness, sodium ions generated from brine are exchanged with calcium and magnesium ions in the water. Activated carbon or metabisulfite is used to remove chlorine.

In some cases, reverse osmosis is applied, and this removes almost all the particulates and organic materials, as well as microorganisms and endotoxins. Electrode ionization, which combines ion exchange membranes and resins, removes the last traces of dissolved ions from water under the influence of a direct electric current. The last stage of the purification is ozone sterilization of water to inactivate residual microorganisms, as ozone is an efficient disinfectant (UV at 254 nm wavelength is then used to break up the spent ozone). The PW generated is then circulated to each point where it is used (point of use—POU).

To obtain WFI, the PW is distilled via several stills. Similarly, WFI is circulated to the POU.

TABLE 9.4   The *US Pharmacopoeia* Specifications for Purified Water and Water for Injection

Variable	Purified Water[a]	Water for Injection
pH	5.0–7.0	5.0–7.0
Conductivity (μS/cm)	≤1.1 at 20 °C	≤1.1 at 20 °C
Total organic carbon (parts per billion)	<500	<500
Microbial	<100 cfu/mL	<10 cfu/100 mL
Endotoxin (EU/mL)	—	0.25

[a]cfu = colony forming unit.

Several important points should be noted:

- PW and WFI are never stagnant; they are recirculated at 1–3 m/s to prevent the growth of microorganisms.
- Pipeworks are designed to be self-draining with a slope of at least 1% over their length to prevent accumulation of water.
- Spray balls are used to continuously wash the PW and WFI storage tanks.
- Plastics such as polypropylene and acrylonitrile butadiene styrene are used for the construction of pretreatment tanks and pipeworks.
- After purification, 316L grade stainless steel is used. This is resistant to corrosion and is electropolished and passivated to reduce roughness, which may act as sites for bacterial growth and future corrosion.
- There should be no deadlegs, that is, areas where water may stagnate. The length of pipe without continuous flow of water should be less than six times the internal diameter of the pipe.
- WFI is circulated hot at about 80 °C as hot water is self-sanitizing.

The quality of PW and WFI is constantly monitored. The FDA has provided guidance for the validation of water systems. The validation program consists of three phases. Phases 1 and 2 are for 2–4 weeks each of continuous sampling and testing of water to establish the effectiveness of the pretreatment and purification and distillation processes. Phase 3 is routine monitoring of the water quality over the remainder of a 1 year period to gauge the influences of seasonal conditions on the water quality.

The specifications for PW and WFI according to the *US Pharmacopoeia* are given in Table 9.4.

9.6.2   Cleaning

Cleaning of product contact surfaces such as reaction and storage vessels is an important aspect of pharmaceutical manufacturing. Residual materials are contaminants and may provide fertile grounds for microorganisms to grow.

TABLE 9.5   Types of Cleaning Agents

Cleaning Agent	Concentration
Acetic acid	100–200 ppm
Peracetic acid	100–200 ppm
Phosphoric acid	1000–2500 ppm
Sodium hydroxide	1500–7500 ppm
Sodium hypochlorite	25–50 ppm
Solubilizing detergents	According to manufacturer's directions

Source: Vos JR, O'Brien RW. Cleaning and validation of cleaning in biopharmaceutical processing: a survey, in Avis KE, Wagner CM, Wu VL, eds., *Biotechnology: Quality Assurance and Validation, Drug Manufacturing Technology Series*, Volume 4, Interpharm Press, Buffalo Grove, IL, 1999.

This is especially the case for biopharmaceutical production, because the soiled materials are normally protein based, and, unlike synthetic drugs, biopharmaceutical drugs are not generally subjected to terminal sterilization.

There are several approaches to cleaning. The favored approach is clean-in-place (CIP), in which cleaning solutions are piped to the vessel under computer control. In cases where CIP is not suitable, clean-out-of-place (COP) is used. This approach is mostly for smaller items. COP may be carried out manually or with automated tanks. A third approach is manual cleaning, although this is prone to human error and is not generally adopted.

Different types of cleaning solutions are used. They include acids, bases, and detergents (Table 9.5).

The effectiveness of cleaning needs to be validated. The types of cleaning agents, concentrations, cleaning cycle, and temperature have to be determined. This is achieved by performing IQ, OQ, and PQ for each piece of equipment that has product contact surfaces. After cleaning, final rinse water samples using PW or WFI are collected. Direct surface sampling using swabs can be used as well. The samples are analyzed for pH, conductivity, microorganism levels, endotoxin, total organic carbon (TOC), residual materials, and other appropriate tests to determine levels of contaminants carried over from previous batches and residuals left by cleaning agents.

Other systems and areas that require cleaning are chromatographic columns and surfaces in the facilities, especially cleanrooms. A rigorous cleaning program has to be implemented to minimize potential product contamination. This includes a limit being set for the maximum carryover of contaminants and validated by the validation process (Exhibit 9.7).

Recently, there is a trend in the use of disposable, or single use, systems to obviate the need for cleaning, both in terms of contamination control and cost saving. Another aspect is operator safety, which can be enhanced through reduction of contact with potentially hazardous chemicals or microorganisms. Setup costs and installation time for disposable systems are cheaper and faster, hence leading to the adoption of such technologies. Disposables may range

Exhibit 9.7   Maximum Allowable Carryover

Equipment is cleaned after a production batch. The maximum allowable carryover (MAC) of materials from one production batch to the next batch is given by the formula

$$MAC = TD \times BS \times SF/LDD$$

where TD = a single therapeutic dose, BS = batch size of next product to be manufactured in the same equipment, SF = safety factor, and LDD = largest daily dose of the next product to be manufactured in the same equipment.

If the therapeutic dose is 100 mg, batch size is 10 kg, largest daily dose is 800 mg, and the safety factor is 1/1000, the MAC is

$$MAC = (100\,mg \times 10,000,000\,mg \times 1/1000)/800\,mg = 1250\,mg$$

Source: Parenteral Drug Association. Technical report No. 29, Points to consider for cleaning validation, *PDA Journal of Pharmaceutical Science and Technology* 52: Nov–Dec Supplement (1998).

from polymer-based syringes to storage containers and bioreactors. In these instances, it is necessary to demonstrate that the systems used do not leach out materials when in contact with the solutions or solvents to be stored or used. Controlled experiments showing realistic case studies are required to convince regulatory authorities that the disposables present no harmful leachables affecting the safety and efficacy of the drugs.

9.6.3   Computer Validation

The pervasiveness of computerized systems within pharmaceutical manufacturing facilities requires that these systems be validated to prevent potential problems from computer software "bugs" and incompatible interfaces between software and hardware. FDA regulations under 21 CFR Part 11 (effective August 20, 1997) spell out the regulatory requirements for electronic signatures and electronic records to ensure that they are trustworthy and reliable (Exhibit 9.8). In February 2003, the FDA undertook to reexamine certain provisions of Part 11 as a result of the response from the pharmaceutical industry. Currently, the FDA enforces a narrow interpretation of Part 11, which applies when electronic records are used in place of paper records. However, Part 11 does not apply when computers are used for producing printouts and the regulated activities are based on a paper system. For systems that predate August 1997, the FDA applies discretion in its enforcement, although these systems must comply with predicate rules effective at the time.

Exhibit 9.8   21 CFR Part 11 Electronic Records; Electronic Signatures—Maintenance of Electronic Records

This regulation is far reaching and contains explicit requirements for computerized systems validation. It "applies to electronic records and electronic signatures that persons create, modify, maintain, archive or transmit." As such, it requires persons to "employ procedures and controls designed to ensure the authenticity, integrity, and, when appropriate, the confidentiality of electronic records, and to ensure that the signer cannot readily repudiate the signed record as not genuine."

The following are examples of some selected requirements:

Section 11.10(a): Validation of systems to ensure accuracy, reliability, consistent intended performance, and the ability to discern invalid or altered records

Section 11.10(b): The ability to generate accurate and complete copies of records in both human readable and electronic form suitable for inspection, reviews, and copying by the agency (FDA)

Section 11.10(d): Limiting system access to authorized individuals

Section 11.10(e): Using secured, computer generated, time-stamped, audit trails

Section 11.50: Signed electronic records to contain information associated with the signing that clearly indicates the printed name of the signer, the date and time of signing, and what the signature means

Overall Approach to Part 11 According to Guidance for Industry Part 11, Electronic Records; Electronic Signatures—Scope and Application: The approach is based on three main elements:

- Part 11 will be interpreted narrowly.
- For those records that remain subject to Part 11, enforcement discretion will be exercised with regard to Part 11 requirements for validation, audit trails, record retention, and record copying in the manner described in the guidance and with regard to all Part 11 requirements for systems that were operational before the effective date of Part 11 (also known as legacy systems).
- The FDA will enforce all predicate rule requirements, including predicate rule record and recordkeeping requirements. (Predicate rules are preexisting regulatory requirements such as GLP, GMP, and GCP guidelines.)

The FDA intends to enforce provisions related to the following controls and requirements:

- Limiting system access to authorized individuals
- Use of operational system checks
- Use of authority checks
- Use of device checks
- Determination that persons who develop, maintain, or use electronic systems have the education, training, and experience to perform their assigned tasks
- Establishment of and adherence to written policies that hold individuals accountable for actions initiated under their electronic signatures
- Appropriate controls over systems documentation
- Controls for open systems corresponding to controls for closed systems bulleted above
- Requirements related to electronic signatures

Source: Food and Drug Administration. *Guidance for Industry Part 11, Electronic Records; Electronic Signatures—Scope and Application.* http://www.fda.gov/cder/guidance/5667fnl. htm [accessed November 15, 2007].

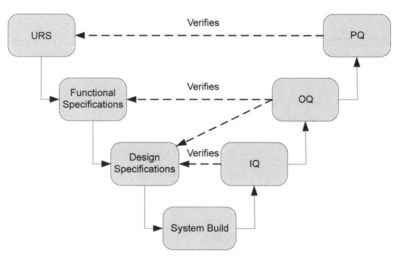

Figure 9.4 Framework for specification and qualification of computerized systems.

Another industry guide for development and testing of computerized systems is the Good Automated Manufacturing Practice by the International Society for Pharmaceutical Engineering. This document sets out the various life cycle stages for software systems design, testing, and validation (Fig. 9.4).

User requirement specifications (URS) for the computerized system are provided by the pharmaceutical firm to the computer systems vendor. The vendor generates functional and design specifications as a basis for designing and coding software for the computerized system. The system is then built, together with all the interfaces to the hardware, and tested by the vendor. After installation of the computerized system, IQ, OQ, and PQ are performed at the pharmaceutical facility to verify that the system is able to meet the URS and design and functional specifications.

Computer validation establishes documented evidence to show that the computerized system will consistently function and meet its predetermined specification and quality attributes with a high degree of assurance. Some of the parameters tested in the validation process include the following:

- IQ: Proper installation of system, correct software version, all parts present and connected correctly, software virus check, integrity of hard disk, availability of source code and manuals
- OQ: Start-up routine, calibration routine, data transfer/backup, data integrity, power failure, auto lock off, human–machine interface, security access/audit trail, system stress test in event of power failure, alarm tests, operator data entry tests
- PQ: System compatibility tests under operational conditions at defined limits.

In accordance with GAMP, there are different validation strategies depending on the categories of software; the degree of validation increases from a general system to a custom system (see Exhibit 9.9).

9.6.4   Process Validation

Processes for the manufacture of drugs must be validated to ensure they consistently produce drugs of the desired specifications. Even in a seemingly straightforward production of a small molecule drug, there are generally many processing steps to achieve the active ingredient. The difficulty of producing proteins and antibodies in biological systems escalates even further, necessitating the control of multiparameters to manage the growth of cells and their reproduction of the desired products.

Process validation entails firstly the definition of both the critical and noncritical parameters. Once they are defined, emphasis can be directed to designing a program to validate these parameters. Some established steps involve the evaluation of process consistency over at least three batches, via the consideration of the processing steps and yield and comparing these with predetermined specifications. Some input parameters that may be considered as critical are temperature, flow rate, and stirring speed, and they are varied and checked against output variables such as yield, purity, and crystallization rate.

Exhibit 9.9 Categories of Software

Category 1—Operating Systems: These are established commercially available operating systems. They are not subject to specific validation; their features are functionally tested and challenged indirectly during testing of the application. Name and version number are documented and verified during IQ.

Category 2—Firmware: Instrumentation and controllers often incorporate firmware. The name, version, and any configuration and calibration for the firmware should be documented and verified during IQ, and functionality tested during OQ.

Category 3—Standard Software Packages: These are commercial, off-the-shelf software packages. The package is not configured, and process parameters may be input into the application. The name and version should be documented and verified during IQ. Compliance to URS should be tested during OQ. Supplier documentation should be assessed and used.

Category 4—Configurable Software Packages: These software packages can be configured according to user requirements. A supplier audit is usually required to confirm software has been developed according to a documented quality system. Validation should ensure software meets URS requirements. Full life cycle validation is needed.

Category 5—Custom (Bespoke) Software: These software packages are developed to meet specific requirements of the user. A supplier audit is usually required to confirm the software has been developed according to a documented quality system. Validation should ensure the software meets URS requirements. Full life cycle validation is needed.

Source: International Society for Pharmaceutical Engineering. *Guide for Validation of Automated Systems: GAMP 4—Good Automated Manufacturing Practice*, ISPE, 2001.

Essentially, the process validation activity is to provide the basis for a robust manufacturing practice.

9.6.5 Analytical Methods Validation

Generally, GMP manufacturers use compendial methods from the *US Pharmacopoeia, British Pharmacopoeia*, or *European Pharmacopoeia* as much as possible for analyses and testing purposes, as these methods have been validated and accepted by regulatory authorities. However, manufacturers are expected to demonstrate that the compendial methods are suitable for the conditions under which the tests are performed.

There are occasions where new analytical methods have to be developed specifically for testing raw materials, intermediates, and finished products that are not covered by compendial methods. In these situations, the analytical methods are required to undergo a validation process to ensure they are suitable. One or more of the following parameters as defined in Exhibit 9.10 must be validated for newly developed analytical methods:

- Specificity
- Accuracy
- Precision
- Repeatability
- Limit of detection
- Limit of quantitation
- Linearity
- Ruggedness
- Robustness

Exhibit 9.10   Analytical Methods Validation

Specificity: Ability to assess unequivocally the analyte in the presence of components that may be expected to be present

Accuracy: Expresses the closeness of agreement between the value that is acceptable, either as a conventional time value or an acceptable reference value, and the value found

Precision: Expresses the closeness of agreement between a series of measurements obtained from multiple sampling of the same homogeneous sample under the prescribed conditions

Repeatability: Expresses the precision under the same operating conditions over a short interval of time

Limit of Detection: Lowest amount of analyte that can be detected in a sample

Limit of Quantitation: Lowest amount of analyte that can be quantitatively determined in a sample with suitable precision and accuracy

Linearity: Ability to obtain test results that are proportional to the concentration of analyte in the sample

Range: Interval between the upper and lower concentration (amounts) of analyte in the sample for which it has an acceptable degree of precision, accuracy, and linearity

Ruggedness: Interval between upper and lower concentration of analyte in the sample for which it has been demonstrated that the analytical procedure has a suitable level of precision, accuracy, and linearity

Robustness: Measurement of its capacity to remain unaffected by small, but deliberate, variations in method parameters; provides an indication of its reliability during normal use

The following table shows the recommended validation requirements for the analytical methods used to characterize the drug molecules.

Validation Parameter	Analytical Method[a]			
	Identification	Impurities (Quantitation)	Impurities (Limit)	Assay
Specificity	+	+	+	+
Accuracy	–	+	–	+
Precision	–	+	–	+
Repeatability				
Intermediate precision				
Reproducibility				
Limit of detection	–	–	+	–
Limit of quantitation	–	+	–	–
Linearity	–	+	–	+
Range	–	+	–	+
Robustness	–	+	–	+

[a]plus sign (+) means evaluation needed; minus sign (–) means evaluation not needed.

Source: Schniepp S, Taylor M, Loffredo D, Vasinko J. *Method Validation: An Overview of Global Standards*, PDA Letter, XLIII, p.1, pp. 20–24 and p. 28 (2007).

Source: International Conference for Harmonization. *Validation of Analytical Procedures: Text and Methodology*, Q2(R1), 2005.

A rationale should be generated to explain and support the reasoning for validating the selected parameters. The use of reference standards during validation helps to reinforce the reliability of the analytical method developed. The conditions of how a test is performed may have a strong influence on the results. These conditions have to be recorded and followed.

9.6.6 Sterilization Processes

Parenteral drug products are required to be sterile. There are principally five different ways to sterilize a product. These are steam, dry heat, radiation, gas, and filtration. Selection of which method to use is based on the product that requires sterilization. For example, protein-based drugs are heat sensitive, so the normal means for sterilizing these products is filtration. The rationale for sterilization validation is to show the reduction in microbial load or destruction of biological indicators.

Steam under pressure at 15 psig (103.4 kPa), 121 °C is a very effective sterilant. Bacterial spores that are resistant to dry heat are killed by steam sterilization. The mechanism is thought to be that the steam causes denaturation of proteins and amino acids within the bacterial cells. An autoclave is a typical steam sterilization device. It is validated taking into account the loading pattern of items in the autoclave chamber and the sterilization cycle used. Often, biological indicators such as *Bacillus stearothermophilus* (also known as *Geobacillus stearothermophilus*) and *Clostridium sporogenes* are used to challenge the effectiveness of sterilization.

Dry heat is used to sterilize and depyrogenate components and drug products. The definition of dry heat sterilization is 170 °C for at least 2 hours and a depyrogenation cycle at 250 °C for more than 30 minutes. Typical equipment includes tunnel sterilizers (force convection, infrared, flame) and microwave sterilizers. An important aspect is the need to ensure air supply is filtered through HEPA filters. Biological indicators such as *Bacillus subtilis* can be used to gauge the performance of sterilization.

Radiation generates high-energy photons, which penetrate microorganisms and cause death through ionization. Commercial radiation sterilization employs gamma-ray radioisotopes such as cobalt-60 and cesium-137. The radiation dose is around 10^3 to 4×10^5 Gy. Radiation may cause degradation in drug products and its effects have to be considered.

Ethylene oxide and hydrogen peroxide are the typical gases for gas sterilization. Their advantage is that they can be used at much lower temperatures than steam sterilization: 27–60 °C for ethylene oxide and 25–40 °C for hydrogen peroxide. Another advantage is that they do not cause damage to the product or the packaging.

For protein-based drugs, filtration via a 0.2 μm filter is an effective way to achieve sterilization. Factors that determine the filtration efficiency include integrity of the filter, pressure, temperature, flow rate, contact time of material with the filter, pH, and viscosity. Validation of filters should include chemical compatibility of the filter with the product and possibility of contaminant from the filters leaching into the product.

The effectiveness of sterilization can be established by culturing samples of the filtrate in a growth medium. Fluid thioglycolate medium and soybean-casein digest medium are normally used. Incubation is 7–14 days at 30–35 °C for fluid thioglycolate medium and 7–14 days at 20–25 °C for soybean-casein digest medium. The absence of microorganism colonies at the end of the growth cycle is an indication of sterility.

Several mathematical functions are used as indicators of microbial destruction. These are D, Z, and F values:

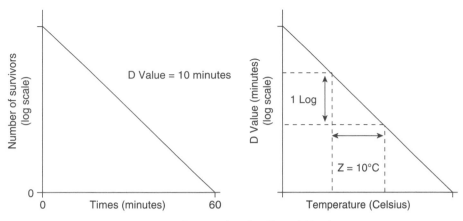

Figure 9.5   Curves showing D and Z values.

D Value: The amount of time, at a given temperature, that is required to reduce the microbial population by one order of magnitude (1 log)

Z Value: The number of degrees of temperature necessary to change the D value by a factor of 10

F Value: The equivalent time, in minutes, at a specific temperature delivered to a product to produce a given sterilization effect at a reference temperature and specific Z value

Both the D and Z values are further illustrated in Fig. 9.5.

9.6.7 Stability Evaluation

The quality of a drug changes over time under the influence of temperature, humidity, and light. It is a requirement that drug products have to be stable during transportation and storage over their projected shelf life. The ICH *Harmonized Tripartite Guideline* Q1A(R2) *Stability Testing of New Drug Substances and Products* (February 2003) sets out the guidelines for testing the stabilities of new drug substances. It should be noted that the test conditions simulating Climate Zones III (hot and dry) and IV (hot and humid tropical) are considered as having the most adverse effects on drug products and as such are acceptable to other Climate Zones.

To evaluate the stability of a drug, stress testing is carried out to determine the effects of environmental conditions on the drug. For example, the effect of temperature is assessed over a range of temperatures in 10 °C intervals, humidity at conditions <75% relative humidity, oxidation and photolysis degradation processes, and hydrolysis of the drug at different pH levels. These evaluations are to assess any changes in the physical, chemical, biological, and microbiological properties of the drug in its container or packaging after it has been transported to various environmental conditions.

TABLE 9.6 Storage Conditions for Evaluating Drug Stability

Study	Storage Condition[a]	Minimum Period Covered by Data at Submission
General Case		
Long term	$25\,°C \pm 2\,°C$	12 months
	60% RH $\pm$ 5% RH	
Intermediate	$30\,°C \pm 2\,°C$	6 months
	60% RH $\pm$ 5% RH	
Accelerated	$40\,°C \pm 2\,°C$	6 months
	60% RH $\pm$ 5% RH	
Drug Substances Intended for Storage in a Refrigerator		
Long term	$5\,°C \pm 3\,°C$	12 months
Accelerated	$25\,°C \pm 2\,°C$	6 months
	60% RH $\pm$ 5% RH	
Drug Substances Intended for Storage in a Freezer		
Long term	$-20\,°C \pm 5\,°C$	12 months

[a]RH = relative humidity.

The ICH has specified storage conditions to evaluate drug stability as part of the submission data for drug approval (Table 9.6).

9.7 THE FDA'S NEW cGMP INITIATIVE

In August 2002, the FDA announced a new initiative to cGMP. The initiative is called *Pharmaceutical cGMPs for the 21st Century: A Risk-Based Approach.* It is a 2 year program, and there are three goals to this initiative:

- To focus the FDA's cGMP requirements on potential risks to public health and channel additional regulatory attention to manufacturing aspects that pose potential risk
- To ensure the regulatory work does not impede innovations in the pharmaceutical industry
- To enhance consistency in approach to assure production quality and safety among the FDA's centers and field groups

A steering committee was formed and the FDA has been working on this initiative with industry groups. The following are the five principles for guiding the initiative:

- Risk-based orientation
- Science-based policies and standards
- Integrated quality systems orientation
- International cooperation
- Strong public health protection

The risk-based approach merges science-based policies and standards with an integrated quality system. This is to ensure that the FDA's resources are directed to address those areas that are considered to have higher risks; for example, companies with previous compliance problems, new companies with unknown history, and processes requiring aseptic procedures.

There are three steps to implementing the risk-based approach:

- The first steps include holding scientific workshops with key stakeholders, re-examining and clarifying the scope and interpretation of Part 11 (see Section 9.6.3), developing technical dispute resolution processes, and harmonizing inconsistencies between the different centers of the FDA.
- The second steps are to utilize new scientific developments and analysis tools to focus on higher risk areas, and include trained product specialists as members of pharmaceutical inspection teams.
- The long-term steps are to develop and implement science-based risk management to regulatory issues and target inspections in risk areas, while encouraging innovations in pharmaceutical companies with proven regulatory history and control. The FDA is also examining how it can facilitate introduction of process analytical technologies to improve manufacturing efficiencies.

The FDA has established a new CDER Division of Compliance Risk Management. It is working with stakeholders to develop a risk model. FDA has released three draft guidance documents: (1) Part 11, *Electronic Records; Electronic Signatures—Scope and Application* (2003), (2) *Comparability Protocols—Chemistry, Manufacturing and Controls Information* (2003), which applies to small molecule and veterinary drugs, and (3) *Protein Drug Products and Biological Products—Chemistry, Manufacturing and Controls Information* (2003). These comparability protocols allow manufacturers to change manufacturing processes, under certain conditions, without submitting a supplement to the FDA for prior approval. Thus, a more streamlined process can be implemented to improve efficiency of production without unduly being burdened by regulatory requirements. Similarly, the ICH has prepared a comparability protocol for biological products under Q5E—*Comparability of Biotechnological/Biological Products Subject to Changes in their Manufacturing Process* (see Exhibit 9.11 for more information on comparability). A parallel track to the risk-based approach is the system approach for GMP inspection. This is discussed in Section 10.3.

Exhibit 9.11   Comparability Studies

Regulatory authorities such as the FDA have prepared guidelines on comparability protocol, *Comparability Protocols—Chemistry, Manufacturing and Controls Information,* to enable manufacturers to follow a plan to establish and substantiate that changes to postapproval processes do not affect drug quality.

The FDA's definition of comparability protocol is "a well-defined, written plan for assessing the effect of specific CMC changes in the identity, strength, purity, and potency of a specific drug product as these factors relate to the safety and effectiveness of the product. A comparability protocol describes changes that are covered under the protocol and specifies the tests and studies that will be performed, including analytical procedures that will be used, and acceptance criteria that will be achieved to demonstrate that specified CMC changes do not adversely affect the product."

The ICH guideline Q5E—*Comparability of Biotechnological/Biological Products Subject to Changes in Their Manufacturing Process*—provides further guidance for the comparability of large molecule drugs, necessitating the requirement to evaluate quality attributes such as physicochemical properties, biological activities, immunological properties, purity, impurities, contaminants, and stability. Bridging clinical and nonclinical data may be required if the information gathered from manufacturing changes are insufficient to show comparability.

The FDA also introduces the concept of process analytical technology (PAT). PAT refers to systems that are used to analyze, monitor, and control manufacturing processes on a continuous basis. The quality attributes and specifications of raw materials, in-process intermediates, and processes are measured in real time and compared with predetermined parameters so that deviations can be rectified in a proactive nature to assure that the end products conform to the level of quality as expected. It is believed that a system based on PAT being implemented in a real-time manner would improve manufacturing efficiency and simultaneously retain or improve the product quality through these interactive measurements and controls.

Some examples of PAT are chemical, physical, microbiological, mathematical, and risk analysis. The FDA suggested the following tools that may be used to manage PAT:

- Multivariate data acquisition and analysis tools
- Modern process analyzers or process analytical chemistry tools

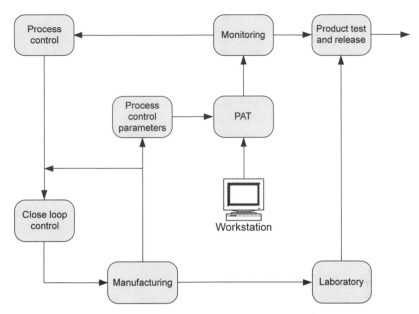

Figure 9.6   Controls in manufacturing process.

- Process and endpoint monitoring and control tools
- Continuous improvement and knowledge management tools

A diagrammatic representation of using PAT and the Laboratory Information Management System (LIMS) controls for the manufacturing of pharmaceuticals/biopharmaceuticals is presented in Fig. 9.6.

9.8   CASE STUDY #9

Risk-Based Approach*

The intention of the risk-based approach introduced by the FDA in 2002 was to modernize the FDA's regulation of the pharmaceutical quality of drugs. This was followed by the EMEA's regulation in recognition of such an important topic. This case study summarizes the progress to date for both the FDA and EMEA in the implementation of the risk-based approach.

*Sources: (1) Food and Drug Administration. *A Risk-Based Approach to Pharmaceutical Current Good Manufacturing Practices (cGMP) for the 21ˢᵗ Century.* http://www.fda.gov/cder/gmp/ [accessed September 8, 2007]. (2) European Medicines Agency. *Implementation of the Action Plan to Further Progress the European Risk Management Strategy: Rolling Two-Year Work Programme* (Mid 2005–Mid 2007). http://www.emea.europa.eu/pdfs/human/phv/37268705en.pdf [accessed September 8, 2007].

FDA: The initiative was originially issued with the following goals:

- Encourage the early adoption of new technological advances by the pharmaceutical industry
- Facilitate industry application of modern quality management techniques, including implementation of quality systems approaches, to all aspects of pharmaceutical production and quality assurance
- Encourage implementation of risk-based approaches that focus both industry and FDA attention on critical areas
- Ensure that regulatory review, compliance, and inspection policies are based on state-of-the-art pharmaceutical science
- Enhance the consistency and coordination of the FDA's drug quality regulatory programs, in part, by further integrating enhanced quality systems approaches into the FDA's business processes and regulatory policies concerning review and inspection activities

Key Accomplishments to Date:

Communication: Several new guidances or draft guidances have been issued to support the goals of the initiative.

Internal Quality Management Systems: Several internal quality management systems have been or are being established.

External Quality Management Systems: The FDA has published a guidance on a quality systems approach to pharmaceutical cGMPs.

International Collaboration: The focus has been on promoting international cooperation directed at assuring drug product quality and consistency of cGMPs.

Implementation of Quality by Design: The focus of this concept is that quality should be built into a product with a thorough understanding of the product and process by which it is developed and manufactured along with a knowledge of the risks involved in manufacturing the product and how best to mitigate those risks.

Process Analytical Technologies: Process analytical technologies (PATs) are to be used as tools for designing and analyzing pharmaceutical development and manufacturing. A number of training programs have been organized to highlight the applicability of PATs and the regulatory pathways.

Pharmaceutical Inspectorate (PI): The PI was established in the ORA to enhance the FDA's overall inspection program.

Conclusion: The FDA continues to facilitate the modernization of regulatory processes related to pharmaceutical products in the 21st century.

EMEA: The European Risk Management Strategy (ERMS) aims to "provide a more coherent approach to the detection, assessment, minimization and communication of risks of medicines in Europe. This should lead to a more proactive approach to safety monitoring of medicines throughout their life-cycle."

The Public Status Report on the Implementation of the European Risk Management Strategy presents the progress made from 2005 to 2007 in this area. Some of these are as follows:

- Implementing the legal tools for monitoring the safety of medicines and for regulatory actions
- Strengthening the spontaneous reporting scheme via electronic reporting
- Launching the European Network of Centres for Pharmacoepidemiology and Pharmacovigilance (ENCePP) project to monitor the conduct of multicenter postauthorization safety studies
- Contributing to the European commission on the conduct of research in pharmacovigilance

9.9 SUMMARY OF IMPORTANT POINTS

1. Through regulations, drugs are manufactured in accordance with Good Manufacturing Practice (GMP). The relevant codes in the United States for small and large molecule drugs are 21 CFR Parts 21 and 211, and 21 CFR Parts 600 and 610, respectively. For the European Union, the requirements for GMP are stipulated in EU Directive 2003/94/EC.
2. With the formation of the International Conference on Harmonization (ICH), the GMP guidelines are presented in Q7. This document has been adopted by the United States, the European Union, and Japan.
3. The following are important elements for GMP:
   - Quality system
   - Trained personnel
   - Suitable facilities
   - Appropriate equipment
   - Documentation and records
   - Materials management
   - Production and in-process controls
   - Identification and labeling
   - Laboratory controls
   - Validation
   - Change controls

4. The following are some important GMP systems:
   - Water system
   - Cleaning practices
   - Computer validation
   - Process validation
   - Test methods validation
   - Sterilization procedures
   - Stability evaluation
5. The risk-based approach introduced by the FDA focuses the pharmaceutical companies' and regulatory authorities' resources into addressing potential risk areas based on the scientific approach. The aim is to ensure pharmaceutical companies adopt new scientific developments and equipment for manufacturing safe, effective, pure, and consistent products.

9.10 REVIEW QUESTIONS

1. Using ICH Q7 as a reference, highlight the important elements of GMP.
2. Discuss the importance of quality system, trained personnel, and validation.
3. Explain the different grades of water in a pharmaceutical setting.
4. List and explain the important parameters for analytical test method validation.
5. How can contamination be controlled in a pharmaceutical plant?
6. Discuss the benefits or otherwise of process analytical technology (PAT).
7. What are the benefits for the implementation of a risk-based approach in GMP manufacturing?

9.11 BRIEF ANSWERS AND EXPLANATIONS

1. Refer to Section 9.4.
2. Quality system provides a global structure and coordination to implement manufacturing policies and procedures in a methodical and controlled manner. It also enables deviations and problems to be addressed in timely ways and reduces recurrence. Trained personnel are necessary to execute manufacturing and test activities by following established procedures and to ensure compliance with GMP and that products are

manufactured to specifications. Validation is conducted such that equipment, computers, processes, and analytical test methods work and function in the specified manner as intended.

3. Purified water (PW) is used for cleaning and preparation of nonsterile drug compounds whereas water-for-injection (WFI) is used for the final rinse and preparation of sterile materials. Refer to Section 9.6.1.

4. Refer to Exhibit 9.10.

5. Contamination is controlled at multiple levels in a pharmaceutical plant. It starts with a clean, well-maintained facility. Raw materials are labeled, checked, and segregated at storage. Equipment is cleaned and tested for cleanliness prior to being used. Personnel are to follow procedures as written in SOPs. Samples and products are clearly labeled and stored. Flows of materials, equipment, personnel, tools, and wastes are controlled.

6. PAT is used to analyze, monitor, and control manufacturing processes on a continuous real-time basis. This process means predetermined characteristics are checked and deviations, if any, are corrected promptly such that product quality is maintained.

7. Refer to Section 9.8. The risk-based approach directs resources to address issues according to risk levels to assure the safety and quality of drug products being manufactured.

9.12 FURTHER READING

Brunkow R, DeLucia D, Green G, et al. *Cleaning and Cleaning Validation: A Biotechnology Perspective*, PDA, Bethesda, MD, 1996.

Carleton FJ, Agalloco JP, eds. *Validation of Pharmaceutical Processes, Sterile Products*, Marcel Dekker, New York, 1999.

Center for Biologics Evaluation and Research. *Guidance for Industry: For the Submission of Chemistry, Manufacturing and Controls and Establishment Description Information for Human Plasma-Derived Biological Products, Animal Plasma or Serum-Derived Products*, FDA, Rockville, MD, 1999.

Center for Biologics Evaluation and Research. *Guidance for Industry: Monoclonal Antibodies Used as Reagents in Drug Manufacturing*, FDA, Rockville, MD, 2001.

Center for Biologics Evaluation and Research. *Points to Consider in the Production and Testing of New Drugs and Biologicals Produced by Recombinant DNA Technology, and the Supplement*, FDA, Rockville, MD, 1985 and 2006.

Food and Drug Administration. *ICH Q10 Pharmaceutical Quality System*, FDA, Rockvill, MD, 2007.

Food and Drug Administration. *Guidance for Industry, 21 CFR Part 11; Electronic Records; Electronic Signatures—Scope and Application*, FDA, Rockville, MD, 2003.

Food and Drug Administration. *Guide to Inspections Validation of Cleaning Processes*, FDA, Rockville, MD, 2006.

Food and Drug Administration. *Points to Consider in the Characterization of Cell Line Used to Produce Biologicals*, FDA, Rockville, MD, 2006.

Gadamasetti KG, ed. *Process Chemistry in the Pharmaceutical Industry*, Marcel Dekker, New York, 1999.

International Conference on Harmonization. Derivation and characterization of cell substrates used for production of biotechnological/biological products, in *Harmonized Tripartite Guideline*, ICH, 2001.

International Conference on Harmonization. Specifications: test procedures and acceptance criteria for new drug substance and new drug products: chemical substances, in *Harmonized Tripartite Guideline*, ICH, 1999.

International Conference on Harmonization. Specifications: test procedures and acceptance criteria for biotechnological/biological products, in *Harmonized Tripartite Guideline*, ICH, 1999.

International Conference on Harmonization. Viral safety evaluation of biotechnology products derived from cell lines of human or animal origin, in *Harmonized Tripartite Guideline*, ICH, 1997.

International Society for Pharmaceutical Engineering. *GAMP Guide for Validation of Automated Systems*, ISPE, 2001.

Medicines Control Agency. *Rules and Guidance for Pharmaceutical Manufacturers and Distributors*, MCA, 1997.

Office of Regulatory Affairs. *Guide to Inspections of Bulk Pharmaceutical Chemicals*, FDA, Rockville, MD, 1994.

Office of Regulatory Affairs. *Guide To Inspections Of Quality Systems*, FDA, Rockville, MD, 1999. http:www.fda.gov/ora/inspect_ref/igs/qsit/qsitguide.htm [accessed April 19, 2002].

Vesper JL. *Documentation Systems, Clear and Simple*, Interpharm Press, Buffalo Grove, IL, 1998.

CHAPTER 10

GOOD MANUFACTURING PRACTICE: DRUG MANUFACTURING

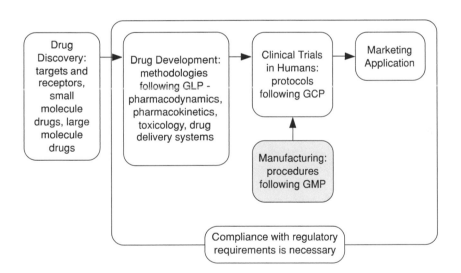

10.1   Introduction   320
10.2   GMP Manufacturing   322
10.3   GMP Inspection   325
10.4   Manufacture of Small Molecule APIs (Chemical Synthesis Methods)   332
10.5   Manufacture of Large Molecule APIs (Recombinant DNA Methods)   340

10.6   Finished Dosage Forms   348
10.7   Case Study #10   352
10.8   Summary of Important Points   355
10.9   Review Questions   356
10.10   Brief Answers and Explanations   356
10.11   Further Reading   357

10.1   INTRODUCTION

There are distinct differences in the manufacture of small and large molecule drugs; the former is mainly based on organic chemical synthesis, while the latter relies on biological systems of recombinant DNA (rDNA) technology. In this chapter, we describe the manufacturing processes for these drugs under GMP environments.

Before the large scale manufacturing of a drug under a GMP quality system is undertaken, there is a development process, which spans the discovery stage and the commercial production of the drug. From the discovery stage, lead compounds are identified. These lead compounds are tested in a number of stages, from laboratory *in vitro* assays to *in vivo* tests, pharmacology, toxicology, and finally into clinical trials as potential drug candidates. An increasing quantity of the drug material is needed as each stage progresses. The demand for material grows from milligrams to grams and kilograms. A development program is phased in to meet this demand by initially producing drugs on a laboratory scale. It then progresses to pilot plant scale to provide more drug material for clinical trials, and finally implements procedures, processes, equipment, and setting up of a manufacturing plant for large-scale commercial production.

Broadly, the drug development program covers the following.

General Items

Raw Materials: Raw materials have a significant impact on the manufacturing process. Issues such as availability and reliability of supply, reactivity, toxicity, handling, and storage have to be considered. Cost is another factor to take into account. Often, trade-offs between costs, manufacturing processes, and yields are considered.

Safety: Production of the requisite drug molecule, called the active pharmaceutical ingredient (API) or bulk pharmaceutical chemical (BPC), may involve materials, solvents, or intermediates that are volatile, toxic, or even explosive. The development program has to determine the appropriate manufacturing processes to ensure that safety is not compromised and the API can be produced and purified to remove impurities and toxic residues.

Reproducibility of Manufacturing Processes: The aim of GMP is to ensure the manufacture of safe, potent, pure, and effective drug in a consistent manner. The development program exists to evaluate procedures and processes that can be implemented in a large-scale manufacturing environment to ensure the drug product conforms to the intended safety, potency, purity, effectiveness, and consistency on a routine basis.

Environmental Factors: In addition to conformance to GMP, the manufacturing plant has to comply with local environmental legislation. This may cover materials transportation, handling, storage, and disposal. The manufacturing plant is set up with systems for controlling gaseous emission, decontamination of solid waste, and treatment of liquid discharge. All these factors are evaluated in the development program.

Small Molecule Drugs

Organic Chemistry Synthesis Route: The production of the API for small molecule drugs requires ingenious and meticulous development of organic synthesis steps. Some drugs may require more than 50 steps to obtain the intended API. For example, there are more than 100 production steps for the manufacture of Roche's AIDS drug enfuvirtide (Fuzeon), approved by the FDA in March 2003; see Exhibit 10.1.

In some cases, drug materials are isolated from natural products. In other cases, natural product extraction constitutes the raw material or intermediate for production of the drug via a semisynthetic route. Methods for chemical reactions, product purification, control parameters, and analytical procedures are developed and they form the basis for the chemistry, manufacturing, and control (CMC) information for regulatory application.

Large Molecule Drugs

Optimization of Protein Synthesis Route: Protein-based drugs are produced using living systems of microbial or mammalian cells. The development

---

Exhibit 10.1   Fuzeon

Fuzeon (enfuvirtide) is an anti-HIV drug. It interferes with the entry of HIV into the CD4 cell. It is a synthetic peptide with 36 amino acids, the N terminus is acetylated while the C terminal forms a carboxamide. The peptide binds to the gp41 subunit of the HIV envelope glycoprotein and prevents the HIV from fusing with CD4.

Source: Food and Drug Administration. http://www.fda.gov/medwatch/SAFETY/2004/oct_PI/Fuzeon_PI.pdf [accessed November 29, 2007].

program commences with selection of the cell line, cloning methods for genes that express the intended protein molecule, and experimentation of conducive growth environments for high yields and determination of effective purification procedures. Similarly, CMC information is submitted for regulatory application.

Drug development work also includes formulation, stability studies, and selection of drug delivery systems, as discussed in Section 5.6. Once the API has been prepared, excipients are added:

- To modify processing properties for the manufacture of finished dosage forms (tablets, capsules, parenterals, etc.)
- As preservatives or buffers to ensure drug stability
- For efficient delivery of the drug to targets

In API manufacture, whether via chemical synthesis, rDNA technology, or extraction from natural products, there are significant changes (physical and chemical) from the starting materials to the API. In the formulation process, however, the quality and specifications of the API are retained. The addition of excipients to produce the drug product in a finished dosage form does not present physical or chemical changes to the API.

It should be noted that GMP regulations are necessary for approved drug products. Regulatory authorities such as the FDA do not expect total GMP compliance for the manufacture of drugs designated for clinical trials. This is recognized, and ICH Q7 GMP Guidance Section 19, "APIs for Use in Clinical Trials," sets out the GMP expectations. It states that "controls used in the manufacture of APIs for use in clinical trials should be consistent with the stage of development of the drug product." As a Sponsor files for an Investigational New Drug (IND), the manufacturing information is presented in the CMC (Section 8.2.2). Initially, for Phases I and II clinical trials, regulatory authorities do not require total GMP compliance in the CMC. However, by Phase III, all quality systems, production processes, and validation issues are expected to have been resolved to enable GMP manufacturing of drug products to be carried out routinely.

10.2 GMP MANUFACTURING

Manufacturing of drugs, whether the API or finished dosage form, is required to comply with GMP regulations (see Chapter 9). Figure 10.1 shows the implementation of GMP concepts in drug manufacture.

The first requirement for GMP manufacture is the availability of trained personnel. Other requirements are as follows:

- Raw materials must conform to specifications.
- Water, in the form of purified water or water-for-injection, must be available as required.

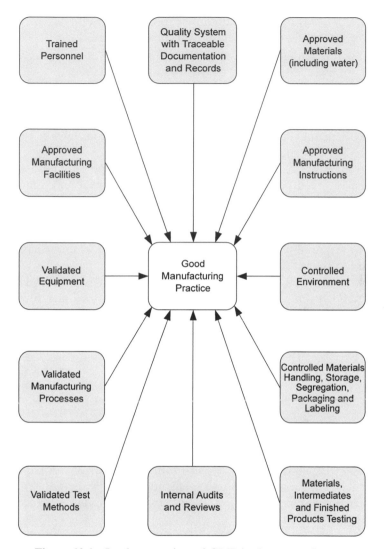

Figure 10.1   Implementation of GMP in drug manufacture.

- An environment should have appropriate controls for temperature, pressure, and relative humidity. For aseptic production, cleanroom conditions monitored for particles and bioburden contamination are necessary.
- Equipment must be validated and maintained with current calibration.
- Processes must be developed and validated to ensure the production of pure and consistent product.
- Operating procedures must be clearly written down, detailing each manufacturing step.
- Approved batch records must be kept for registering all relevant information during the manufacturing process.

- Quality records must be maintained to document all tests pertaining to the raw materials, intermediates, and finished product.
- A validated cleaning procedure must be in place to ensure reaction vessels and containers do not carry contaminants.

All these items must be in place before the manufacturing process begins. In the manufacturing process, deviations from specified conditions and processes may happen. For example, the pH for the reactions may be outside the range stated in the procedure or the reaction process may produce more heat, leading to a greater temperature rise than is programmed. These deviations have to be resolved before the following steps can proceed. Similarly, raw materials, intermediates, and products that are outside specifications require out-of-specifications investigations. When resolved satisfactorily, the materials, intermediates, and products are released. Change controls and corrective actions are required when investigations show failure in controls that need rectification.

QC tests are carried out according to validated analytical methods or established methods from pharmacopoeias: *US Pharmacopoeia* and *British Pharmacopoeia*. Exhibit 10.2 lists some of the QC analytical methods performed on drug intermediates and products.

Exhibit 10.2   Selected Analytical Methods

High Performance Liquid Chromatography (HPLC): This is a separation method for characterizing or determining the purity of a drug material. The material is passed through a chromatographic column with solid matrix, which binds the material and separates the material according to its physicochemical properties.

Sodium Dodecyl Sulfate–Polyacrylamide Gel Electrophoresis (SDS-PAGE): This method is used to separate proteins based on molecular weights. SDS is added to the proteins to produce a net negative charge. Under an electric field, the negatively charged proteins migrate to the anode. Smaller molecules migrate longer distances and are separated from larger molecules. Bromophenol blue is used as a color marker to show the progress of electrophoresis. Dithiothreitol (DTT) or mercaptoethanol is added to the sample to disentangle disulfide linkages of proteins in a "reduced" condition.

Isoelectric Focusing: This is an electrophoretic method in which the proteins are separated based on their charge characteristics. This is accomplished by the proteins moving through a medium with a pH gradient.

The protein stops at the point where the pH equals the protein's iso-electric point (pH where the protein has no net charge).

Capillary Electrophoresis (CE): The CE instrument consists of a source/sample vial, a destination vial and a small capillary filled with electrolyte joining the two vials. A voltage is applied and separates the sample according to size and charge, which is detected by UV absorbance.

Quantitative Polymerase Chain Reaction (QPCR): This method can determine the amount of DNA or RNA in a sample. Using the QPCR method, DNA can be amplified many times, allowing minute quantities to be assessed.

Spectroscopy: Drug compounds absorb visible, infrared, and UV radiation at frequencies that are characteristic of the compounds. Quantitative measurements can be calculated from the absorbance readings at specific frequencies or wavelengths.

Circular Dichroism: This method is used to determine the enantiomers in racemic mixtures. The isomers rotate polarized light in different directions depending on their chiral characteristics.

Atomic Spectroscopy: This method is used to determine the concentration of an element in a drug substance. The intensity of the emission lines of the element measured at specific wavelengths shows its concentration.

Mass Spectroscopy: This is based on measurement of the ratio of mass to number of positive or negative charges of the substance to be analyzed. The pattern generated is characteristic of the drug substance. One method is the use of matrix-assisted laser desorption ionization–time of flight (MALTI-TOF) mass spectroscopy. This gives partial sequences of peptide fragments. From these, the protein identity can be revealed through a database search.

Limulus Amebocyte Lysate (LAL) Test: This test is used to detect the presence of endotoxins in the drug substance. It relies on the coagulation reaction between the endotoxin and the blood of a horseshoe crab.

NMR and ELISA: These methods are discussed in Chapters 3 and 4.

10.3   GMP INSPECTION

Regulatory authorities inspect GMP facilities to ensure drugs are manufactured according to GMP requirements. *The Compliance Program Guidance Manual for FDA Staff: Drug Manufacturing Inspections Program 7356.002* (February 2002) states the strategy for inspection as follows:

• Evaluating through factory inspections, including the collection and analysis of associated samples, the conditions and practices under which drugs and drug products are manufactured, packed, tested, and held, and
• Monitoring the quality of drugs and drug products through surveillance activities such as sampling and analyzing products in distribution

The FDA carries out the inspections once every 2 years. The FDA has adopted a systems approach to GMP inspection. A GMP facility is divided into six systems:

• *Quality System:* This consists of procedures and specifications to assure the overall compliance for the facility. Quality control, change control, batch release, internal audits, and quality records are part of the quality system.
• *Facilities and Equipment System:* This includes (1) buildings and facilities along with maintenance; (2) equipment IQ, OQ, calibration, maintenance, cleaning, and validation of cleaning processes; and (3) utilities such as HVAC, compressed gases, steam, and water systems.
• *Materials System:* This is concerned with the segregation and storage of raw materials, components, and finished products, inventory control, and distribution of finished products.
• *Production System:* This includes manufacturing processes, sampling and testing, batch records, and process validation.
• *Packaging and Labeling System:* This includes control and issuance of labels, packaging operations, and validation of these operations.
• *Laboratory Control System:* This includes laboratory test methods, stability program, and analytical method validation.

The FDA carries out two types of inspections: surveillance inspections and compliance inspections. Surveillance inspections are the biennial inspections. Compliance inspections are to follow-up on previous corrective actions. Compliance inspections also include "For Cause Inspections," which are inspections to audit a specific problem that has come to the FDA's attention, for example, product recall and industry or public complaints. There are two options of inspections: (1) the Full Inspection Option and (2) the Abbreviated Inspection Option.

The Full Inspection Option is a surveillance or compliance inspection that is thorough and gives the FDA a deep understanding of the cGMP program in a manufacturing facility. This type of inspection is conducted when the FDA has little knowledge about the facility, such as a new facility or where the facility has a history of noncompliance or when the FDA has doubt about the facility's quality system. The Full Inspection Option audits at least four systems in the facility, one of which must be the quality system.

The Abbreviated Inspection Option is a surveillance or compliance inspection. It is a shortened inspection. This is performed when the facility has a satisfactory cGMP compliance record and there are no product problems or there has been little change since the last inspection. At least two systems are audited, including the quality system.

Figure 10.2 shows in simplified form the flow of an FDA inspection and the actions taken. Exhibit 10.3 summarizes the FDA guidelines for the inspection of manufacturers of bulk pharmaceutical chemicals (BPCs and APIs), biotechnology, and finished dosage form drugs. Inspections of biopharmaceutical manufacturing facilities are carried out by Team Biologics, which is a partnership between the Office of Regulatory Affairs and the Center for Biologics Evaluation and Research (CBER) of the Food and Drug Administration (FDA). Team Biologics consists of personnel with skills and experience in biopharmaceutical manufacturing to ensure critical areas are inspected.

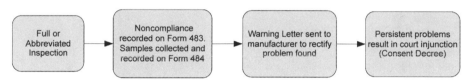

Figure 10.2 FDA inspection and action process.

Exhibit 10.3   FDA Guides to Inspections

Guide to Inspections of Bulk Pharmaceutical Chemicals (BPC)—1991

Buildings and Facilities: Contamination/Cross Contamination, Water System/Water Quality, Aseptic/Sterile Processing

Equipment: Multipurpose Equipment, Equipment Cleaning and Use Log, Equipment Located Outdoors, Protected Environment, Cleaning of Product Contact Surfaces

Raw Materials

Containers, Closures, and Packaging Components

Production and Process Controls: Mother Liquors, In-Process Blending/Mixing, Validation of Process and Control Procedures, Reprocessing, Process Change, Impurities

In-Process Testing: Packaging and Labeling of Finished BPC

Expiration Dating or Reevaluation Dating

Laboratory Controls

Stability Testing

Reserve Samples
Batch Production Records

Biotechnology Inspection Guide Reference Materials and Training Aids—1991

Cell Culture and Fermentation
    Master Cell Bank and Working Cell Bank
        Origin
        Characterization and History Qualifying Tests
        Storage Conditions and Maintenance
    Media
        Raw Materials
        Bovine Serum
        Sterilization
Culture Growth
    Inoculation and Aseptic Transfer
    Monitoring of Growth Parameters and Control
    Containment and Containment Control
Ascites Production
    Mouse Colony
        Characterization and Control of the Mouse Colony
        Animal Quarters/Environmental Controls
    Manufacturing Processes
        Animal Identification
        Tapping Procedure
        Storage and Pooling of Ascites
        Purification
Extraction, Isolation, and Purification
    Process Types
    Process Validation
        Documentation
        Validation
        Follow-up Investigations
    Process Water/Buffers/WFI
    Plant Environment
Cleaning Procedure
    Detailed Cleaning Procedure

Sampling Plan
Analytical Method/Cleaning Limits
Processing and Filling
Processing
In-Process Quality Control
Filling
Lyophilization
Laboratory Controls
Training
Equipment Maintenance/Calibration/Monitoring
Method Validation
Standard/Reference Material
Storage of Labile Components
Laboratory SOPs
Testing
Quality
Identity
Protein Concentration/Content
Purity
Potency
Stability
Batch-to-Batch Consistency
Environmental Coverage
Environmental Assessment

Guide to Inspections of Dosage Form Drug Manufacturer

Organization and Personnel
Buildings and Facilities
Equipment
Components and Product Containers
Production and Process Controls: Critical Manufacturing Steps, Equipment Identification, In-Line and Bulk Testing, Actual Yield, Personnel Habits
Sterile Products: Personnel, Buildings, Air, Environmental Controls, Equipment, WFI, Containers and Closures, Sterilization, Laboratory Controls, Production Records

> **Exhibit 10.4   GMP Consent Decree**
>
> In May 2002, Schering-Plough was fined US\$500 million for GMP violations by the FDA under the consent decree scheme. The issue centered on the GMP violations of the manufacturing facilities in New Jersey and Puerto Rico. A total of 13 inspections were carried out by the FDA from 1998 to 2002. The noncompliances were related to the facilities, quality assurance, manufacturing, equipment, laboratories, and labeling.
>
> In addition to the US\$500 million that Schering-Plough had to pay, it also had to settle about US\$500,000 for inspection costs, recall several products, suspend or discontinue certain products, and revamp its quality system to ensure future compliance.
>
> *Source*: Food and Drug Administration. *The Food & Drug Letter*, Issue No. 653, June 2002.

If the inspector believes the cGMP has been violated, Form FDA-483 is used to record the observations. Samples may be taken by the FDA inspector for analysis. In this case, Form FDA-484 is issued to the manufacturer for the receipt of samples. A normal practice for the manufacturer is to take more samples for internal analysis and compare them with the FDA data when required.

After inspection, the inspector prepares a detailed Establishment Inspection Report (EIR). This is the FDA's primary record for the inspection. Time is given to the manufacturer to respond to the deficiencies found and recorded on Form FDA-483. Failure to comply with satisfactory resolution of the deficiencies found will result in the FDA sending out a Warning Letter notifying the manufacturer to comply. If the manufacturer is unable to resolve the deficiency after the deadline set by the FDA, the FDA may proceed to prosecute the manufacturer with an injunction. The injunction is a court order called Consent Decree, and the manufacturer may be required to cease operations until the problem is rectified (see Exhibit 10.4).

In Europe, inspections are conducted by member states on behalf of the European Union. For drugs approved under the centralized procedure, inspections are coordinated by the European Medicines Agency (EMEA; refer to Sections 7.3 and 8.3). For countries that are members of the Pharmaceutical Inspection Cooperation Scheme (PIC/S; see Section 7.13), there is mutual recognition of inspections performed by members.

Some typical problems found in GMP inspections are the following:

- *Out-of-Specifications:* There are insufficient investigations to determine the root cause of problems and issues are not closed in a timely manner.
- *Product Sterility:* The tests performed are superficial and not validated.
- *Environment Monitoring:* Personnel are not monitored, there is inadequate monitoring, microorganisms are not monitored, there is no

identification of contaminants, and alert limits for contaminants are set too high.

- *Raw Materials, Components, and Finished Product:* There is no audit procedure or an insufficient audit procedure exists; test methods lack validation.
- *Training:* There is no training plan, training is not documented, and investigation of problems in manufacturing does not lead to retraining of involved staff.
- *Materials:* There is no segregation of materials.
- *Calibration:* There is a lack of scheduling and lapsed calibration validity.
- *Documentation:* Manufacturing steps are not signed off, changes are not explained, and obsolete copies of documents are used.
- *Process:* Personnel are not following set procedures; procedures are not validated.
- *Internal Audit and Review:* Infrequent internal audits are performed, superficial audit is insufficient to reveal problems, and there is no follow-up on issues observed at internal audits.
- *Management:* There is a lack or insufficient commitment on GMP issues.

The following are examples of some of the typical violations leading to drug product recall according to the FDA:

- cGMP deviations
- Subpotent products
- Product failed *US Pharmacopoeia* dissolution test requirements
- Product failed endotoxin/pyrogen tests
- Presence of contaminants in product
- Label mix-up on the product
- Stability data do not support expiration date
- Product lacks stability
- Product failed content uniformity
- Product failed pH test requirements

In early 2007, the EMEA reported the analysis of GMP deficiencies in its document entitled *Good Manufacturing Practice: An Analysis of Regulatory Inspection Findings in the Centralised Procedure*. It detailed the deficiencies reported in 435 inspections of manufacturers of medicinal products and starting materials in the European Union and third countries in the period 1995 to 2005. Altogether there were 9465 deficiencies, of which 193 were critical (2%), 989 major (10%), and 8283 (88%) others. The top 20 deficiencies are presented in Table 10.1.

TABLE 10.1   Deficiencies of Manufacturers of Medicinal Products and Starting Materials in the EU and Developing Countries During 1995–2005

Ranking	Category of GMP Deficiency	Number
1	Contamination, microbiological—potential for	112
2	Documentation—quality system elements/procedures	102
3	Regulatory issues: unauthorized activities	66
4	Design and maintenance of premises	59
5	Regulatory issues: noncompliance with marketing authorization	55
6	Sterility assurance	53
7	Documentation—manufacturing	50
8	Documentation—specification and testing	46
9	Equipment validation	43
10	Design and maintenance of equipment	36
11	Personnel issues: duties of key personnel	35
12	Supplier and contractor audit and technical agreements	34
13	Contamination, chemical/physical—potential for	33
14	Process validation	33
15	Environmental monitoring	25
16	Personnel issues: hygiene/clothing	25
17	Investigation of anomalies	22
18	In-process controls—control and monitoring of production operations	18
19	Line clearance, segregation and potential for mix-up	18
20	Personnel issues: training	17

Source: EMEA Inspections. *Good Manufacturing Practice: An Analysis of Regulatory Inspection Findings in the Centralized Procedure*, Doc. Ref. EMEA/INS?GMP/23022/2007 [accessed August 28, 2007].

The new initiative from the FDA, *Pharmaceutical Current Good Manufacturing Practices (cGMPs) for the 21st Century: A Risk Based Approach*, will have a major effect on the conduct of GMP inspections (see Section 9.7). In early 2003, the FDA identified three categories of potentially higher-risk drug manufacturing sites for prioritizing inspections:

- Sites making sterile drugs
- Sites making prescription drugs
- Sites of new registrants not previously inspected by the FDA

10.4   MANUFACTURE OF SMALL MOLECULE APIs (CHEMICAL SYNTHESIS METHODS)

10.4.1   Conventional Synthesis Techniques

The manufacturing process for a small molecule API is shown in Fig. 10.3. Typically, the chemical reactions are performed in large reaction vessels. For

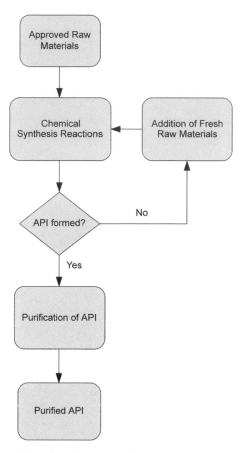

Figure 10.3   Chemical synthesis of small molecule APIs.

commercial production of an API, the reaction vessel can typically range from 1000 to 20,000 L in volume (Fig. 10.4).

The reaction vessel is normally made of glass-lined stainless steel with a jacket for heating and cooling and consists of:

- Charge-hole for addition of solid raw materials
- Metered-pump input for liquids
- Supply of gases as reactants or inert blanket
- Stirrer for mixing the raw materials
- Condenser unit for solvent reflux
- Vents with filters for gas emission or depressurization
- Transfer line for discharge/separation of reactants/products
- Probes for measuring the temperature, pH, and pressure
- Sampling ports for withdrawals of samples for analysis

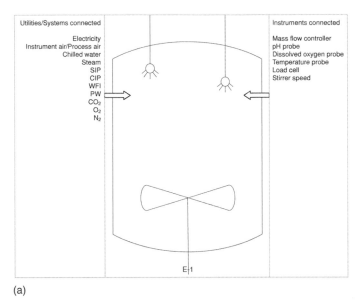

(a)

(b)

Figure 10.4   Reaction vessel for the manufacture of an API: (a) schematic drawing and (b) photo courtesy of Vega Grieshaber KG, Germany.

Production of the API begins with the selection of a synthetic route, as determined in the development program. Raw materials are added into a reaction vessel. These raw materials as reactants are heated or cooled in the reaction vessel (normal range is from −15 to 140 °C; purpose-built vessels are needed for extreme reactions that require lower or higher temperature controls or pressurization of reaction processes). The chemical synthesis reactions are monitored and controlled via sensor probes (pH, temperature, and pressure) with in-process feedback controls for adjustments and alarms when necessary. Samples are withdrawn at defined intervals for analysis to determine the reaction progress. Catalysts, including enzymes, may be added to speed up and direct the reaction along a certain pathway.

It is important to maintain a uniform reaction environment within the vessel chamber by using a stirrer to agitate and mix the reactants. Gaseous discharge is vented through filters to the outside environment. As the reactions may generate substantial heat (exothermic) and pressure, or may even be potentially explosive, special precautionary features are designed into the vessel.

At the end of the synthesis reactions, the product can be pumped to another vessel or container via transfer lines. If the chemical reactions proceed to completion with negligible trace quantities of impurities, the next stage of production may commence in the same reaction vessel with addition of fresh raw materials. This process is called telescoping.

The finished product is centrifuged and purified via a number of processes, including filtration, fractional distillation, condensation, crystallization, and chromatographic separation techniques. The purified API is tested and then it is ready to be formulated into the finished dosage form, as discussed in Section 10.6. Exhibit 10.5 illustrates some of the typical reagents for API manufacture and Exhibit 10.6 presents selected chemical reactions as examples of the

---

Exhibit 10.5   Typical Reagents for API Manufacture

Solvents: Water (purified water or water-for-injection grade); toluene, methanol, ethanol, ether, acetate, dimethyl sulfoxide, tetrahydrofuran, hexane, cyclohexane, dichloromethane, acetonitrile, acetone

Oxidizing Agents: Hydrogen peroxide, chromic acid, potassium permanganate, manganese dioxide, ozone

Reducing Agents: Hydrogen, lithium aluminum hydride, sodium borohydride, di-isobutyl aluminum hydride, iron metal

Acids: Sulfuric acid, hydrochloric acid, phosphoric acid, methanesulfonic acid, acetic acid, formic acid

Bases: Sodium hydroxide, ammonia, triethylamine, pyridine, butyl lithium, sodium hydride, α-methylbenzylamine

Halogenation Reagents: Halogens, N-bromo- and N-chlorosuccinimide, thionyl chloride, phosphorus oxychloride

Alkylating Agents: Dimethyl sulfate, methyl iodide, methyl tosylate

Sulfur Reagents: Thiols and sulfides, hydrogen sulfide, sodium sulfide, sodium thiocyanate, thiourea, sodium metabisulfide

Phosphorus Reagents: Phosphorus halides

Boron Reagents: Diborane, boron trifluoride, dialkyl borinates, aryl boronic acids.

Source: Lee S, Robinson G. *Process Development: Fine Chemicals from Grams to Kilograms*, Oxford Chemistry Series, Oxford University Press, Oxford, 1995.

Exhibit 10.6   Selected Chemical Reactions as Examples for API Manufacture

Halogenation

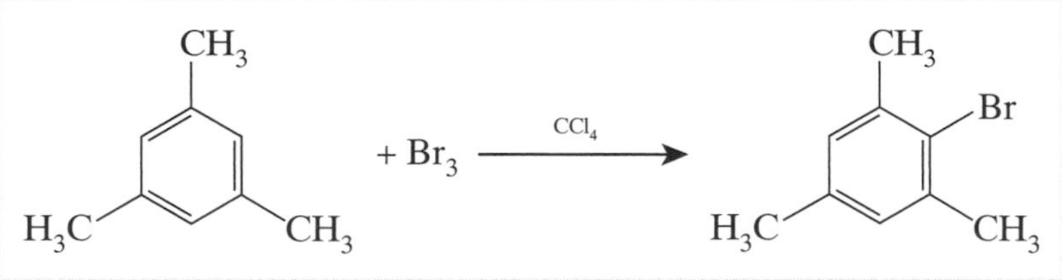

Alkylation

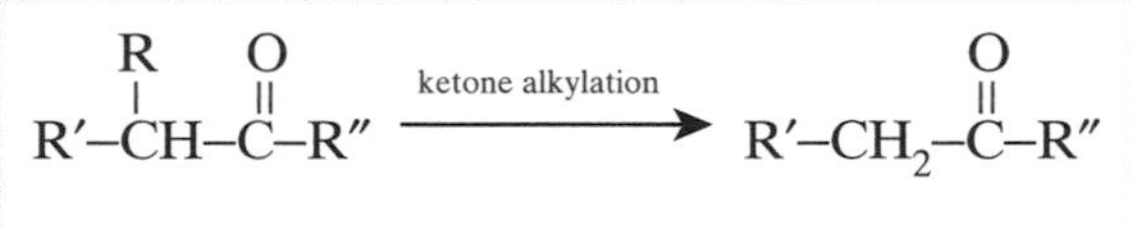

Acylation

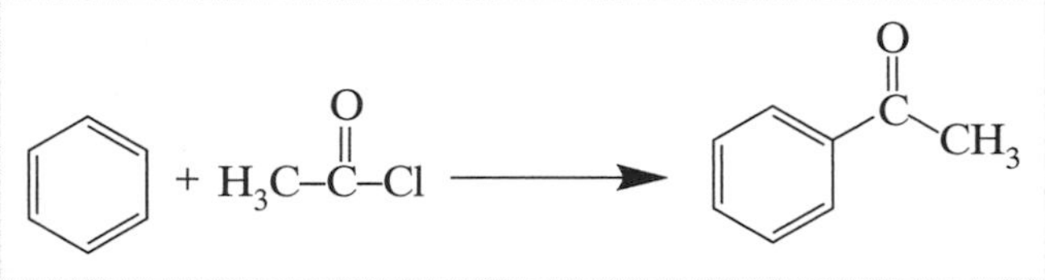

Grignard Reaction

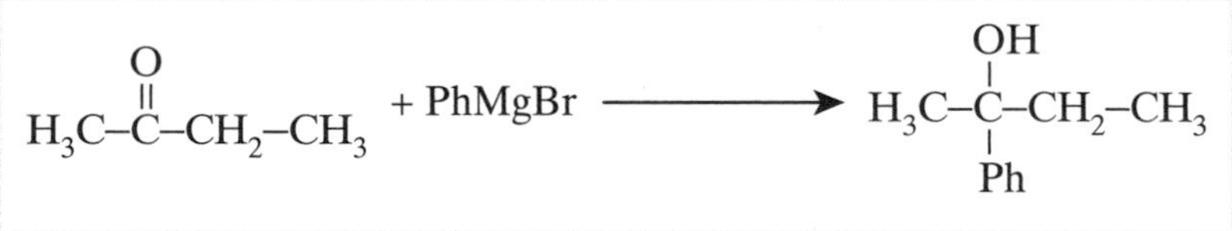

Source: Hornback JM. *Organic Chemistry*, Brooks/Cole, Belmont, CA, 1998.

synthesis processes for drug manufacture. Purification processes for drug materials are described in Exhibit 10.7.

The production of the API and finished dosage form is required to comply with GMP regulations discussed in Chapter 9 and Section 10.2. The quality system, quality control, and validation of equipment and processes have to be developed and adhered to in the manufacturing process. Proper records and documentation are required to be kept in the forms of batch records,

Exhibit 10.7   Purification of an API

Filtration/Fractional Distillation/Condensation: Filters are used to remove solid particles from a solvent. The use of $0.2\,\mu$m filters can remove microbial contamination. Filtered solutions can be fractionally distilled and condensed to obtain the API.

Crystallization: Crystallization is used to separate the API from its solvent and impurities, or to separate racemic mixtures in solution. Crystallization occurs from a supersaturated solution. Important conditions are the temperature, concentration, stirring rate, and heating and cooling rate. Seeding with the desired API can assist in providing nucleation sites for the preferential crystallization of the API.

Source: Carstensen, JT, *Advanced Pharmaceutical Solids, Drugs and the Pharmaceutical Sciences*, Vol. 110, Chapter 6, pp. 89–106, Marcel Dekker, New York, 2001.

Exhibit 10.8   Synthesis of Paclitaxel (Taxol)

An introduction to Taxol (Bristol-Myers Squibb) is presented in Exhibit 3.4. The active pharmaceutical ingredient (API) is paclitaxel. The chemical name is $5\beta,20$-epoxy-$1,2\alpha,4,7\beta,10\beta,13\alpha$-hexahydroxytax-11-en-9-one 4,10-diacetate 2-benzoate 13 ester with (*2R, 3S*)-*N*-benzoyl-3-phenylisoserine.

Early production of 1 kg of paclitaxel required extraction from about 13,000 kg of the Pacific yew tree bark. This process was refined, and paclitaxel is now produced by a semisynthetic route. The starting material, 10-deacetyl baccatin III (10-DAB), is obtained from the needles of *Taxus baccata* (European yew) or *T. wallichiana* (Himalayan yew). The yield is around 1000 kg of needles to produce 1 kg of 10-DAB.

Source: Cabri W, Di Fabio R. *From Bench to Market: The Evolution of Chemical Synthesis*, Oxford University Press, Oxford, 2000.

test records, and manufacturing procedures. Reaction vessels and associated equipment must be calibrated, validated, and cleaned to acceptable levels before being used; this is especially the case for multiproduct plants, where more than one API is manufactured.

As an example, we present in Exhibit 10.8 the synthesis of paclitaxel (Taxol, Bristol-Myers Squibb), an important anticancer drug for breast and ovarian cancer and Kaposi sarcoma. It illustrates the complexity in the synthesis of drug molecules.

10.4.2   Chiral Synthesis Techniques

Production of synthetic drug often gives rise to racemic mixtures of API enantiomers; that is, they are mirror images of each other (see Section 3.6). The first problem is that the production process generates equal amounts of the enantiomers and only one enantiomer is active. In this case, only half the yield is effective for drug application. Another problem is that often one isomer is effective and the other may be benign or create undesirable side effects when interacting with biological receptors that are chiral themselves. Examples of the differing effectiveness of chiral drugs are presented in Exhibit 10.9.

Manufacturing of chiral drugs has become increasingly important; first to improve potency, second to improve yield, and third to extend the patent life for approved drugs based on racemic mixtures. Stereoselective synthetic methods are used to produce chiral drugs. There are three basic methods being applied:

- Enzyme and nonenzyme catalysts
- Chiral building blocks
- Chiral auxiliary

Enzyme and Nonenzyme Catalysts: By nature, enzymes themselves are chiral and they catalyze a variety of chemical reactions with stereoselectivity. These reactions include oxidation, reduction, and hydration. Examples of enzymes are oxidases, dehydrogenases, lipases, and proteases. Metoprolol, an adrenoceptor-blocking drug, is produced using an enzyme-catalyzed method.

Nonorganic and organometallic catalysts are also used to channel the reactions toward the chiral synthesis pathway. The drug called levodopa, (*S*)-3,4-dihydroxyalanine, is an effective drug against Parkinson's disease. It is stereoselectively manufactured using catalysts such as rhodium or ruthenium complexes.

Chiral Building Blocks: Some drugs are made using chiral building blocks to generate the required chiral center in the drug. The introduction of chiral centers ensures that the reaction proceeds in the desired direction. The preparation of enalapril, an ACE inhibitor, is an example of the use of chiral building blocks.

Chiral Auxiliary: A chiral auxiliary is an intermediate formed by the attachment of a pure enantiomer to an achiral substrate. The attachment, called a chiral auxiliary, restricts the approach of reactants to react in specific ways to produce the chiral molecule. The antibacterial drug aztreonam is synthesized using the chiral auxiliary method.

Exhibit 10.9   Examples of Chiral Drugs

Thalidomide: The (*R*)-enantiomer is a sedative and the (*S*)-enantiomer is teratogenic (i.e., causes fetal deformity).

Propranolol: An antihypertensive drug: the (*S*)-enantiomer is 130-fold more potent than the (*R*)-enantiomer, a β-adrenoceptor antagonist.

Dexetimide: Dexetimide has 10,000-fold more affinity for the muscarinic acetylcholine receptor than its enantiomer, levetimide.

Dextropropoxyphene: The (*2R,3S*)-enantiomer marketed as Darvon is an analgesic, whereas the (*2S,3R*)-enantiomer called Novrad is an antitussive.

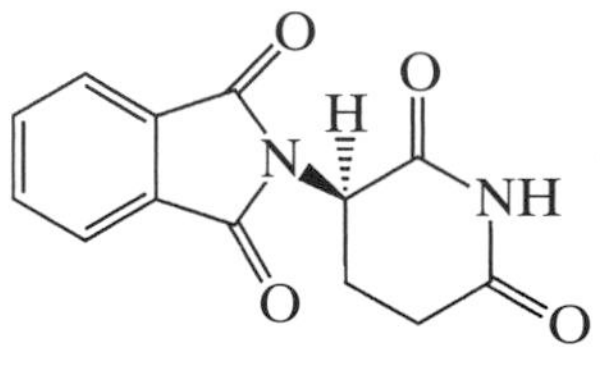

(*R*)-Thalidomide 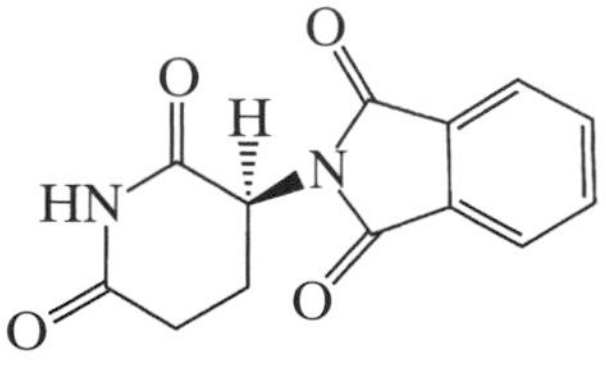   (*S*)-Thalidomide

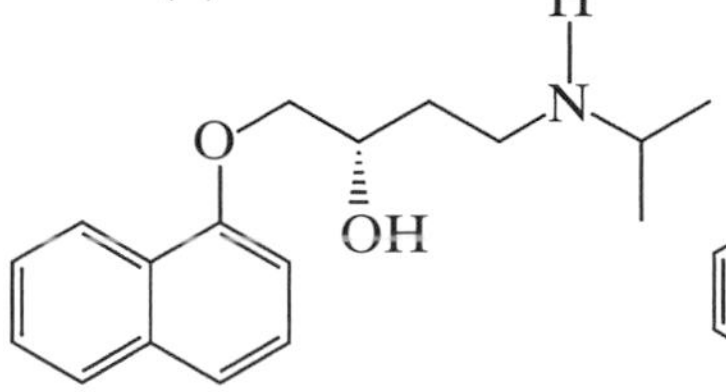

(*S*)-Propranolol 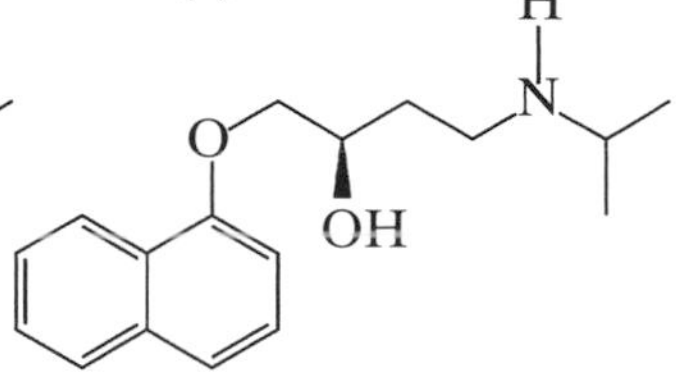   (*R*)-Propranolol

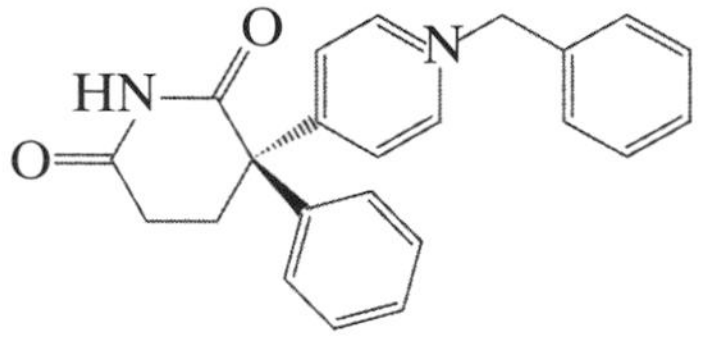

(*S*)-(+)-Dexetimide 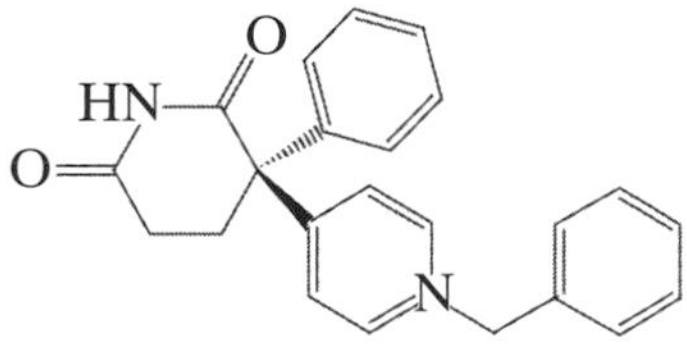   (*R*)-(−)-Levetimide

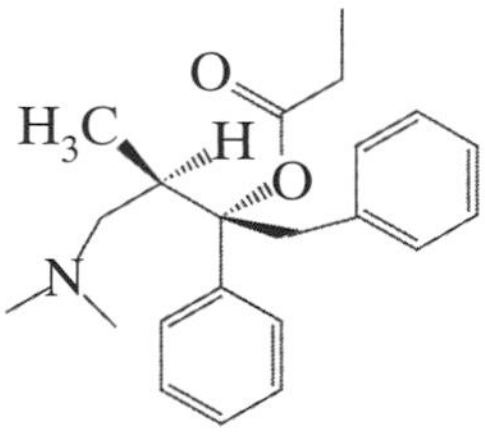

Darvon 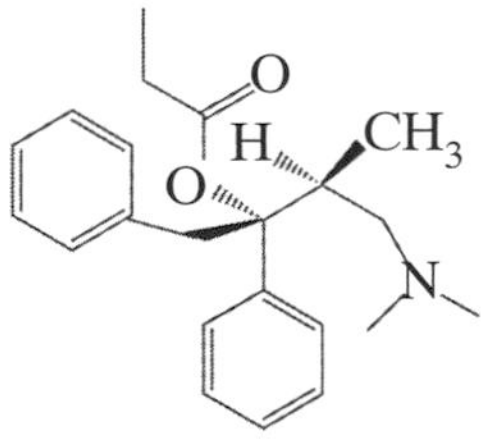   Novrad

10.5 MANUFACTURE OF LARGE MOLECULE APIs (RECOMBINANT DNA METHODS)

The manufacturing process for a typical large molecule protein-based API is shown in Fig. 10.5. A description of the plasmid vector is given in Appendix 4. The production of these APIs uses "factories" that are living cells in the form of cell lines, which can grow indefinitely under appropriate conditions. As discussed in Section 4.3.3, monoclonal antibodies are conventionally produced using hybridoma cell lines, but now cell culture methods are becoming more important. Other protein-based drugs are produced using a variety of cell lines: from bacterial or fungal to insect and mammalian cell lines.

Currently, about equal numbers of the approved protein drugs are derived from microbial and mammalian cells, although more drugs are expected to be produced from mammalian cell lines in the future. There are pros and cons for each type of cell as production "factories" for the protein drugs.

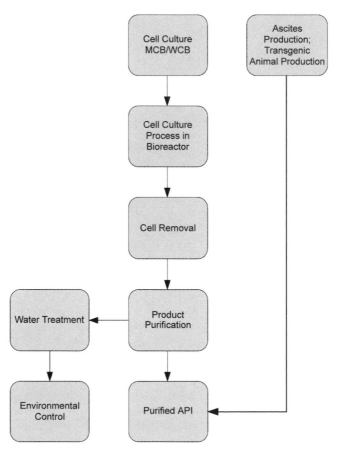

Figure 10.5 Cell culture production of biopharmaceutical drug.

- *Microbial Cells (e.g., E. coli Cells)*

  Advantages: Cells grow rapidly on relatively inexpensive media; the fermentation technology for growing microbial cells is well established.

  Disadvantages: The protein expressed is accumulated within the cell matrix (intracellular); the protein does not undergo posttranslational modifications (resulting in proteins that may be structurally different or less useful to humans); the presence of lipopolysaccharides (pyrogens—microbial substances that cause fever) is likely to contaminate the product; and there is a need for more extensive chromatographic purification.

- *Mammalian Cells (e.g., Chinese Hamster Ovary (CHO) and Baby Hamster Kidney (BHK) Cell)*

  Advantages: Posttranslational modification of protein product can be performed; extracellular expression of proteins requires less complex purification processes.

  Disadvantages: Cells have complex nutritional requirements; the production cost is higher; cells grow more slowly and are susceptible to physical damage; they require specially designed bioreactors.

The recombinant technique involves transfecting cells with DNA that codes for the production of the intended protein. The process of transfection into a plasmid is shown in Exhibit 10.10 (also refer to Appendix 4). Once transfected, clones that express the protein of interest are selected and used for production.

A master cell bank (MCB) is set up, which forms the first generation of these clones (see Exhibit 10.11 about the guidelines for testing and characterizing cell lines and cell banks). They are stored in hundreds of vials in liquid nitrogen freezers (cells in the log phase in growth medium added with 5–10% dimethyl sulfoxide (DMSO) as a cryoprotectant and stored at −150°C or below) to preserve them indefinitely. From the MCB, a vial is taken and cells are grown to produce a working cell bank (WCB), and they too are maintained in liquid nitrogen freezers. Subsequently, a vial is taken from the WCB for each batch of production. This two-tier system of MCB and WCB can supply production needs indefinitely and also ensures the consistency and fidelity of protein production. For example, 200 vials each of the MCB and WCB for a production rate of 10 batches per year will last 4000 years.

Cells from the WCB are cultured initially in flasks, which contain nutrient medium. The medium may contain the following:

- Amino acids
- Vitamins A, D, E, and K
- Ionic salts (Na^+, K^+, Mg^{2+}, Ca^{2+}, Cl^-, SO_4^{2-}, PO_4^{3-}, HCO^{3-})
- Glucose (as a source of energy)
- Organic supplements (proteins, peptides, nucleoside, citric acid, lipids, cholesterols)
- Hormones, growth factors, antibodies, and antibiotics

Exhibit 10.10   Recombinant DNA Techniques—Genetic Engineering

The first step is the isolation of DNA genes that code for the production of the desired protein. The next stage is the insertion of these genes (foreign DNA) into a vector, or carrier. Common vectors used are the bacteriophage (a virus) and the bacterial plasmids, which are circular bacterial DNA.

Both the foreign DNA and plasmid are cleaved by enzymes called restriction endonucleases. They are mixed and then joined together using another enzyme called ligase.

The plasmid with the inserted DNA genes is transformed into bacterial cells or transfected into mammalian cells. Methods used include electroporation, microinsertion, or chemical mediated transfection. When these cells are cultured in medium with nutrients, they grow and divide. In the growth process, the foreign genes express proteins within the cell (intracellular) or outside the cell (extracellular). At the end of the growth cycle, the cells are killed and the proteins are extracted and purified as the protein drug material. The process is illustrated below, where the cells act as "factories" for producing the protein of interest.

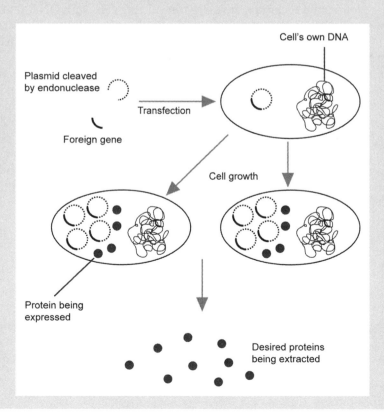

Exhibit 10.11   Guidelines for Testing and Characterizing Cell Lines/Banks

Cell Line: One needs to know the origin, source, and history of the cells. Records must be kept for the cultivation of cells, medium used, genetic manipulation, selection criteria, isolation methods, identification, characteristics, and tests for endogenous and adventitious agents.

> *Human Cell Lines:* Characteristics of donor, tissue or organ of origin, ethnic and geographical origin, age, sex, and general physiological conditions
>
> *Animal Cell Lines:* Species, strains, breeding conditions, tissue or organ of origin, geographical origin, age, sex and general physiological conditions.
>
> *Microbial Cell Lines:* Species, strain, genotype, phenotype, pathogenicity, toxin production, and other biohazard information

Cell Bank: Records must be kept of the cell banking method and procedure.

> *Characterization and Testing of Cell Banks:* Test for adventitious agents, endogenous agents, and molecular contaminants (toxins, antibiotics) to confirm identity, purity, and suitability for manufacturing use
>
> *Evaluation of Stability:* Consistency of the coding sequence of the expression construct in generating the product
>
> *Karyology and Tumorigenicity:* May be required to test for safety of cell line; for those products with no cells, karyology and tumorgenicity not necessary but demonstration of residual host cell DNA is required; for those products with presence of live cells, karyology and tumorgenicity required

Examples of characterization testing of CHO MCB include sterility, mycoplasma, identity, retrovirus, adventitious agents, and bovine and porcine virus.

Source: International Conference on Harmonization. *Derivation and Characterisation of Cell Substrates Used for Production of Biotechnology/Biological Products*, Q5D, July 1997. http://www.ich.org/LOB/media/MEDIA429.pdf [accessed September 28, 2007].

In certain cases, serum (fetal bovine serum—FBS) is added to promote the growth of cells. However, the bovine spongiform encephalopathy (BSE) problem has necessitated tight control on the quality of FBS (refer to Exhibit 10.12). This increases production and downstream processing costs. For new cell lines being developed, serum-free and protein-free media are used to circumvent the possibility of virus contamination from animal sources and the variation that may arise from use of serum from animal herds.

Most cells grow well under a fermentation process at a pH of around 7.0–7.4. However, as cells grow, CO_2 is produced. To maintain optimal growth, the media are often buffered with, for example, phosphate buffered saline. Cells go through different phases of growth (Fig. 10.6). The cell viability, density, and consumption of nutrients are constantly monitored (Fig. 10.7).

When the cells grow to a certain density (number of cells per milliliter, around $1–10 \times 10^6$ cells/mL) and have an acceptable viability (normally >90% survival rate), they are inoculated into larger vessels called bioreactors. There may be several steps for growth in different size bioreactors before a final production bioreactor is used, which may be as large as 10,000–20,000 L.

Cells grow under different conditions that have been optimized in the development stage. Some cells prefer to anchor onto solid substrates. In this case, microcarrier beads or hollow fibers are used to provide attachment for

---

Exhibit 10.12   Bovine Spongiform Encephalopathy

Bovine spongiform encephalopathy (BSE or mad cow disease) is a progressive neurological degenerative disease in cattle. It is caused by a mutated protein called a prion. BSE was first reported in the United Kingdom in 1986. Creutzfeldt-Jakob disease (CJD) is a rare disease that occurs in humans. Evidence to date indicates it is possible for humans to acquire CJD after consuming BSE-contaminated cattle products.

A number of measures have been taken to contain BSE. Thousands of cattle have been culled and there are controls prohibiting the feeding of mammalian proteins to ruminant animals (cows, sheep, and goats). There are also surveillance programs set up to monitor CJD in humans.

The FDA and European regulatory authorities have strongly recommended that drug manufacturers not use materials derived from ruminant animals in countries where BSE has been reported. Manufacturers of protein drugs that require fetal bovine serum (FBS) for cell growth use FBS from countries such as the United States, Australia, and New Zealand—countries considered safe from BSE. As precautionary measures, newly developed cell lines for production of protein drugs are focusing on serum-free and protein-free growth medium.

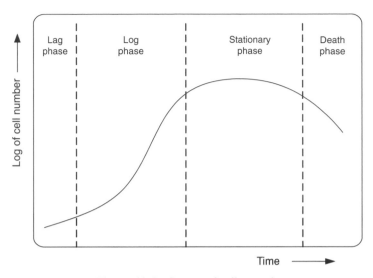

Figure 10.6   Stages of cell growth.

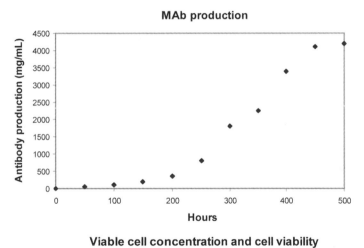

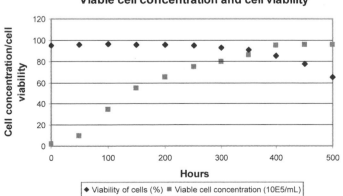

Figure 10.7   Monitoring of cell growth over time.

Exhibit 10.13   Cell Culture Methods

Suspension Process (Stirred/Sparged Tank)

Advantages: Easy to operate in batch and fed-batch; easy to obtain cell sampling to determine cell concentration and viability; easy to scale up

Disadvantages: Sensitive to shear force from stirrer; foaming when serum is added to media

Anchorage-Dependent Process (Microcarriers, Hollow Fiber)

Advantages: Established technology; higher product titer than suspension process

Disadvantages: Cells only grow when attached to solid substrates; cells require additional attachment factor; not easy to scale up; more cleaning validation issues and high disposal cost; more difficult to operate than suspension process

the cells. Some other cells grow best in suspension within the media. Yet, in other cases, continuous supply and harvesting of the cells are optimal; these are the perfusion techniques. The advantages and disadvantages of these methods are described in Exhibit 10.13. Cells are monitored for growth, viability, consumption of nutrient, discharge of metabolites, and use of oxygen and carbon dioxide.

As cells grow, proteins are secreted. At the end of the growth cycle, the proteins are harvested. For cells that produce intracellular proteins, the cell membranes are lysed to free the proteins. Yield of unpurified proteins may range from milligrams to 1–5 g/L. The proteins are then purified, normally via several stages through chromatographic columns (Exhibit 10.14), resulting in about half the amount from the cell culture process as purified product. The purified protein as an API is tested to ensure it meets specifications. Once it passes the requisite specifications, it is ready to be processed into the finished form (see Section 10.6).

Cells grow from a single vial (1–5 mL) to thousands of liters in bioreactors. Many generations of cell division and growth are involved. It is important that in this process, which may last from weeks to months, the cells do not mutate and faithfully express the intended protein. Cells at the completion of production are collected as samples and grown further as an end of production cell bank (EPCB). Analysis conducted on the products from the EPCB is to verify that the protein is produced properly, even after many generations of cell reproductions.

Exhibit 10.15 presents the production of etanercept (Enbrel).

Exhibit 10.14   Purification Using Chromatographic Techniques

Chromatographic separation relies on the affinity of binding between different components of the API in liquid and the solid matrix column. The API is separated from the impurities by percolating the liquid through chromatographic columns filled with solid phase matrices. The matrices are made of different materials and separate the components on the basis of physicochemical properties such as charge, size and shape, hydrophobic and hydrophilic characteristics, complex formation with certain ions or metals, and interaction with dyes.

Ion Exchange: Separation is based on selective, reversible adsorption of charged molecules to an immobilized ion exchange group of opposite charge. An ion exchanger consists of an insoluble porous matrix to which charged groups have been covalently bound. Anion exchanger group: DEAE, diethylaminoethyl; QAE, quaternary aminoethyl; Q, quaternary ammonium. Cation exchanger group: carboxymethyl, sulfopropyl.

Affinity: Product to be purified binds to an affinity ligand that is coupled to a matrix. The ligand is specific for a particular type of protein/peptide molecule, or group of such molecules. The targeted product binds to the ligand under specific conditions of high or low ionic strengths and at a certain pH. After the unbound impurities are removed, the product can be eluted by using a gradient of increasing or decreasing ionic strength or by changing the pH.

Hydrophobic: Proteins and peptides differ from one another in their hydrophobic properties. Salt solutions are used to mediate the binding of molecules to a hydrophobic matrix substituted with a hydrophobic ligand.

Reversed Phase: This technique is based on hydrophobic regions on the surface of proteins and the hydrophobic groups covalently attached to the surface of the matrix. Organic solvents are required for elution. It is suitable for peptides and proteins up to 2.4×10^4 Da.

Gel Filtration: Separation is in accordance with size. Large molecules elute in void volumes and are eluted earlier. Small molecules penetrate pores of the matrix and elute later because of the increase in path length.

Exhibit 10.15 Etanercept (Enbrel)

Etanercept (Enbrel, Immunex) is a dimerized protein molecule that consists of a portion of the tumor necrosis factor receptor coupled to the Fc portion of human IgG1 (see Section 4.3). It has 934 amino acids and a molecular weight of approximately 150 kDa.

Etanercept is produced by recombinant technology in a Chinese hamster ovary mammalian cell expression system. The WCB is grown in a proprietary media system. The cells are cultured initially in flasks and then inoculated into the bioreactor vessel. The product is purified in a number of chromatographic steps, followed by viral inactivation and viral filtration steps.

The finished form is a sterile, lyophilized powder formulated with trimethamine, mannitol, and sucrose as excipients.

Source: Food and Drug Administration, Center for Biologics Evaluation and Research. *Chemistry, Manufacturing and Controls Review, BLA 980286, TNFR:Fc, Immunex*, FDA, Rockville, MD, 1998. Immunex, *Enbrel*, http:///www.immunex.com/search/searchresults.jsp [accessed January 10, 2003].

10.6 FINISHED DOSAGE FORMS

10.6.1 Examples of Different Dosage Forms

A finished dosage form is a drug formulated with an API and excipients in a form that is suitable for administering to patients. The following are some of the reasons for preparing finished dosage forms:

- An exact quantity of the effective drug is incorporated into the formulation.
- Drug products are easier to handle and this increases compliance in taking or administering the drug.
- Preservatives and stabilizers can be added to the API to improve shelf life and result in less stringent storage conditions.
- Taste, color, and odor of the API can be masked by additions of excipients.
- Delivery vehicles (Section 5.6) can be used to provide more specific targeting of the drug to receptors.
- Extended drug effect can be maintained with controlled release formulations.
- Different types of delivery mechanisms can be achieved for effective drug action (e.g., intravenous injection, inhalation, or sublingual application).

TABLE 10.2   Finished Dosage Form Application

Route of Administration	Finished Dosage Form
Oral	Tablets, capsules, solutions, syrups, gels, powders
Sublingual	Tablets, lozenges
Parenteral	Solutions, suspensions
Topical	Ointments, creams, pastes, powders, lotions, solutions, aerosols
Inhalational	Aerosols, sprays
Rectal	Solutions, ointments, suppositories
Vaginal	Solutions, ointments, tablets, suppositories
Urethral	Solutions

Source: Ansel HC. *Introduction to Pharmaceutical Dosage Forms*, 3rd ed., Lea & Febiger, London, 1981, p. 48.

Table 10.2 lists the finished dosage forms for various routes of drug administration. The choice of which finished dosage form to administer to a patient depends on a number of factors. These factors include the nature of the disease, time required for onset of drug action, age of patient, site of intended receptor, and health status of patients. In general, where possible, drug manufacturers provide several dosage forms for an API to enable it to be applied in different ways for achieving reliable and effective therapy.

Solids: Solid dosage forms are the most common means for presenting the drug product for patient administration. Most APIs are in crystallized or powder forms. They are ground to predetermined sizes using mills or pulverizers. The APIs and excipients are then mixed using blenders or tumblers.

Tablets are manufactured through a compression process. Excipients such as binders, lubricants, colorants, flavorings, and disintegration modifiers are added. The production process has to ensure that tablets have the required mechanical strength and do not crumble. Tablets may be coated or uncoated. Uncoated tablets consist of granules of API and excipients compressed into tablets. Various substances are applied to coat tablets, from sugar to waxes, gums, plasticizers, and flavorings. Effervescent tablets contain acids or carbonates, which, when mixed with water, release carbon dioxide to disperse the drug materials. Release modifiers are added to tablets to alter the time and duration of drug release. Enteric coatings are used to protect drugs from being dissolved in the acid environment of the stomach.

Capsules consist of shells for enclosing drug materials. Hard gelatin shells are made of gelatin, sugar, and water. Soft gelatin shells have additional glycerin or sorbitol to soften the wall. Powder or liquid can fill the capsule shells. Hard capsules consist of two prefabricated shell sections. Drug API and excipients in solid form or paste are placed in one section and it is then capped with the other section. Soft capsules are mainly filled with liquids and sealed in one operation. There are modified-release capsules for delayed-release or sustained-release application. Another type is the specially formulated shells that are resistant to acid in stomach, for drug release in the intestine.

An important specification for solid dosage form manufacture is the dissolution factor. The product is formulated and manufactured such that it will have the specified dissolution profile for maximum effectiveness.

Liquids: The liquid dosage form comes in several categories: solution, emulsion, and suspension. They are prepared by dissolving the API in solvents such as water (purified water), alcohol, glycerin, or glycol. Flavorants, colorants, antioxidants, preservatives, and agents for stabilizing, emulsifying, and thickening are often added to the solution to prepare the required liquid formulation. Important criteria in the manufacture of liquid dosage form are the uniformity of dispersion and the effectiveness of preservatives.

Parenterals: The most important criterion for parenterals is that they have to be sterile for injection or infusion administration. Excipients are added to make parenterals isotonic with blood, improve solubility, and control pH of the solution. The solvent vehicles include water-for-injection, sterile sodium chloride, potassium chloride, or calcium chloride solution, and nonaqueous solvents such as alcohol, glycol, and glycerin. Preservatives, antioxidants, and stabilizers are normally added to enhance the properties of the drug product.

Manufacturing is performed in cleanroom conditions. Sterilization processes in the form of heat, steam, gas, or radiation are applied to ensure microorganisms are destroyed in the drug product. For protein-based drugs that can be damaged by the normal sterilization processes, the product is manufactured under aseptic conditions. Both sterility and pyrogen tests are performed to ensure parenteral drug products are safe to be injected.

Inhalants: Inhalants are pressurized dosage forms whereby powder or liquid drug substances are delivered in fine dispersions of aerosols or sprays by propellants. Typically, the particle size of the API is in the range of 2–20 μm for delivery to the respiratory system, and larger sizes for topical applications. Excipients such as preservatives, stabilizers, and diluents are added. The manufacturing of inhalants involves filling the container with the API, propellant, liquid (if required), and excipients. The container is capped with a valve assembly and actuator for control of dosage. Production issues are the effectiveness of preservatives, container leakages, and control of dosage delivery.

Ointments and Creams: Ointments are applied to the skin for topical treatment or to be absorbed into the blood system for delivery to target areas. They are semisolid preparations obtained by mixing the API with selected ointment bases depending on intended use. These bases include petrolatum, paraffin, mineral oil, lanolin, and glycols. Preservatives are often added to ensure the ointments will maintain the recommended shelf life.

Creams are less viscous than ointments. They are dispersions of the API in emulsions. Both oil-in-water and water-in-oil emulsions have their applications.

10.6.2 Packaging and Labeling

The finished dosage forms are packaged into blister packs, bottles, vials, syringes, aerosol containers, or tubes. Nowadays, packaging has tamper-proof designs to ensure the integrity of the packaging. Labeling of the packaging is in accordance with information submitted to regulatory authorities. Exhibit 10.16 describes the FDA regulations for packaging and labeling of intermediates, APIs, and finished dosage forms.

10.6.3 Cold Chain

The need to transport temperature-sensitive raw materials and products, such as cell line, medium, large molecule drugs, and vaccines, means that some form of control during transportation is needed. For example, a working cell bank for the production of proteins may be transported in liquid nitrogen ($-196\,°C$) and that of protein and vaccines in dry ice ($-78\,°C$) in order to protect the integrity of the materials. Data loggers are used to record the temperature

---

Exhibit 10.16   Packaging and Labeling of APIs and Intermediates (ICH (1997) Guide for API: Good Manufacturing Practice), Including Finished Dosage Forms

- Written procedures for receipt, identification, quarantine, sampling, examination, testing, release, and handling of packaging and labeling materials
- Records of shipment and packaging
- Containers suitable for intended use; not reactive, additive, or absorptive to intermediates or API and must protect contents from deterioration and contamination
- Access to labels limited to authorized personnel
- Reconciliation of quantities of labels issued and used
- Procedures to ensure correct packaging and labels are used
- Labeling operations should prevent mix-ups
- Examination of containers and packages to ensure use of correct labels
- Transport materials with seals that will alert recipient of the possibility of alteration if seal has been breached

during the transit to provide evidence that the goods remained under the desired temperature and their functional integrity is not compromised.

10.7   CASE STUDY #10

Generics and Biogenerics*

Generics: From earlier chapters we know that generics are copies of "innovator" or "branded" drugs when their patents have expired. These generics are small molecule drugs and they are formulated with the same active ingredients, dosages, and routes of administration as the innovator drugs. Typically, they are sold at a fraction of the price charged by the pharmaceutical companies for their patented products. The price disparity of patented drugs and generics is due to the recovery of costs by the pharmaceutical companies for expenses and risks taken in the discovery and development of the patented drugs, including preclinical tests and clinical trials. By the time the patented drugs are approved by regulatory authorities, pharmaceutical companies typically have less than 10 years to recoup their investments. In contrast, companies producing the generics do not have to invest in drug discovery nor the extensive preclinical tests and clinical trials that are mandatory to bring a drug to market. They do, however, need to demonstrate bioequivalence, which costs miniscule amounts compared the expenses invested by the innovator pharmaceutical companies.

Generics came into effect through the Drug Price Competition and Patent Term Restoration Act of 1984 in the United States, generally known as the Hatch–Waxman Act, which paved the way for the entry of generics as the US Congress sought to lower drug prices for consumers by encouraging generic competition. Since then, many generics have appeared as soon as the patents of innovative drugs expired. In the United States, generics approval is via the ANDA process in accordance with the Federal Food, Drug and Cosmetic Act (FD&C Act), Section 505(j) (see Exhibit 8.2). An important criterion is that generics must show bioequivalence to the patented drug; that is, they must be absorbed into the body at a similar rate and extent.

Sources: (1) European Medicines Agency. *Guidance for Users of the centralized Procedure for Generics/Hybrid Applications*, October 2006. http://www.emea.europa.eu/pdfs/human/euleg/22541106en.pdf [accessed September 12, 2007]. (2) Woodcock J. *Statement before the Subcommittee on Health, Committee on Energy and Commerce*, May 2007. http://www.fda.gov/ola/2007/policy05022007.html [accessed September 13, 2007]. (3) European Medicines Agency. *Guideline on Biosimilar Biological Medicinal Products Containing Biotechnology-Derived Proteins as Active Substance: Quality Issues*, June 2005. (4) Woodcock J, et al. Opinion: the FDA's assessment of follow-on protein products: a historical perspective, *Nature Reviews Drug Discovery*, April 2007. (5) EPO, by any other name, editorial, *Nature* 449:259 (2007). (6) Covic A, Kuhlmann M. Biosimilars: recent developments, *International Urology and Nephrology* 39:261–266 (2007).

In the European Union, generics can be applied through an abridged procedure or centralized procedure according to Regulation 726/2004, Article 3(3). The applicant does not have to provide results of preclinical tests and clinical trials if there is evidence to demonstrate the product is a generic copy of an approved reference drug. The generic is defined as a medicinal product that:

- Contains the same qualitative and quantitative composition in active substance as the reference medicinal product
- Possesses the same pharmaceutical form as the reference medicinal product
- Has a bioequivalence that has been demonstrated by appropriate bioavailability studies

Biogenerics: Biogenerics, in some cases referred to as biosimilars and follow-on biologics, are by definition copies of off-patent biopharmaceuticals/biologics. The lack of consensus on the terms—biogenerics, biosimilars, and follow-on biologics—amply demonstrates the uncertainty in this area. The problem stems from the fact that biologics, by their very nature, are not as well defined as small molecule drugs and as such they are difficult to characterize precisely. They are sensitive to processing and storage conditions (e.g., pH, pressure, temperature, mixing speed), in addition to the nutrients and growth factors added; all of which affect how the cells grow and express the proteins. Furthermore, proteins are known to undergo posttranslational modifications (glycosylation, sulfation, acetylation, phosphorylation, etc.) that affect how the amino acids fold and interact in their 3D structures. Currently, manufacturers of protein-based drugs have to demonstrate comparability of products if they make changes to their processes. This is to show that products remain as pure, safe, and effective irrespective of the changes, and they are under control by the manufacturers.

Thus, arguments abound as to what would constitute similarity in these protein drugs to claim them as biogenerics, considering that these proteins have thousands of atoms with molecular weight often in excess of 40 kDa with complex constructs, compared to small molecules of nominally less than 0.5 kDa with relatively simple structures.

This debate is particularly pertinent as some of these first generation biopharmaceuticals produced by rDNA technology are coming off-patent, as shown in Table 10.3, and the generics industry aims to tap into this market, while at the same time regulatory authorities are under pressure to work toward making drugs more affordable to patients.

Although a number of assays and technologies are available to characterize and test protein molecules, such as peptide mapping, protein sequencing, carbohydrate analysis, electrophoresis, ELISA, and mass spectroscopy, they are not as definitive as the methods used for small molecule drugs. Hence, the test for similarity is not as well defined in the case of proteins. However, as

TABLE 10.3   Biologics Products Facing Biogenerics Challenge

Product Class	Leading Brands	Company	EU Patent Expiry	US Patent Expiry
Erythropoietin alpha	Epogen/Procrit// Eprex/Erypo	Amgen/J&J	Expired	2012
Erythropoietin beta	NeoRecormon	Roche/Wyeth/ Chugai	Expired	Expired
Interferon-β1-a	Avonex, Rebif	Biogen Idec, Serono	2012	2008 and 2013
Interferon-β1-b	Betaferon	Bayer (Schering)	Expired	Expired
Granulocyte colony-stimulating factor	Neupogen	Amgen	Expired	2013
Interferon-α-2b	Intron A	Schering Plough	Expired	Expired
Interferon-α-2a	Roferon-A	Chiron	2007	2012
Soluble TNF-α receptor	Enbrel	Amgen/Wyeth	2010	2009
TNF-α antibody	Remicade	Centocor, Schering Plough and Tanabe	2010/2011/ 2012	2011
CD20 antibody	MabThera/ Rituxan	Genentech/ Roche	2013	2015
ErbB2 receptor antibody	Herceptin	Genentech/ Roche	2014	2014
EGFR antibody	Erbitux	BMS	2010	2015
VEGF antibody	Avastin	Roche	2019	2017

Source: Ledford H. News Feature, The same but different, *Nature* 449:274–276 (2007).

technology progresses, this notion of similarity and that processes have great influence on structures is being challenged. Examples provided are the human growth hormones (hGHs) from different sources, which have all been approved and appeared to be similar in treatment efficacy: they are produced by Pharmacia and Ferring using standard *Escherichia coli*, by Eli Lilly and Novo Nordisk using a special strain of *E. coli*, and by Serono through expression in mouse cell line.

The regulatory framework is slowly changing. In the European Union, a guideline for biosimilars has been prepared for consideration: *Guideline on Biosimilar Biological Medicinal Products Containing Biotechnology-Derived Proteins as Active Substance*. The EMEA acknowledged that biosimilars are unlike generics and that such biopharmaceuticals would require stringent testing before marketing authorization. For example, for products such as recombinant EPO, a safety study and two randomized, double-blinded, placebo-controlled trials would have to be performed.

In the United States, although there is no guideline as yet, the FDA is working toward guidance under the FD&C Act for public comments. It appears the use of Section 505(b)(2), which calls for certain preclinical and clinical data, although not as extensive as for innovator drugs, may allow the FDA to review and approve biogenerics in conjunction with the PHS Act. Currently, both the EMEA and the FDA are treating each case of approval for biogenerics on a case-specific basis. In fact, in 2006, both agencies have approved the first recombinant, Omnitrope (somatropin), a human growth hormone for long-term treatment in pediatrics for growth failure and replacement therapy in adults with growth hormone deficiency. Omnitrope was approved by comparing with Gonotropin, a product approved in 1985. The approval was via the abridged procedure in the European Union and Section 505(b)(2) in the United States.

The FDA stressed its commitment to review case-specific biogenerics by considering the following:

- Evidence of integrity and consistency of the manufacturing process
- Conformance of manufacturing standards to existing regulations, if any
- Demonstration of a product's consistency with appropriate reference standards or comparators (using relevant assays), including comparative pharmacokinetic and pharmacodynamic data
- The extent to which the existing body of clinical data and experience with the approved product can be relied on

It is conceivable that, in the not too distant future, biogenerics will be approved with applicants supplying certain preclinical data and clinical trial information to prove comparability and bioequivalence. At the same time, the regulatory authorities will rely on their experience and expertise in approving innovator biologics to guide them as they evaluate biogenerics.

10.8 SUMMARY OF IMPORTANT POINTS

1. Small molecule drugs are produced using organic synthesis processes whereas large molecule drugs are derived mainly from living cells, such as microbial and mammalian cells.
2. Regulatory authorities inspect GMP facilities to ensure compliance to GMP. The FDA carries out surveillance and compliance inspections. A system-based approach is adopted: quality, facilities and equipment, materials, production, packaging and labeling, and laboratory control. Deficiencies are reported on Form FDA-483, which may lead to a warning letter and consent decree if unresolved.
3. Typically, small molecule drugs are produced using reactions such as oxidation–reduction, acid–base, halogenation, alkylation, and substitution.

The resulting drug substances, called active pharmaceutical ingredients (APIs), are recovered and purified from solvents. More recent methods aim to isolate chiral compounds to improve drug–target interactions.

4. Most large molecule drugs are produced in microbial or mammalian cell culture systems using rDNA techniques. Foreign genes that express the desired drug molecules are inserted into plasmid vectors, which are then introduced into the microbial or mammalian cells and grown in nutrient-rich media. The drug molecules are recovered through a series of chromatographic purification steps. Since the end products are sensitive to environmental factors, they are not amenable to final sterilization. As such, aseptic processes have to be employed to prevent contamination.

5. Purified drug substances are mixed with excipients into finished dosage forms: solids, liquids, parenterals, inhalants, and ointments and creams, then packaged and labeled and shipped for distribution.

6. As novel drugs come off patents, generics are produced and increasingly compete with novel drugs. Biogenerics (a misnomer term, see Section 10.7), however, due to the complexity of the drug molecules, are closely reviewed by regulatory authorities and only a handful have been approved based on individual cases.

10.9   REVIEW QUESTIONS

1. Differentiate the production processes for small and large molecule drugs.
2. Describe the system-based approach of FDA inspections, with reference to the surveillance and compliance inspection program.
3. Explain selected reaction synthesis and purification steps for small molecule drugs.
4. Discuss the importance of testing and characterizing cell lines and cell banks.
5. Why is there a need to remove bovine serum from cell culture media?
6. Distinguish the characteristics of the different types of columns for protein purification.
7. Explain what is meant by biogenerics and the regulatory status with respect to approving these drugs.

10.10   BRIEF ANSWERS AND EXPLANATIONS

1. Refer to Sections 10.1, 10.4, and 10.5: the organic synthesis route is used for small molecule drugs and living cells for large molecule drugs.

2. Refer to Section 10.3.

3. Refer to Section 10.4 and Exhibits 10.5–10.9.

4. Because living cells are complex, it is necessary to establish their history and assure that they are free of components and viruses that may have deleterious effects on the drug molecules and humans when these cells are administered with such drug molecules.

5. Refer to Exhibit 10.12.

6. Refer to Exhibit 10.14.

7. Refer to Section 10.7. The industry consensus is that biogenerics will be a reality in the near future, resulting from improvements in technology, which would be able to provide more precise characterization of proteins, and the continued pressure to drive down rising healthcare costs.

10.11   FURTHER READING

British Pharmacopoeia Commisssion, Stationery Office Books, 2006.

Butler M, ed. *Cell Culture and Upstream Processing*, Taylor & Francis, New York, 2007.

Cabri W, Di Fabio R. *From Bench to Market: The Evolution of Chemical Synthesis*, Oxford University Press, Oxford, UK, 2000.

Campbell MK, Farrell SO. *Biochemistry*, 5th ed., Thomson Brooks/Cole, Belmont, CA, 2006.

Channarayappa. *Molecular Biotechnology: Principles and Practices*, Universities Press, Hyderabad, India, 2007.

FDAnews. *Surviving an FDA inspection*, Washington Business Information, Virginia, 2001.

Food and Drug Administration. *Biotechnology Inspection Guide Reference Materials and Training Aids*, FDA, Rockville, MD, 2006.

Food and Drug Administration. *Guide to Inspections of Bulk Pharmaceutical Chemicals*, FDA, Rockville, MD, 2006.

Food and Drug Administration. *Guide to Inspections of Dosage Form Drug Manufacturers*, FDA, Rockville, MD, 2006.

Food and Drug Administration. *The Compliance Program Guidance Manual for FDA Staff: Drug Manufacturing Inspections Program 7356.002*, FDA, Rockville, MD, 2007.

Gosling JP, ed. *Immunoassays: Practical Approach*, Oxford University Press, Oxford, UK, 2000.

International Conference on Harmonization. *Derivation and Characterization of Cell Substrates Used for Production of Biotechnological/Biological Products*, ICH, 1997.

International Conference on Harmonization. *Specifications: Test Procedures and Acceptance Criteria for Biotechnological/Biological Products*, ICH, 1999.

International Conference on Harmonization. *Specifications: Test Procedures and Acceptance Criteria for New Drug Substances and New Drug Products: Chemical Substances*, ICH, 1999.

International Conference on Harmonization. *Viral Safety Evaluation of Biotechnology Products Derived from Cell Lines of Human or Animal Origin*, ICH, 1997.

King FD, ed. *Medicinal Chemistry Principles and Practice*, Royal Society of Chemistry, Cambridge, UK, 1999.

Krogsgaard-Larsen P, LilJefors T, Madsen U, eds. *Textbook of Drug Design and Discovery*, 3rd ed., Taylor & Francis, London, 2002.

Lee S, Robinson G. *Process Development: Fine Chemicals from Grams to Kilograms*, Oxford Science Publications, Oxford, UK, 1995.

Repic O. *Principles of Process Research and Chemical Development in the Pharmaceutical Industry*, Wiley, Hoboken, NJ, 1998.

Thomas G. *Medicinal Chemistry, An Introduction*, Wiley, Chichester, UK, 2000.

US Pharmacopoeia 29 and National Formulary 24, The Official Compendia of Standards, 2006.

Walsh G. *Biopharmaceutical: Biochemistry and Biotechnology*, Wiley, Chichester, UK, 1998.

Welling PG, Lasagna L, Banakar UV, eds. *The Drug Development Process: Increasing Efficiency and Cost Effectiveness*, Marcel Dekker, New York, 1996.

CHAPTER 11

FUTURE PERSPECTIVES

11.1	Past Advances and Future Challenges	360
11.2	Small Molecule Pharmaceutical Drugs	360
11.3	Large Molecule Biopharmaceutical Drugs	362
11.4	Traditional Medicine	364
11.5	Individualized Medicine	366
11.6	Gene Therapy	366
11.7	Cloning and Stem Cells	367
11.8	Old Age Diseases and Aging	369
11.9	Lifestyle Drugs	371
11.10	Performance-Enhancing Drugs	373
11.11	Chemical and Biological Terrorism	376
11.12	Transgenic Animals and Plants	376
11.13	Antimicrobial Drug Resistance	379
11.14	Regulatory Issues	380
11.15	Intellectual Property Rights	381
11.16	Bioethics	382
11.17	Concluding Remarks	384
11.18	Case Study #11	387
11.19	Further Reading	389

Drugs: From Discovery to Approval, Second Edition, By Rick Ng
Copyright © 2009 John Wiley & Sons, Inc.

11.1 PAST ADVANCES AND FUTURE CHALLENGES

Drug discovery and development underwent astounding changes in the last decade. These have been fueled by advances in many areas, especially cell and molecular biology, recombinant DNA technology, genomics, proteomics, biochemical and chemical informatics, as well as laboratory equipment and automation. In tandem with these advancements, there were changes in regulatory requirements, in order to approve drugs in an efficient manner to benefit those in need of the medication. Great strides have also been made in the harmonization of regulations to adopt some form of international regulatory standards, in the hope of lowering regulatory costs and expediting approvals. New and more efficient manufacturing technologies and processes have enabled purer and more potent drugs to be produced.

Against this backdrop of advances, the pharmaceutical industry also faces unprecedented challenges in many areas. The pipelines for new drugs are drying up, in spite of huge investments allocated for research and development. Drugs are not being discovered and developed fast enough to fill the pipeline. The prospects of generating more blockbuster drugs are not encouraging. Some new technologies, such as combinatorial chemistry and high throughput screening, have yet to live up to expectations of speeding up drug discovery. Clinical trials have become more complex, lengthy, and expensive. In addition, there are ethical, social, political, and intellectual property issues that need due consideration. There are many more diverse questions, such as gene therapy, cloning, intellectual property, sustainable biodiversity, transgenic production systems, bioterrorism, cost of treatment, and quality of life, that challenge the industry to face them squarely.

In this chapter, we discuss some of these issues and the likely course of events that may unfold in the decades ahead.

11.2 SMALL MOLECULE PHARMACEUTICAL DRUGS

There are two distinct routes to the discovery of small molecule drugs: (1) from natural products and (2) from rational design.

11.2.1 Drugs from Natural Products

Proponents of drugs from natural products argue that natural products provide a vast diversity of chemical compounds. These compounds with myriad chemical compositions and structures serve as reservoirs for many pharmacologically active lead compounds to be discovered.

As described in Exhibit 3.2, there are now regulations enacted to protect the environment with respect to natural product collection. Bioprospecting from natural habitats has to take into account the 1993 *Convention on Biologi-*

Exhibit 11.1   Marine Bioprospecting

There are 34 fundamental phyla of life: 17 occur on land and 32 in the sea (including some overlaps). More chemical diversity is found among marine life forms. Most of these are from invertebrate organisms—sponges, tunicates, and mollusks. Some of the compounds from marine life forms are extremely potent, given that these organisms have to defend themselves from attacks in vast volumes of water that dilute the compound.

The range of climatic conditions, from tropical waters to cold arctic ocean, shallow continental shelves to great ocean depths with high pressure and low oxygen content, means that there is a potentially huge supply of life forms with extensive biodiversity.

Source: Willis RC. Nature's pharma sea, *Modern Drug Discovery* 5:32–38 (2002).

cal Diversity: the sovereign rights of nations over their biological resources have to be respected.

Hitherto, the normal source of collections has been from terrestrial habitats. However, marine bioprospecting represents a vast untapped area that is likely to be intensified. Exhibit 11.1 describes the diversity and life forms in this habitat.

It can be envisioned that laboratory equipment and assay systems will continue to play crucial roles in natural product drug discovery. Although high throughput technologies have not yet delivered more approved drugs, there will continue to be a push for even higher density screening throughputs. Further miniaturization of liquid dispensing and more specific and sensitive assay systems will continue to be developed. Larger compound libraries and more comprehensive databases will provide another natural progression to widen the boundaries of chemical diversities, with the hope that drug candidates will be found within these boundaries.

11.2.2   Drugs from Rational Design

Combinations of technologies such as X-ray crystallography, NMR, bioinformatics, computational chemistry, and combinatorial chemistry have yet to realize their full potential to design safe and effective drugs with high success rates. The use of microarrays and proteomics to identify targets that cause diseases will help to define the focus on these targets. A variety of technology platforms are then used to simulate drug compound–target interactions with the aim of interrupting or diverting disease pathways.

An often-quoted limitation of rational drug design is the lack of biodiversity and chemical space (the various possible chemical compositions in nature, estimated to be as high as 10^{100} different compounds) in the libraries of compounds examined. The use of advanced software, together with artificial intelligence for simulation, is an important tool to extend the chemical space and provide a greater diversity of chemical structures for use as scaffolds to test for potential drug candidates. There are likely to be more useful rules such as Lipinski's rule (see Section 3.3.4) being implemented to test the scaffolds and functional groups to be attached. The development of intelligent, information-rich systems will be a key to the success of rational design technological platforms. Combinatorial chemistry with chiral selectivity will help to design and generate potent drugs more expeditiously. Increasing contributions will be expected to come from siRNA and systems biology as these two fields of studies develop.

Imatinib mesylate (Gleevec, Novartis), zanamivir (Relenza, GlaxoSmith-Kline), and oseltamivir (Tamiflu, Roche) are examples of drugs (Exhibits 3.11 and 3.7) that show the successful contributions of rational drug design. For example, the X-ray structure of angiotensin converting enzyme (ACE) has been reported (see Exhibit 11.2), and this may pave the way for more effective ACE inhibitors to be developed for the treatment of hypertension and heart disorders.

11.3   LARGE MOLECULE BIOPHARMACEUTICAL DRUGS

The biopharmaceutical industry for producing large molecule protein-based drugs has grown exponentially in the past 25 years (Exhibit 4.1). The knowledge base of life sciences doubles every 14 months. Molecular biology has contributed to many types of proteins being expressed in different cell systems. Manufacturing processes for producing consistent, pure, and potent biopharmaceuticals are now widely available.

To date, most approved protein-based drugs are for therapeutic or replacement therapies. They are recombinant versions of natural proteins such as insulin and erythropoietin. Their characteristics and functions are relatively well defined and known. The next phase of biopharmaceuticals, such as antibodies and vaccines, is more complex and requires more tests and characterizations. Controls for the reliability, contamination, and fidelity of expression systems will be high on the agenda in the coming decade.

The high cost of biopharmaceuticals has been due in part to the stringent requirements of aseptic manufacturing processes, process control, and stability issues of the proteins. Some efforts are likely to be directed at developing more robust protein expression systems, better control parameters for efficient and reliable manufacturing processes, and more stable formulations. Production of biopharmaceuticals using transgenic plants and animals may also lower manufacturing costs (see Section 11.12). Most forecasts predict that we are at the threshold of seeing many more effective and potent biopharmaceuticals for a variety of treatments.

Exhibit 11.2   X-Ray Structure of Angiotensin Converting Enzyme

Angiotensin converting enzyme (ACE) is an important enzyme for the regulation of blood pressure. It exists in two forms: the somatic form has 1277 amino acids, while the sperm cell form has 701 amino acids. The somatic form consists of two domains: the carboxy-terminal (C) domain and the amino-terminal (N) domain. The sperm cell form consists of only the C domain. Studies have shown that the C domain is the dominant angiotensin converting site for controlling blood pressure and cardiovascular functions.

The structure of testicular ACE (sperm cell form) was determined using X-ray crystallography. It has 27 helices, six short strands of β-structure, and six glycosylation sites. There are two highly homologous but not identical domains: the C terminal is responsible for the conversion of angiotensin I to angiotensin II and affects blood pressure, whereas the N terminal is involved in hemoregulation.

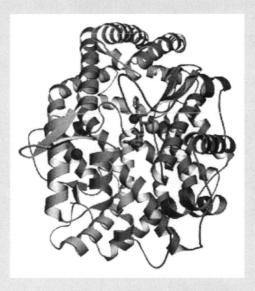

The structure with an inhibitor, lisinopril, revealed the exact nature of the binding of the drug to the active site. With this detailed analysis, more efficacious drugs are expected to be developed to bind to the active site (C terminal, such as RXPA380) and control blood pressure.

Sources: (1) Natesh R, Schwager SL, Sturrock ED, Acharya KR. Crystal structure of the human angiotensin-converting enzyme-lisinopril complex, *Nature* 421:551–554 (2003). Used with permission. (2) Turk B. Targeting proteases, successes, failures and future prospects, *Nature Reviews Drug Discovery* 5:785–799 (2006).

11.4   TRADITIONAL MEDICINE

There has been a revival in the use of traditional medicine (TM) in the last decade. TM is also called complementary or alternative medicine (CAM). It is likely that this revival trend will continue when there are still unmet needs to be filled by "Western" drugs. The quandary with Western drugs is that pharmaceutical firms have to ensure a reasonable return on shareholders' investment. With many drug discovery and development programs costing billions of dollars, most pharmaceutical firms concentrate on blockbuster drugs with large market potential. The way is thus left open for the treatment of many diseases by other means such as TM.

Another perceived view about Western drugs is the adverse effects. Western medicine is sometimes viewed as a "sledgehammer" method rather than the more holistic approach using TM. The costs of Western medicine can be prohibitive as well, especially for patients in countries where there are no comprehensive healthcare systems. For example, treatment with pegylated interferon for hepatitis C or antibody for colon cancer can cost tens of thousands of dollars per year. Limitations of Western drugs in some cases, such as resistance to antibiotics, also highlight problems and pave the way for people to consider TM. See Exhibit 11.3 on how GlaxoSmithKline and

---

Exhibit 11.3   AIDS Drugs in Developing Countries

GlaxoSmithKline, under pressure for not doing more to help the AIDS sufferers in South Africa, relented and handed over the rights to the manufacture of its AIDS drugs to a South African firm. GlaxoSmithKline waived its royalty rights and the South African drug firm will pay 30% of net sales to government organizations fighting HIV and AIDS. The drugs are the antiretroviral drugs zidovudine (AZT, Retrovir) and lamivudine (3TC, Lamivir), and a combination drug of the two, Combivir. This action will help to supply badly needed drugs to developing countries; it also sets up the precedent of differential strategies for the manufacture and supply of drugs in the world.

Source: Reuters, October 8, 2001.

Bristol-Myers Squibb licensed its AIDS drug, Reyataz, free of charge to two generic manufacturers in South Africa and India. This decision will make available the generic form of the drug to sub-Saharan Africa AIDS victims at affordable prices. Reyztaz was first marketed in the United States in 2003 and its patent expires in 2017.

Source: United Press International, February 15, 2006.

Bristol-Myers Squibb made available their AIDS drugs to developing nations.

For TM to be accepted into mainstream medical treatment, a likely scenario is the application of scientific methodologies and controls for TM development, evaluation, and production. Many of the tools for high throughput screening (HTS) and assay systems can be used to test the efficacy of TM, similar to the irrational approach of screening natural products. Pharmacology studies have to be conducted in accordance with Good Laboratory Practice (GLP).

Some regulatory authorities have foreseen the future impact of TM and set up appropriate guidelines. The European Union has legislation for traditional herbal products. Another example is the Therapeutic Goods Administration of Australia, which has set up a complementary medicine section that controls the regulatory practices for TM.

The conditions for growing and harvesting of traditional herbs are sources of variability and contamination. Good Agricultural Practice (GAP) is required to minimize contamination, such as heavy metals and fertilizers, to the raw herbs and also to improve agricultural methods for ensuring consistent levels of active ingredients. Factors such as climatic variations and processing conditions may also affect the quality of TM. Better characterization methods, as well as Good Manufacturing Practice (GMP), are needed for the scientifically based development and production of reliable, consistent, and efficacious TM.

Exhibit 11.4 shows the strategy drawn up by the World Health Organization for TM.

Exhibit 11.4   The WHO Traditional Medicine Strategy, 2002–2005

There are four objectives:

Policy: Integrate TM/CAM with national health care systems, as appropriate, by developing and implementing national TM/CAM policies and programs

Safety, Efficacy, and Quality: Promote the safety, efficacy, and quality of TM/CAM by expanding the knowledge base on TM/CAM, and by providing guidance on regulatory and quality assurance standards

Access: Increase the availability and affordability of TM/CAM, as appropriate, with an emphasis on access for poor populations

Rational Use: Promote therapeutically sound use of appropriate TM/CAM by providers and consumers

Source: World Health Organization. *WHO Traditional Medicine Strategy*, WHO, 2002–2005. http://whqlibdoc.who.int/hq/2002/WHO_EDM_TRM_2002.1.pdf [accessed October 9, 2007].

11.5 INDIVIDUALIZED MEDICINE

The challenge is that, one day, drugs will be tailor-made for individuals and adapted to each person's genetic makeup. In this way, the drugs will be used optimally and adverse events will be minimized, if not eliminated. Environment, diet, age, lifestyle, and state of health can all influence a person's response to drugs, but an understanding of an individual's genetic makeup is thought to be the key to creating personalized drugs.

An individual's blood samples can be collected and analyzed. Through the study of single nucleotide polymorphisms and pharmacogenomics, the genes that cause diseases can be pinpointed. The results will show the individual's disease condition or predisposition to some diseases. In this way, treatment or preventive measures can be prescribed.

It is envisaged that cheaper and more affordable drugs may also result as drugs are designed for specific groups of people. Clinical trials may be shortened considerably and save millions of dollars. Large patient population groups to trial a drug may no longer be required. These drugs can also be introduced faster into the market to treat people in need of these tailor-made medications.

Exhibit 11.5 shows an example of how differences in genetic make-up affect the effectiveness of a drug on individuals.

11.6 GENE THERAPY

Currently, there is still a gap for the potential of gene therapy to be fulfilled. Gene therapy clinical trials have been conducted for diseases such as severe combined immunodeficiency disease (SCID, "bubble baby" syndrome), sickle cell anemia, cystic fibrosis, familial hypercholesterolemia, and Gaucher disease.

The aim of gene therapy is to supply healthy genes to replace those that are missing or flawed. One key to the success of gene therapy is the vectors that are used to transport the genes (see Section 4.6). Another critical success factor is the understanding of the effects and functions of genes. Of the estimated 30,000 human genes, we know the functions of relatively few. Although some diseases such as sickle cell anemia and cystic fibrosis are caused by single genes, there are other diseases that may be the result of multiple gene disorders, and the relationships of these genes have to be studied.

There are two types of gene therapy: somatic cell and germ cell gene therapy. Somatic cells are nonreproductive cells and, as such, somatic cell gene therapy affects the individual only. The change in gene is not passed on to the next generation. Germ cell gene therapy involves changes to the reproductive cells, the sperm and egg, with the result that the new genes are passed on to future generations. Although most researchers support research on somatic cell gene therapy, there are differences in opinion concerning germ cell gene therapy. For example, the following are some the ethical questions:

Exhibit 11.5   Genetic Variations

Single nucleotide polymorphisms (SNPs) are DNA sequence variations among individuals. Research is under way to find out about specific SNPs (or sets of SNPs) that are associated with various medical conditions and to study the differences in SNP patterns among various human populations. It is hoped that knowledge of SNPs will improve medical treatment by enabling prediction of disease risk and response to therapies.

Source: Coronini R, et al. Decoding the literature on genetic variation, *Nature Biotechnology* 21:21–29 (2003).

In one study, researchers from Vanderbilt University in Nashville, Tennessee (USA), compared responses to a β-blocker called atenolol among 34 patients. All patients had genetic variations affecting one of the building blocks of the receptor that binds to β-blocker drugs, which affected the way the receptor responded to the binding of the drug.

Thirteen had one type of genetic variation, Gly389, and 21 had another variation, called Arg389.

Patients with the Arg389 variant achieved a significantly lower resting blood pressure and heart rate with the drug than did Gly389 patients, suggesting that the drug was more effective for them. However, this finding held true only at rest, and not during exercise.

Source: Sofowora GG, et al. A common β_1-adrenergic receptor polymorphism (Arg389Gly) affects blood pressure response to β-blockade, *Clinical Pharmacology & Therapeutics* 73:366–371 (2003).

- If gene therapy can remedy missing or faulty genes, why can it not be applied to germ cell gene therapy to stop the fault from passing onto future generations?
- Gene therapy is costly. Who decides which patient receives the therapy? Who pays for the treatment?

Exhibit 11.6 describes a recent gene therapy trial that resulted in unexpected outcomes, which the regulatory authorities have to consider. A recent report of gene therapy for treatment of Alzheimer's disease is also included.

11.7 CLONING AND STEM CELLS

We can divide cloning into therapeutic cloning and reproductive cloning. Therapeutic cloning is synonymous with stem cell research. Under proper control and environment, embryonic stem cells can potentially be directed to grow and develop into different tissues that are invaluable for replacing

Exhibit 11.6   Gene Therapy Trials

SCID Gene Therapy Trial: Infants with severe combined immunodeficiency disease (SCID, bubble boy syndrome) have a gene defect that leads to a complete lack of white blood cells. Without treatment, these infants die from complications of infectious diseases during the first few years of life. The only treatment currently approved for this condition is a bone marrow transplant.

Gene therapy offers another potential avenue to "fix" the defective gene. The therapy itself is by no means straightforward. In a French gene therapy trial, three boys with SCID were treated using retrovirus-based gene therapy. They later developed cancer and one died of leukemia. Reviews showed that the retrovirus inserted near oncogenes and promoted development of cancer.

Source: Branca MA. Gene therapy: curse or inching towards credibility? *Nature Biotechnology* 23:519–521 (2005).

Recent Gene Therapy Study: In a recent gene therapy study, a gene for the expression of a protein called neprilysin was introduced into transgenic mice. Neprilysin regulates amyloid levels, which are implicated in Alzheimer's disease (Exhibit 11.8). The results showed a 50% reduction in the levels of amyloid.

Source: Brown M. Gene therapy success for Alzheimer's? *BioMedNet*, May 2003. http://news.bmn.com/news/story?day=030501&story=2 [accessed May 20, 2003].

damaged or diseased tissues and organs. Reproductive cloning is the replication of another living being with genes from only one individual. An example is the cloning of Dolly the sheep in 1996 (Exhibit 11.7).

As discussed in Section 4.7, stem cells have the potential to treat medical conditions beyond the scope that can be offered by drugs alone. However, there are many scientific and ethical hurdles to overcome. On the scientific front, stem cell research activities will intensify over the next decade. These challenges can broadly be divided into (1) determining how to develop stem cells into specific tissues and (2) implanting these tissues into the body without rejection by the recipient's immune system. On the ethical front, it is expected that there will be more debates on the ethical issues of stem cell research. Most scientists consent to therapeutic cloning (stem cell research) but not reproductive cloning. The ethical issue of stem cell research concerns harvesting cells from embryos that are a few days old. This action destroys the embryos. Some questions are:

Exhibit 11.7   Dolly the Sheep (1996–2003)

In February 1997, the Scottish scientist Ian Wilmut and colleagues at the Roslin Institute announced the birth of a cloned sheep called Dolly in July 1996. They had removed the nucleus from the egg cell of a sheep and replaced it with the nucleus from an adult sheep. Dolly was born from a surrogate mother sheep and is an exact clone of the adult sheep, unlike offspring from the reproductive process, in which the offspring inherits the genes from both parents.

Dolly suffered from premature arthritis in 2002 and had to be put down in February 2003 at the age of 6½, because of progressive lung disease common in older sheep. It is not known whether Dolly's premature death is related to cloning; her life was about half the normal sheep lifespan of 12 years.

- What is the legal and religious status of the embryos?
- Who has the right to give informed consent?
- Under what conditions can this consent be given?

In November 2003, the members of the Europe Parliament voted to approve embryonic stem cell research, using techniques similar to that adopted for cloning Dolly the sheep, although severe restrictions were put in place. For US scientists, however, the US legislation meant that they were only allowed to performed research using 12 existing sources of the embryonic stem cells and were not allowed to create any new sources.

11.8   OLD AGE DISEASES AND AGING

We are living in an aging society. The United Nations estimates that the world will have two billion people over the age of 60 by 2050. With the aging population, there are old age diseases that we have to face and treat with more effective therapies. Hypertension, stroke, Alzheimer's disease, heart disease, type II diabetes, Parkinson's disease, and osteoporosis are some examples (Exhibit 11.8).

For some of these diseases, such as hypertension and heart disease, drugs such as ACE inhibitors and beta-blockers are available for treatment. For some other diseases, such as Alzheimer's disease, more effective drugs have yet to be discovered. For stroke, two late stage (Phase III) trials of NXY-059 and desmoteplase failed to meet the trial criteria. Other clinical trials in progress for ischemic stroke are presented in Table 11.1.

It is also expected that, in the coming decades, there will be more research on aging. Exhibit 11.9 describes some recent findings concerning aging.

Exhibit 11.8   Old Age Diseases

Hypertension: Hypertension, or high blood pressure, is the elevation of arterial blood pressure. For an adult, a systolic pressure above 140 mmHg or diastolic pressure above 90 mmHg is considered hypertension. The cause for hypertension may be due to narrowing or hardening of blood vessels, kidney diseases, or other unknown origins. Some commonly used drugs for the treatment of hypertension are angiotensin converting enzyme (ACE) inhibitors, angiotensin II receptor blockers, and calcium antagonists (calcium-channel blockers). The mechanism is such that ACE inhibitors block the enzyme from hydrolyzing angiotensin I to angiotensin II; angiotensin II receptor blockers stop the effects of angiotensin, while calcium antagonists are used to reduce heart rate and relax blood vessels.

A 5 year randomized clinical trial in the United States found that in spite of the availability of hypertensive medication, awareness promotions, and guidelines, only a third of all the hypertensive patients have their blood pressure under effective control due to noncompliance in medication. More tailored behavioral management intervention may help to improve compliance and achieve better control.

Source: Bosworth HB, et al. The take control of your blood pressure study: study design and methodology, *Contemporary Clinical Trials* 28:33–47 (2007).

Stroke: A stroke occurs when there is an interruption of blood supply to the brain. An ischemic stroke occurs when a clot prevents blood flow in the brain. A hemorrhagic stroke is when there is a rupture of a blood vessel in the brain. In either case, the brain cells in the affected area die. This area is called an infarct. Medical treatment is required to arrest the damage. More effective treatment can be administered within 6 hours of the onset of stroke. A stroke may result in weakness, paralysis, impairment of speech and memory, or even death. Medical treatment includes the use of anticoagulants to treat stroke victims.

Alzheimer's Disease: This disease is due to the accumulation of β-amyloid protein in the brain. The protein is believed to trigger brain degeneration through cell death of the neurons. Alzheimer's disease is characterized by loss of memory and intellectual performance, and slowness in thought. In the United States, a class of drugs called cholinesterase inhibitors is approved to treat Alzheimer's disease. Both Europe and the United States have approved a drug called memantine for treatment of Alzheimer's disease.

Source: Food and Drug Administration, FDA News. *FDA Approves Memantine (Namenda) for Alzheimer's Disease.*   http://www.fda.gov/bbs/topics/NEWS/2003/NEW00961.html [accessed September 27, 2007].

Diabetes: Refer to Exhibit 4.13 for a description of diabetes.

Parkinson's Disease: Parkinson's disease is a progressive neurological disorder. It is due to the degeneration of neurons in the part of the brain that controls movement. The degeneration of neurons causes a decrease in the level of dopamine, a neurotransmitter chemical necessary for the proper transmission of signals. Patients experience tremors in limbs, rigidity, difficulty in movements, and loss of facial expression. Recent clinical trials of a new drug, called glial derived neurotrophic factor (GDNF), which controls dopamine-producing nerve cells, showed conflicting results. In 2005 the Sponsor, Amgen, decided to terminate the Phase II trials due to potential possibility that the drug may harm patients.

Source: Amgen. *Following Complete Review of Phase 2 Trial Data Amgen Confirms Decision to Halt GDNF Study; Comprehensive Review of Scientific Findings. Patient Safety, Drove Decision.* http://wwwext.amgen.com/media/media_pr_detail.jsp?year=2005&releaseID=673490 [accessed September 27, 2007].

Osteoporosis: Osteoporosis is the loss of structural bony tissue and gives rise to brittleness in bones. This may lead to fractures of hips, spines, and wrists. Osteoporosis can begin at a young age if a person does not receive enough calcium and vitamin D. A person reaches maximum bone strength between 25 and 30 years of age; after that, the bone strength decreases by about 0.4% per year. After menopause, bone strength reduces by about 3% per year. Drugs such as estrogen, calcitonin, alendronate, raloxifene, and risedronate are approved for the treatment of postmenopausal osteoporosis.

11.9 LIFESTYLE DRUGS

As society becomes more affluent, there are demands for lifestyle drugs, to treat non-life-threatening conditions, or even to make a person feel more confident or look better. It is likely that there will be a proliferation of lifestyle drugs in the future.

The following are major areas for lifestyle drugs:

- Obesity treatment
- Aging: enhance muscular tone and youthful vitality, antiwrinkle
- Memory enhancement
- Sexual dysfunction

TABLE 11.1   Clinical Trials for Ischemic Stroke

Agent	Mechanism	Sponsor	Stage
Viprinex (ancrod)	Fibrinolytic	Neurobiological Technologies	Phase III (two trials)
Albumin	Free radical scavenger and other mechanisms	National Institute of Neurological Disorders and Stroke (NINDS)	Phase III
Magnesium (given prehospital)	NMDA receptor blockage and other mechanisms	NINDS	Phase III
Citicholine	Phosphatidylcholine precursor	Grupo Ferrer	Phase III
AX200 (G-CSF)	Various neuroprotection and restoration	Syngis Pharma	Phase II
Microplasmin	Fibrinolytic	ThromoGenics	Phase II
Erythropoietin and human chorionic gonadotropin	Neural stem cell proliferation and differentiation	Stem Cell Therapeutics	Phase II
Viagra (sildenafil)	Phosphodiesterase type 5 inhibitor	Mort and Brigitte Harris Foundation	Phase I
ReN001	Neural stem cell treatment	Reneuron	IND submitted

Source: Garber K. Stroke treatment—light at the end of the tunnel? *Nature Biotechnology* 25:838–840 (2007).

- Smoking cessation
- Hair loss therapy

We have witnessed lifestyle drugs in the form of orlistat (Xenical, Roche) prescribed for obesity management (Exhibit 2.11), and growth hormones have been promoted for enhancement of muscle tone and youthful vitality (Exhibit 4.11). Botulinum toxin (Botox) is being injected as an antiwrinkle treatment (Exhibit 11.10), and a vitamin A derivative (Retin-A gel, Tretinoin) is prescribed for the treatment of facial wrinkles. Drugs such as tacrine and donepezil, which work by attacking enzymes that break down acetylcholine, have been approved for boosting memory. Sildenafil (Viagra, Pfizer) is used to treat sexual dysfunction (Exhibit 3.18).

Currently, bupropion (Zyban, GlaxoSmithKline) is the only approved drug for helping cease cigarette smoking, and finasteride (Propecia, Merck) has been approved by the Food and Drug Administration (FDA) for the treatment of male baldness.

Based on the current trend, we can only expect that more lifestyle drugs will be approved in the coming decades.

Exhibit 11.9   Aging

Aging is another research area that challenges scientists to understand and perhaps devise means to slow the process. An increasing number of scientists believe that aging is due to the prolonged process of oxidative damage. It has been found that oxygen radicals attack cell proteins and membranes, with the mitochondria being the most susceptible.

An antioxidant enzyme, superoxide dismutase (SOD), breaks down oxygen radicals and renders them harmless. Fruit flies and rats with mutated genes for SOD expression live longer than normal by as much as 40%. Can this antioxidant really prolong human lifespan? There are no proven data to show that consuming copious amounts of antioxidant will help. One possible reason is that the human body can only accept a certain level of antioxidant; excessive amounts are excreted. A recent article by noted scientists in the field of aging has warned against the false belief of slowing aging by consuming hormones or antioxidants.

Sources: 1. Olshansky SJ, Hayflick L, Carnes BA. No truth to the fountain of youth, *Scientific American* June:92–95 (2002). (2) Hopkin K. Your new body—making Methuselah, *Scientific American* September:32–37 (1999).

Exhibit 11.10   Botox

Botox is a toxin produced by the bacterium *Clostridium botulinum*. When Botox is injected into facial tissues, it is absorbed by the nerve endings of muscle fibers. Nerve transmissions are interrupted and consequently the muscle relaxes. The relaxed muscle is then no longer effective to pull the facial lines to show the wrinkles.

Treatment with Botox is temporary. Once the nerve endings return to normal, the muscle will resume its contractual pull on the wrinkles.

11.10   PERFORMANCE-ENHANCING DRUGS

In this competitive world, especially in the sports field, athletes try their best to outperform each other. Unfortunately, some athletes resort to the use of performance-enhancing drugs to attain a competitive edge. Table 11.2 is a list of banned performance-enhancing drugs published by the World Anti-Doping Association (WADA) and the International Olympic Committee (IOC). This list is effective from January 1, 2007.

Despite measures such as regular screenings and threats of suspensions, some athletes continue to take risks and consume these banned drugs. It is clear that to wrestle with cases of banned drugs requires better detection

TABLE 11.2   List of Drugs Banned in Sports

Substances Prohibited at All Times		Specific Substances and Derivatives
Anabolic agents	Anabolic androgenic steroids (AAS)	
	Exogenous AAS	1-Androstenediol, 1-androstenedione, bolandiol, bolasterone, boldenone, boldione, calusterone, clostebol, danazol, dehydrochlormethyltestosterone, desoxymethyltestosterone, drostanolone, ethylestrenol, fluoxymesterone, formebolone, furazabol, gestrinone, 4-hydroxytestosterone, mestanolone, mesterolone, metenolone, methandienone, methandriol, methasterone, methyldienolone, methyl-1-testosterone, methylnortestosterone, methyltrienolone, methyltestosterone, mibolerone, nandrolone, 19-norandrostenedione, norboletone, norclostebol, norethandrolone, oxabolone, oxandrolone, oxymesterone, oxymetholone, prostanozol, quinbolone, stanozolol, stenbolone, 1-testosterone
	Endogenous AAS	Androstenediol, androstenedione, dihydrotestosterone, prasterone, testosterone and their metabolites and isomers
	Other anabolic agents	Clenbuterol, tibolone, zeranol, zilpaterol
Hormones and related substances		Erythropoietin (EPO), growth hormone (hGH), insulin-like growth factors (e.g., IGF-1), mechano growth factors (MGFs), gonadotrophins (LH, hCG—prohibited in males only), insulin, corticotropins
Beta-2 agonists		All beta-2 agonists including isomers, except formosterol, salbutamol, salmeterol, and terbutaline
Agents with antiestrogenic activity		Aromatase inhibitors (including anastrozole, letrozole, aminoglutethimide, exemestane, formestane, testolactone), selective estrogen receptor modulators—SERMs (including raloxifene, tamoxifen, toremifene), clomiphene, cyclofenil, fulvestrant
Diuretics and other masking agents		Diuretics, amiloride, bumetanide, canrenone, chlorthalidone, etacrynic acid, furosemide, indapamide, metolazone, spironolactone, thiazides, triamterene

TABLE 11.2 *Continued*

Substances Prohibited in Competition	Specific Substances and Derivatives
Stimulants	Adrafinil, adrenaline, amfepramone, amiphenazole, amphetamine, amphetaminil, benzphetamine, benzylpiperazine, bromantan, cathine, clobenzorex, cocaine, cropropamide, crotetamide, cyclazodone, dimethylamphetamine, ephedrine, etamivan, etilamphetamine, etilefrine, famprofazone, fenbutrazate, fencamfamin, fencamine, fenetylline, fenfluramine, fenproporex, furfenorex, heptaminol, isometheptene, levmethamphetamine, meclofenoxate, mefenorex, mephentermine, mesocarb, methamphetamine, methylenedioxyamphetamine, methylenedioxymethamphetamine, *p*-methylamphetamine, methylephedrine, methylphenidate, modafinil, nikethamide, norfenefrine, norfenfluramine, octopamine, ortetamine, oxilofrine, parahydroxyamphetamine, pemoline, pentetrazol, phendimetrazine, phenmetrazine, phenpromethamine, phenterimine, 4-phenylpiracetam, prolintane, propylhexedrine, selegiline, sibutramine, strychnine, tauminoheptane
Narcotics	Buprenorphine, dextromoramide, diamorphine (heroin), fentanyl and derivatives, hydromorphone, methadone, morphine, oxycodone, oxymorphone, pentazocine, pethidine
Cannobinoids	
Glucocorticosteroids	

Source: World Anti-Doping Agency. *The World Anti-Doping Code, The Prohibited 2007 List— International Standard.* http://www.wada-ama.org/rtecontent/document/2007_List_En.pdf [accessed September 10, 2007].

technology together with more stringent monitoring and legal control. Undoubtedly, stamping out banned drugs in the sports arena will be very difficult and protracted, as evidenced by the recent cases of doping scandal at the Torino 2006 Winter Olympics and the 2007 Tour de France.

11.11   CHEMICAL AND BIOLOGICAL TERRORISM

The anthrax case in the United States (11 infected, 5 died) in late 2001, ricin (a potent poison that inhibits protein synthesis) found in early 2003 in the United Kingdom and France, and sarin (an organophosphate nerve gas) poisoning in Japan in 1995 (11 deaths, 5500 people affected) highlighted that terrorism with biological and chemical materials is real. Both chemical and biological terrorism can cause tremendous medical, social, commercial, legal, and political upheavals and problems. In late 2002, the Russian authorities used a gas based on opiate fentanyl to secure the release of hostages held by Chechen rebels in a Moscow theatre.

Governments in many countries are collaborating to examine ways to improve response preparedness in the event of chemical and biological terrorism. Both the European Medicines Agency (EMEA) and the FDA have implemented counterterrorism strategies. These include streamlining of regulatory approvals for vaccines and therapeutics for diseases that could result from the use of biological weapons, such as anthrax, botulism, smallpox, plague, tularemia, and hemorrhagic fevers. A list of potential toxic chemicals has also been prepared. Other antiterrorism solutions are to improve the networks of health reports so that outbreaks can be detected early and precautionary measures taken within a short time. Effective diagnostic tests to ascertain the nature of the chemical and biological threat would need to be developed. In addition, stockpiling antibiotics and vaccines may help to mitigate terrorist attacks, although this is a complex matter because of the multitude of microorganisms and chemicals that may be deployed. Sources of funding would also be needed to develop therapies and antidotes to presumed agents of terrorism.

In June 2002, the FDA amended its regulations for the approval of certain drugs based on animal efficacy data. For those drugs that are intended to protect or treat individuals exposed to lethal or disabling toxic substances or organisms, marketing approval may be granted based on evidence of effectiveness from appropriate animal studies when human efficacy studies are not ethical or feasible. Under this "animal efficacy rule," the FDA approved pyridostigmine bromide for US military personnel. This drug increases survival rate after exposure to Soman nerve gas poisoning. Another medication approved is a lotion called Reactive Skin Decontamination Lotion, which is a liquid decontamination lotion for topical application to remove or neutralize chemical warfare agents and T-2 fungal toxin. Exhibit 11.11 provides some basic information about sarin and anthrax.

11.12   TRANSGENIC ANIMALS AND PLANTS

The production of drugs under GMP conditions is costly, especially for protein-based drugs, which require aseptic processes. Manufacturers have looked to

Exhibit 11.11   Bioterrorism Agents

Anthrax: Anthrax is a toxin with three separate components: a protective antigen (PA), an edema factor (EF), and a lethal factor (LF).

The LF is the most disruptive to cellular functions and disables intracellular signaling molecules. It prevents macrophages from releasing tumor necrosis factor (TNF) and interleukin cytokines, although the production of TNF and cytokines in the macrophages is not impeded. The host's immune system is compromised and is unable to eliminate the anthrax bacillus.

Ultimately, the macrophages die, releasing the enormous built-up stores of TNF and cytokines, triggering a septic shock-like collapse of multiple organ systems.

Scientists at the Harvard Medical School have prepared a recombinant form of a receptor. The idea is to use this cloned receptor as a decoy and mop up the anthrax molecules in circulation. This has been confirmed in laboratory experiments, but ongoing studies are needed.

Currently, the FDA approved treatment for anthrax is ciprofloxacin, doxycycline, and penicillin.

Sources: (1) Stubbs MT. Anthrax X-rayed: new opportunities for biodefence, *Trends in Pharmacological Sciences* 23:539–541 (2002). (2) Center for Disease Control and Prevention. *Anthrax Treatment*. http://www.bt.cdc.gov/agent/anthrax/faq/treatment.asp [accessed September 27, 2007].

Sarin: Pure sarin is a colorless, odorless, volatile, and highly lethal compound. It inhibits the enzyme action of cholinesterase, causing the production of excessive amounts of acetylcholine, which in turn affects the central nervous system.

Diazepam and pralidoxime iodide are prescribed for victims affected by sarin.

transgenic animals and plants as possible "factories" for the production of cost-effective protein-based drugs.

To produce protein-based drugs, DNA genes that code for the expression of the desired protein are inserted into animals or plants. These animals or plants treat the foreign DNA as part of their own genome. As the animals or plants grow, the protein is expressed. Most of the proteins are collected in milk or in eggs of animals, and in fruits or tubers of plants. The proteins are then extracted, purified, and formulated as the protein-based drugs. There are several potential issues. First, animals or plants may produce proteins that have different protein sequences, structures, and glycosylation patterns than the proteins from human origin. This would render the drug less effective. Second,

Exhibit 11.12   Edible Vaccines

Researchers from the University of Maryland in Baltimore (Maryland), the Boyce Thompson Institute for Plant Research in Ithaca (New York), and Tulane University in New Orleans (Louisiana) have performed the first human trial of edible vaccine. Potatoes were genetically engineered to produce a diarrhea-causing toxin secreted by the bacterium *Escherichia coli*.

Of the 14 volunteers for the Phase I study, 11 were given the transgenic potatoes containing the toxin as vaccine and three had ordinary potatoes. Blood and stool samples were collected from the volunteers to evaluate the vaccine's ability to stimulate both systemic and intestinal immune responses. Ten of the 11 volunteers who ingested the transgenic potatoes had fourfold rises in serum antibodies at some point after immunization, and six of the 11 developed fourfold rises in intestinal antibodies. The potatoes were well tolerated and no one experienced serious adverse effects.

Research on other edible vaccines is in the pipeline: (1) potatoes and bananas that might protect against Norwalk virus, a common cause of diarrhea, and (2) potatoes and tomatoes that might protect against hepatitis B.

The WHO met in January 2005 and set up guidelines for plant-derived vaccines, including the disposal of waste materials. The meeting concluded that the present guidelines on traditional vaccine development and evaluation can be applied to edible vaccines.

Sources: (1) MolecularFarming.com. *Molecular Farming of Edible Vaccines*. http://www. molecularfarming.com/ediblevaccine.html [accessed February 24, 2003]. (2) World Health Organization. *Report: WHO Informal Consultation on Scientific Basis for Regulatory Evaluation of Candidate Human Vaccines from Plants*, January 2005. http://www.who.int/ biologicals/Plant%20Vaccine%20Final%20Mtg%20Repor%20Jan.2005.pdf [accessed September 27, 2007].

other biological materials from the animals or plants that are potential contaminants for humans may be present, requiring stringent steps for their removal. Third, transgenic animals or plants have to be separated from natural animals and plants to prevent cross-contamination. Fourth, both the animals and plants have to be kept under close surveillance to ensure they are free of diseases.

Recent work on the potential of edible vaccines is described in Exhibit 11.12. In August 2006 the European Union approved for marketing A-Tryn, a recombinant form of human antithrombin (anticoagulant), which is produced in the milk of genetically engineered goats, showcasing the transgenic production of drugs.

11.13 ANTIMICROBIAL DRUG RESISTANCE

Antimicrobials, including antibiotics, are drugs used to destroy or slow the growth of microorganisms in our body. The mechanisms of action of these drugs on the microbes can be classified as follow:

- Interference with DNA synthesis
- Interference with protein synthesis
- Interference with cell wall synthesis
- Interference with cell membrane permeability
- Inhibition of enzyme(s) of the microorganism

Unfortunately, through misuse or overuse of antimicrobial drugs, such as suboptimal dosage and noncompliance in administrating the medication, the microbes are incompletely eradicated. As microbes survive, they grow and reproduce by dividing every few hours. Soon they evolve, mutate, and adapt to the new environment, where the progeny become resistant to antimicrobials. This is a natural selection process where the microbes with the resistance genes survive and pass them on to future generations.

For example, methicillin-resistant *Staphylococcus aureus* (MRSA) is a strain of bacteria that has become resistant to methicillin. In healthy individuals, the *S. aureus*, though present, does not cause active infection. But in the immune suppressed and the elderly, *S. aureus* infection can result in morbidity and mortality. Worse still is the fact that MRSA is spread in healthcare institutions and community centers such as hospitals, medical centers, nursing homes, childcare centers, gymnasiums, and confined living quarters. The terms HA-MRSA and CA-MRSA refer to hospital-associated and community-associated MRSA.

Enterococci bacteria, although less common than *S. aureus*, can infect hospitalized patients, complicate diseases, and prolong hospital stays. A particular strain that is vancomycin resistant (VRE) can be fatal and accounts for one-third of the infections in intensive care units.

Hygiene and sanitation play an important role in the transmission of microbe infections. At the same time, new and more effective antimicrobials are being developed:

- Dalfopristin binds to bacterial ribosome and inhibits protein synthesis.
- Linezolid stops protein synthesis in ribosome by inhibiting formation of initiation complex.
- Daptomycin causes membrane depolarization in bacteria and prevents membrane transport.
- Oritavancin binds to normal cell wall precursors and inhibits cell wall synthesis.
- Platensimycin inhibits a pathogen's ability to synthesize fatty acids, which are essential components of cell membranes (in development).

11.14   REGULATORY ISSUES

Regulatory requirements are dynamic. They are introduced or amended as circumstances change. The amendment by the FDA to the animal efficacy rule (Section 11.11) and the initiative for a new risk-based approach to cGMP (Section 9.7) are examples of dynamic responses to changing environment and conditions. In Europe, the recognition and adoption of regulatory controls for traditional medicine is another positive development. The formation of the International Conference on Harmonization (ICH) in harmonizing drug regulations, such as the ICH Q7 document, has helped to establish a common denominator for regulatory requirements of GMP for many countries. The other common technical documents and guidelines also assist in streamlining regulatory processes. Table 11.3 is a list of drugs published by the WHO, which

TABLE 11.3   Pharmaceuticals: Restrictions in Use and Availability

Monocomponent Products

Acetylsalicylic acid	Adalimumab
Amphetamine	Anagrelide
Aristolochic acid	Astemizole
Benfluorex	Benzbromarone
Bicalutamide	*Camelia sinesis*
Celecoxib	Cisapride
Codeine	Danazol
Ephedra	Epoetin alfa
Levacetylmethadol	Loratadine
Muromonab-CD3	Natalizumab
Nefazodine hydrochloride	Nevirapine
Nimesulide	Oseltamivir
Paroxetine	Phenylpropanolamine
Promethazine	Repaglinide
Rofecoxib	Salmeterol
Valdecoxib	Vanlafaxine hydrochloride

Combination Products

Abacavir, Lamivudine, Tenofovir	Atazanavir-Ritonavir
Benziodarone and benzbromarone-allopurinol	Didanosine, Lamivudine, Tenofovir
Paracetamol-dextropropoxyphene	Rifampicin and Pyrazinamide

Group Products

Cyclooxygenase-2 inhibitor	Dietary supplements
Herbal medicines	Nonsteroidal anti-imfammatory drugs (NSAIDs)
Statins	

Source: World Health Orga nization. *Pharmaceuticals: Restrictions in Use and Availability*. http://www.who.int/medicines/publications/restrictedpharm2005.pdf[accessed October 3, 2007].

Exhibit 11.13   FDA's Oversight on Gene Therapy and Cancer Vaccine

The FDA has not yet approved for sale any human gene therapy product. However, gene-related research and development is continuing to grow and the FDA is very involved in overseeing this activity. Since 1989, the FDA has received about 300 requests from medical researchers and manufacturers to study gene therapy and to develop gene therapy products. Presently, the FDA is overseeing approximately 210 active Investigational New Drug (IND) gene therapy studies.

Several gene therapies have received orphan drug designation; these are treatments for cystic fibrosis, Gaucher disease, and metastatic brain tumor.

Source: Food and Drug Administration, Center for Biologics Evaluation and Research. *Cellular & Gene Therapy*. http://www.fda.gov/cber/gene.htm [accessed September 21, 2007].

While the FDA is adopting a cautious approach to cancer vaccine, such as DCVax-Brain, the Swiss Institute of Public Health has conditionally allowed the use of this vaccine by patients. DCVax consists of a patient's dendritic cells that have been pulsed with antigens derived from a tumor cell lysate prepared from surgically resected glioblastoma (brain cancer) tissue. It was developed by a company in the United States but has not yet been approved by the FDA.

Source: Fox JL. Uncertainty surrounds cancer vaccine review at FDA, *Nature Biotechnology* 25:827–828 (2007).

have restrictions in use and availability because of reasons such as being banned, withdrawn, restricted in use, or not approved by certain governments in different countries.

As gene therapy and stem cell research progress, we can expect more regulatory requirements to be developed to ensure proper safeguards are implemented. Similarly, xenotransplantation and control of biopharmaceutical products will experience specific regulatory controls as new advances are made. Exhibit 11.13 presents the FDA's current oversight on gene therapy and its cautious approach to cancer vaccine.

11.15   INTELLECTUAL PROPERTY RIGHTS

Intellectual property rights (IPRs) are an important asset for the pharmaceutical industry. Most successful companies have suites of patents to protect their IPRs. The effect of IPR protection is amply demonstrated by the Prozac case. Fluoxetine (Prozac, Eli Lilly) was a blockbuster drug for many years. During

Exhibit 11.14   Prilosec's Legal Battle

The AstraZeneca patent on omeprazole expired in October 2001. AstraZeneca went to court to seek extended patent protection for a special formulation of omeprazole. The special formulation consists of a subcoating layer inserted between the core of the drug's active ingredient and the outer coating. The subcoating is formulated to protect the drug from being broken down quickly by the harsh acids in the stomach.

In October 2002, a US federal judge ruled that three generic companies had infringed on AstraZeneca's patent. However, a fourth company that has its own patent for coating the drug was cleared to market the drug in generic form.

Source: Debaise C. *Wall Street Journal*, October 12, 2002.

the first half of 2001, when Prozac was still protected by patent, the sales were US$1.3 billion. Within one year of patent expiration, generics from other companies were released to the market. The sales of Prozac were reduced to US$380 million for the first half of 2002.

Pharmaceutical companies are adopting strategies to protect their products. These range from new claims for the drugs, reformulations, or isolation of the effective enantiomer (see Section 3.6). Product life-cycle management of drugs and IPRs protection are strategies that pharmaceutical firms will focus on more closely in the coming decades. This is especially the case where pharmaceutical firms try to exclude the encroachment of generics for as long as possible. We describe in Exhibit 11.14 the legal battle between AstraZeneca, the manufacturer of Prilosec, and other companies trying to manufacture omeprazole generics. A brief description of the FDA's rules for generics is presented in Exhibit 11.15 to explain how the prevailing system works.

It should be noted that at times pharmaceutical firms have to waive their IPRs for a certain segment of the market due to commercial or political issues. This is shown in Exhibit 11.16. Another recent case is the decision by Novartis to exit from the Indian market when its intellectual property right for Gleevec was not recognized by the Indian government.

11.16   BIOETHICS

Bioethics is a discipline that deals with the moral issues of biological research and its application in medicine. The United Nations Educational, Scientific and Cultural Organization (UNESCO) in its 2005 Universal Declaration on Bioethics and Human Rights addresses ethical issues related to medicine, life sciences, and associated technologies as applied to human beings. The aim of the declaration is to "provide a universal framework of principles and procedures

Exhibit 11.15   The FDA's Rule on Generics

To encourage generic production, the FDA allows submission of an Abbreviated New Drug Application (ANDA) by a generic manufacturer before the expiration of the patented drug. For the FDA to commence review of the ANDA, the generic drug applicant must certify that the patent for the existing drug is invalid and the generic product does not infringe it. The generic drug applicant must also notify the patent holder that it has filed the ANDA. If the patent holder files an infringement suit against the generic applicant within 45 days of the ANDA notification, the patent holder is given a one-time "stay" of a generic drug's entry into the market for resolution of a patent challenge unless, before that time, the patent expires or is determined to be invalid or not infringed. This 30 month stay gives the patent holder time to assert its patent rights in court before a generic competitor is permitted to enter the market.

Source: Center for Drug Evaluation and Research. *FDA Generic Drugs Final Rule and Initiative*, FDA, Rockville, MD, 2003. http://www.fda.gov/oc/initiatives/generics/default.htm [accessed September 21, 2007].

Exhibit 11.16   Sleeping Sickness in Africa

Sleeping sickness is caused by the presence of parasitic protozoans (*Trypanosoma gambiense* or *T. rhodesiense*) in the blood. The parasite causes drowsiness, lethargy, and eventually death if the patient is untreated. As many as 55 million people in 36 African countries are exposed to this disease. In the late 1990s, the WHO persuaded companies such as Bristol-Myers Squibb, Aventis, and Bayer to waive their patent rights and donate drugs for treating sleeping sickness to Africans. In addition, these companies agreed to contribute US$5 million per year for 5 years for monitoring, treatment, and research and development of drugs for the treatment of sleeping sickness.

Source: Wickware P. Resurrecting the resurrection drug, *Nature Medicine* 8:908–909 (2002).

to guide States in the formulation of their legislation, policies or other instruments in the field of bioethics."

In many areas, bioethics will continue to pose vexing questions that communities and governments must face. In September 2007, the Human Fertilization and Embryology Authority (HFEA) of the United Kingdom approved the research of hybrid embryos, which involves the insertion of human DNA into animal (cow or rabbit) cells where the genetic material has been removed. Proponents argue that the research is critical to discovering treatments for genetic diseases such as Alzheimer's and Parkinson's, as there is a lack of human eggs for research. Opponents, however, dispute the ethicality of hybrid embryos, claiming research would lead to the formation of modified humans from traces of animal genes that may remain in the cell, even though the embryo is only allowed to grow for up to 2 weeks. Where should the boundary of medical research be drawn? Any suggestion is sure to provoke arguments, but necessary to be defined.

11.17   CONCLUDING REMARKS

A reflection on the last 100 years shows the tremendous progress made in drug discovery and development. With better sanitation and healthcare, life expectancy has increased from an average in the 50s at the beginning of the 20th century, to an average in the 80s for developed nations today. However, there are startling differences in the life expectancy and disease types in rich and poor nations and these require the attention of governments and the WHO. Table 11.4 shows the leading causes of death in rich and poor countries. These poorer nations are still affected by diseases that are well controlled, eradicated, or treated in the developed countries.

The drug market continues to expand, even during times of economic downturn. With the introduction of many new technologies and processes, we can expect more effective and specific drugs in the decades ahead. However, a worrying trend has also appeared. It is the declining number of new drugs approved in the past 5 years, in spite of the billions of dollars spent on research and development (see Fig. 1.2). In 2006, the FDA approved 18 new small molecule drugs—new molecular entities (NMEs) or new chemical entities (NCEs). In the same period, only 4 large molecule drugs were approved under the biologics license applications (BLAs) (Fig. 11.1). An insight into drug regulations in the future can be seen from the FDA's long-term objectives:

- Sustain availability of safe and effective new and generic products by improving rapid, transparent, and predictable science-based review of marketing applications.
- Increase the number of safe and effective new products available to patients, including products for unmet medical and public health needs, emerging infectious diseases, and counterterrorism.
- Improve the safe and effective use of medical products with better information technology and effective risk/benefit communication.
- Prevent harm from products by increasing the likelihood of detection and interception of substandard manufacturing processes and products.

TABLE 11.4   Leading Causes of Death in the World by Broad Income Group, 2002

Cause	Number of Deaths (in millions)
High Income Countries	
Coronary heart disease	1.34
Stroke and other cerebrovascular diseases	0.77
Trachea, bronchus, lung cancers	0.46
Lower respiratory infections	0.34
Chronic obstructive pulmonary disease	0.30
Colon and rectal cancers	0.26
Alzheimer's and other dementias	0.22
Diabetes mellitus	0.22
Breast cancer	0.15
Stomach cancer	0.14
Total	*4.20*
Middle Income Countries	
Stroke and other cerebrovascular diseases	3.02
Coronary heart disease	2.77
Chronic obstructive pulmonary disease	1.57
Lower respiratory infections	0.69
HIV/AIDS	0.62
Perinatal conditions	0.60
Stomach cancer	0.58
Trachea, bronchus, lung cancers	0.57
Road traffic accidents	0.55
Hypertensive heart disease	0.54
Total	*11.51*
Low Income Countries	
Coronary heart disease	3.10
Lower respiratory infections	2.86
HIV/AIDS	2.14
Perinatal conditions	1.83
Stroke and other cerebrovascular diseases	1.72
Diarrheal diseases	1.54
Malaria	1.24
Tuberculosis	1.10
Chronic obstructive pulmonary disease	0.88
Road traffic accidents	0.53
Total	*16.94*

Source: World Health Organization. *The Top 10 Causes of Death (2002), March 2007.* http://www. who.int/mediacentre/factsheets/fs310/en/[accessed September 23, 2007].

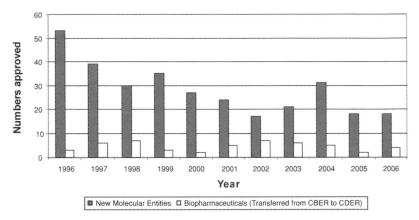

Figure 11.1 New molecular entities and biopharmaceutical approvals by the FDA.

- Improve the infrastructure for problem detection and product information dissemination, to strengthen consumer protection and take timely, effective risk management actions with all FDA-regulated products.

Pharmaceutical firms have to reexamine their strategies to devise means to increase their drug pipelines for continuous streams of products. The high failure rates of Investigational New Drugs during clinical trials (see Exhibit 5.8) necessitate the development of better assay systems and animal models that correlate closely with human pharmacodynamics and pharmacokinetics. The study of pharmacogenomics will be crucial to address this issue.

In addition, the pharmaceutical industry is no longer isolated from society as a whole. It has to factor in consideration for social, political, and ethical issues. The attitudes and concerns of society have to be taken into account. The cost of drugs and their availability to developing nations are other important aspects for consideration. Furthermore, pharmaceutical firms need to be proactive and vigilant to comply with changes in regulations and government policies. New areas of research, such as gene therapy and stem cells, have opened up many ethical issues to be resolved. Production of protein-based drugs using transgenic animals and plants raises potential topics for ethical and political debates.

These challenges require investment, commitment, and ingenious solutions, as shown in Fig. 11.2. Measures such as licensing, alliance, mergers, acquisitions, and outsourcing are needed to help companies and focus on core competencies. Ultimately, pharmaceutical firms and research organizations, together with government authorities, international organizations, and representatives from society, have to collaborate and address these challenges. The results will be the development of novel and efficacious drugs and therapies to treat patients and improvements in the quality of life.

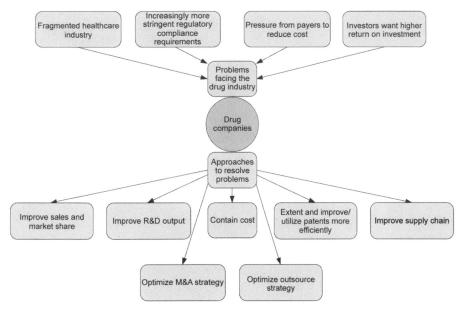

Figure 11.2   Challenges facing the pharmaceutical companies. (*Source*: Adapted from Belsey, MJ. Drug developer strategies to boost competitiveness, *Nature Reviews Drug Discovery*, 6:265–266 (2007).)

11.18   CASE STUDY #11

Alzheimer's Disease

There is still no effective drug for the treatment of Alzheimer's disease. To understand the strategy used to develop a drug for its treatment, a necessary condition rests with an understanding of the cause of the disease.

The key features of Alzheimer's disease are the accumulation of β-amyloid (Aβ) protein outside neurons in the brain and the presence of tangled proteins, called tau, within neurons. It is found that Aβ provides the initial insult to set off the disease.

Recent work shows further understanding:

- The amyloid precursor protein (APP) protrudes from the neuron membrane.
- Both proteases, β-secretase and γ-secretase, cut out Aβ from APP, using aspartic acid as a catalyst in a two-stage process.
- There are three genes believed to be involved in the cutting of the amyloids: one gene encodes for the APP while two encode the enzymes β-secretase and γ-secretase.
- The aggregation of Aβ outside the neuron somehow alters tau inside the neuron, perhaps through a mechanism that involves the kinase proteins.

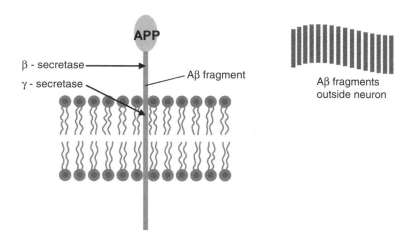

From this knowledge, several probable solutions for developing drugs are being attempted to find a treatment for this disease.

Small Molecule Drug Solution

- Develop a protease inhibitor: there is no known inhibitor for β-secretase yet; an inhibitor of γ-secretase is known and a trial is currently in progress.
- Design a drug to prevent the aggregation of Aβ outside the neuron.
- Devise antagonist to block kinases, thus mitigating the effect of Aβ on tau.

Large Molecule Drug Solution

- Provide vaccine to act prophylatically: use Aβ or its fragments as an antigen to elicit antibody response. A conducted trial showed antibodies were generated as predicted but for unknown reasons patients suffered encephalitis; other vaccines are contemplated which may use different parts or fragments of Aβ to evoke an immune response and to reduce the encephalitis adverse reaction (see Exhibit 4.6).
- Generate therapeutic antibodies to target Aβ for destruction, similar to the example of Herceptin's mechanism of action; but it may be difficult to deliver large protein molecules across the blood–brain barrier of the tightly packed endothelial cells.
- Use gene therapy to surgically implant "healthy" genes with nerve growth factor to cure the three errant genes.

With concerted efforts it is hopeful that better and more effective medication for the treatment of Alzheimer's disease will be available in the not too distant future.

11.19 FURTHER READING

Adams GP, Weiner LM. Monoclonal antibody therapy of cancer, *Nature Biotechnology* 23:1147–1157 (2005).

American College of Physicians. *New Antibiotic: When to Use and When Not to Use.* http://www.acponline.org/ear/vas2002/new_antibiotics.htm [accessed October 18, 2007].

Atun RA, Sheridan D, eds. *Innovation in the Biopharmaceutical Industry*, World Scientific, Singapore, 2007.

Burnett JC, et al. The evolving field of biodefence: therapeutic development and diagnostics, *Nature Reviews Drug Discovery* 4:281–297 (2005).

Cohen JC, Illingworth P, Schuklenk U, eds. *The Power of Pills: Social, Ethical and Legal Issues in Drug Development, Marketing, and Pricing*, Pluto, London, 2006.

Dibner MD. Biopharmaceuticals in 2005 and beyond, *BioExecutive International* January:2–8 (2005).

Dove A. Uncorking the biomanufacturing bottleneck, *Nature Biotechnology* 20:777–779 (2002).

Drews J. Strategic trends in the drug industry, *Drug Discovery Today* 8:411–420 (2003).

Food and Drug Administration, Office of Management Budget Formulation and Presentation. *Human Drugs Performance Goals.* http://www.fda.gov/oc/oms/ofm/budget/2008/Detail/2B-HumanDrugsPerf.htm *Biologics Performance Goals.* http://www.fda.gov/oc/oms/ofm/budget/2008/Detail/2C-BiologicsPerf.htm [accessed September 29, 2007].

Fleming E, Ma P. Drug life-cycle technologies, *Nature Reviews Drug Discovery* 1:751–752 (2002).

Franz S. Screening the right candidate, *Nature Reviews Drug Discovery*. http://www.nature.com/drugdisc/nj/articles/nrd1082.html [accessed October 6, 2007].

Gershell L, Atkins JH. A brief history of novel drug discovery technologies, *Nature Reviews Drug Discovery* 2:321–327 (2003).

Gibbs WW. A diabetes switch? *Scientific American* June:16 (1999).

Happier, hornier, hairier, *Nature Biotechnology* 21:1 (2003).

Harvey AL, ed. *Advances in Drug Discovery Techniques*, Wiley, Hoboken, NJ, 1998.

Hopkin K. Your new body—making Methuselah, *Scientific American* September:32–37 (1999).

Jones D. Anticoagulants, the clot thickens, *Nature Reviews Drug Discovery* 2:251 (2003).

Jurgen D. *In Quest of Tomorrow's Medicine*, Springer-Verlag, New York, 1999.

Larru M. Adult stem cells: an alternative to embryonic stem cells? *Trends in Biotechnology* 19:487 (2001).

Larvol BL, Wilkwerson LJ. *In silico* drug discovery: tools for bridging the NCE gap, *Nature Biotechnology* 16:33–34 (1998).

Natesh R, Schwager SL, Sturrock ED, Acharya KR. Crystal structure of the human angiotensin-converting enzyme–lisinopril complex, *Nature* 421:551–554 (2003).

Olshansky SJ, Hayflick L, Carnes BA. No truth to the fountain of youth, *Scientific American* June:92–95 (2002).

Pence GE. *Classic Cases in Medical Ethics*, McGraw-Hill, New York, 2004.

Roden DM, George AL Jr. The genetic basis of variability in drug responses, *Nature Reviews Drug Discovery* 1:37–44 (2002).

Schwartz L, Preece PE, Hendry RA. *Medical Ethics, A Case-Based Approach*, Saunders, Edinburgh, 2002.

Seeberger PH, Blume T. *New Avenues to Efficient Chemical Synthesis: Emerging Technologies—Proceedings 06.3*, Springer-Verlag, Berlin, 2007.

Stubbs MT. Anthrax X-rayed: new opportunities for biodefence, *Trends in Pharmacological Sciences* 23:539–541 (2002).

Ulrich R, Friend SH. Toxicogenomics and drug discovery: Will new technologies help us produce better drugs? *Nature Reviews Drug Discovery* 1:84–88 (2002).

Walsh G. Biopharmaceutical benchmarks 2006, *Nature Biotechnology* 24:769–776 (2006).

Walters WP, Namchuk M. Designing screens: how to make your hits a hit, *Nature Reviews Drug Discovery* 2:259–266 (2003).

Walters WP, Stahl MT, Murcko MA. Virtual screening—an overview, *Drug Discovery Today* 3:160–178 (1998).

Wechsler J. The push for generics challenges manufacturers, *Pharmaceutical Technology* July:26–34 (2003).

Willis RC. Nature's pharma sea, *Modern Drug Discovery* 5:32–38 (2002).

World Anti-Doping Agency and International Olympic Committee. *Prohibited Classes of Substances and Prohibited Methods 2003*, WADA, 2003.

World Health Organization. *WHO Traditional Medicine Strategy*, WHO, 2002. http://www.who.int/medicines/library/trm/trm_strat_eng.pdf [accessed March 20, 2003].

APPENDIX 1

HISTORY OF DRUG DISCOVERY AND DEVELOPMENT

A1.1 Early History of Medicine   391
A1.2 Drug Discovery and Development in the Middle Ages   394
A1.3 Foundation of Current Drug Discovery and Development   394
A1.4 Beginnings of Modern Pharmaceutical Industry   395
A1.5 Evolution of Drug Products   396
A1.6 Further Reading   397

A1.1 EARLY HISTORY OF MEDICINE

Drug discovery and development has a long history and dates back to the early days of human civilization. In those ancient times, drugs were not just used for physical remedies but were also associated with religious and spiritual healing. Sages or religious leaders were often the administrators of drugs. The early drugs or folk medicines were derived mainly from plant products and supplemented by animal materials and minerals. These drugs were most probably discovered through a combination of trial and error experimentation and observation of human and animal reactions as a result of ingesting such products.

Although these folk medicines probably originated independently in different civilizations, there are a number of similarities, for example, in the use

Drugs: From Discovery to Approval, Second Edition, By Rick Ng
Copyright © 2009 John Wiley & Sons, Inc.

of the same herbs for treating similar diseases. This is likely to be a contribution by ancient traders, who in their travels might have assisted the spread of medical knowledge.

Folk medicines were the only available treatments until recent times. Drug discovery and development started to follow scientific techniques in the late 1800s. From then on, more and more drugs were discovered, tested, and synthesized in large-scale manufacturing plants, as opposed to the extraction of drug products from natural sources in relatively small batch quantities. After World War I, the modern pharmaceutical industry came into being, and drug discovery and development following scientific principles was firmly established.

Although pharmaceutical drugs are now widely used worldwide, many ethnic cultures have retained their own folk medicines. In certain instances, these folk medicines exist side by side and are complemented by pharmaceutical drugs.

The following are some snapshot examples of how drugs were discovered from the early human civilizations.

A1.1.1 Chinese Medicine

Traditional Chinese medicine (TCM) is believed to have originated in the times of the legendary emperor Sheng Nong in 3500 BC. The dynasty system and meticulous recording have helped to preserve the TCM scripts of old China. Some important medical writings are *Shang Han Lun* (Discussion of Fevers), *Huang Di Nei Jing* (The Internal Book of Emperor Huang), and *Sheng Nong Ben Cao Jing* (The Pharmacopoeia of Sheng Nong—a legendary emperor). Exhibit A1.1 relates a legend about the discovery of a herb for treating injuries.

The Chinese pharmacopoeia is extensive. Some of the active ingredients from Chinese herbs have been used in Western drugs; for example, reserpine

---

Exhibit A1.1   A Legend about San Qi

Chinese legend describes the story of the emperor Sheng Nong, who one day tried to kill a snake by beating it. The snake returned a few days later, apparently none the worse after the beating. He beat it again and left it mortally injured. Again, the snake returned several days later. This time, after the beating, he observed that the snake crawled back into the bush and ate a plant material. This plant is now called San Qi (*Panax notoginseng*) and is used for treating external injuries. It is an ingredient for the well-known TCM herbal formula known as Yunnan Bai Yao.

Source: Reid D. *Chinese Herbal Medicine*, Shambhala Publications, Boston, 1996.

from *Rauwouofia* for antihypertensive and emotional and mental control, and the alkaloid ephedrine from *Mahuang* for the treatment of asthma.

A1.1.2 Egyptian Medicine

Ancient papyrus provided written records of early Egyptian medical knowledge. The Ebers papyrus (from around 3000 BC) provided 877 prescriptions and recipes for internal medicine, eye and skin problems, and gynecology. Another record, from the Kahun papyrus of around 1800 BC, detailed treatments for gynecological problems. Medications were based mainly on herbal products such as myrrh, frankincense, castor oil, fennel, sienna, thyme, linseed, aloe, and garlic.

A1.1.3 Indian Medicine

The Indian folk medicine, called Ayurvedic medicine, can be traced back 3000–5000 years and was practiced by the Brahmin sages of ancient times. The treatments were set out in sacred writings called Vedas. The material medica are extensive and most are based on herbal formulations. Some of the herbs have appeared in Western medicines, such as cardamom and cinnamon. Susruta, a physician in the fourth century AD, described the use of henbane as antivenom for snakebites.

A1.1.4 Greek Medicine

Some of the Greek medical ideas were derived from the Egyptians, Babylonians, and even the Chinese and Indians. Castor oil was prescribed as a laxative; linseed or flaxseed was used as a soothing emollient, laxative, and antitussive. Other treatments include fennel plant for relief of intestinal colic and gas, and asafetida gum resin as an antispasmodic. The greatest Greek contribution to the medical field is perhaps to dispel the notion that diseases are due to supernatural causes or spells. The Greeks established that diseases result from natural causes. Hippocrates, the father of medicine, at about 400 BC is credited with laying down the ethics for physicians. Exhibit A1.2 describes the mythology of Asclepius, the Greek God of Medicine.

A1.1.5 Roman Medicine

As great administrators, the Romans instituted hospitals, although these were used mainly to cater to the needs of the military. Through this work, organized medical care was made available. The Romans also extended the pharmacy practice of the Greeks. Dioscorides and Galen were two noted physicians in Roman days. Dioscorides's *Materia Medica* contains descriptions of treatments based on 80% plant, 10% animal, and 10% mineral products.

Exhibit A1.2   Asclepius: Greek God of Medicine

In Greek mythology, Asclepius, the god of medicine, studied medicine under Chiron. He excelled over Chiron, and his medical skills were reputed to be able to bring back the dead. This incurred the wrath of Pluto, the god of the underworld, and the envy of other gods. They complained to Zeus, who also thought that he alone should have the power of life and death. Zeus slew Asclepius with a thunderbolt. However, Asclepius's daughters, Panacea and Hygeia, survived and carried on to tend to the sick.

Source: Leadbetter   R.   *Asclepius.*   http://www.pantheon.org/articles/a/asclepius.html [accessed May 31, 2001].

A1.2   DRUG DISCOVERY AND DEVELOPMENT IN THE MIDDLE AGES

The Middle Ages, from around AD 400 to 1500, witnessed the decline of the Roman influence. This was also the time when plagues scourged many parts of Europe. Diseases such as bubonic plague, leprosy, smallpox, tuberculosis, and scabies were rampant. Many millions of people succumbed to these diseases.

A1.2.1   The Early Church

There are some references to herbs in the Bible. However, the Church's main contribution to medicines is the preservation and transcription of Greek medical manuscripts and treatises. This enabled the knowledge developed in the ancient times to be continued and later used in the Renaissance period.

A1.2.2   Arabian Medicine

Through trade with many regions, the Arabians learned and extended medical knowledge. Their major contribution is perhaps the knowledge of medical preparations and distillation methods, although the techniques were probably derived from the practices of alchemists. Avicenna, around AD 900–1000, recorded a vast encyclopedia of medical description and treatment. Another noted physician was Rhazes, who accurately described measles and smallpox.

A1.3   FOUNDATION OF CURRENT DRUG DISCOVERY AND DEVELOPMENT

The Renaissance period laid the foundation for scientific thoughts in medicinal preparations and medical treatments. There were many advances made in

> **Exhibit A1.3   Edward Jenner's Smallpox Vaccine**
>
> In the late 1700s, Jenner heard that people who worked with cattle and had caught the cowpox disease (a mild disease related to smallpox) were immune and never caught smallpox. In 1796, he proceeded to inoculate a boy using the fluid from the blister of a woman with cowpox. The boy developed cowpox. Two months later, Jenner inoculated the boy with fluids from the blister of a smallpox sufferer. The boy became immune and did not get smallpox.
>
> Through his work, Jenner invented vaccination and saved many lives. However, in today's regulatory control, Jenner's method would not have been approved.
>
> *Source*: BBC. *Medicine Through Time, Edward Jenner (1749–1823)*. http://www.bbc.co.uk/education/medicine/nonint/indust/dt/indtbi2.shtml [accessed May 25, 2002].

anatomy, physiology, surgery, and medical treatments, including public health care, hygiene, and sanitation.

A1.3.1   Smallpox

In 1796, Edward Jenner successfully experimented with smallpox inoculations (Exhibit A1.3). This paved the way for the use of vaccination against some infectious diseases.

A1.3.2   Digitalis

In the late 1700s, William Withering introduced digitalis, an extract from the plant foxglove, for treatment of cardiac problems.

A1.3.3   Scurvy

John Hunter (1768) noted that scurvy was caused by the lack of vitamin C. He prescribed the consumption of lemon juice to treat scurvy.

A1.3.4   Rabies

Louis Pasteur (1864) discovered that microorganisms cause diseases, and he devised vaccination against rabies. This was achieved through the use of attenuated rabies virus.

A1.4   BEGINNINGS OF MODERN PHARMACEUTICAL INDUSTRY

Despite the advances made in the 1800s, there were only a few drugs available for treating diseases at the beginning of the 1900s:

- *Digitalis:* Extracted from a plant called foxglove, digitalis stimulates the cardiac muscles and was used to treat cardiac conditions.
- *Quinine:* Derived from the bark of the Cinchona tree, quinine was used to treat malaria.
- *Ipecacuanha:* Extracted from the bark or root of the Cephaelis plant, ipecacuanha was used to treat dysentery.
- *Aspirin:* Extracted from the bark of willow tree, aspirin was used for the treatment of fever.
- *Mercury:* This was used to treat syphilis.

More systematic research was being performed to discover new drugs from the early 1900s.

Paul Ehrlich used an arsenic compound, arspheamine, to treat syphilis. Gerhard Domagh found that the red dye Prontosil was active against streptococcal bacteria. Later, French scientists isolated the active compound to be sulfanilamide, and this gave rise to a new range of sulfa drugs against hosts of bacteria.

A1.4.1 Penicillin

In 1928, Alexander Fleming discovered that *Penicillium* mold was active against staphylococcus bacteria. Ernst Chain rediscovered this fact some 10 years later, when he collaborated with Howard Florey. By 1944, large-scale production of penicillin was available through the work of Howard Florey and Ernst Chain. This work foreshadowed the commencement of biotechnology, where microorganisms were used to produce drug products. A description of the discovery and large-scale manufacturing of penicillin is given in Exhibit A1.4.

A1.5 EVOLUTION OF DRUG PRODUCTS

In the early days, until the late 1800s, most drugs were based on herbs or extraction of ingredients from botanical sources.

The synthetic drugs using chemical methods were heralded at the beginning of the 1900s, and the pharmaceutical industry was founded. Many drugs were researched and manufactured, but mostly they were used for therapeutic purposes rather than completely curing the diseases.

From the early 1930s, drug discovery concentrated on screening natural products and isolating the active ingredients for treating diseases. The active ingredients are normally the synthetic version of the natural products. These synthetic versions, called new chemical entities (NCEs), have to go through many iterations and tests to ensure they are safe, potent, and effective.

Exhibit A1.4   The Development of Penicillin

In the 1930s, Howard Florey and Ernst Chain worked with a team of scientists at Oxford University in Britain. Ernst Chain discovered an earlier paper by Alexander Fleming on the antibacterial properties of penicillin.

The Florey–Chain team's investigation showed that penicillin interferes with the cell wall of bacteria. Bacteria cells ruptured instead of continuing to grow. In 1938, their animal test, on eight mice given lethal doses of infectious bacteria, showed stunning results. The four mice with penicillin survived, whereas four controls with no medication died. Their first human patient who suffered from infection showed early improvement with penicillin, but died subsequently when the stock of penicillin was exhausted.

The team worked on the technology for large-scale production of penicillin. Commercial quantities were available before the end of World War II and saved millions of lives, especially soldiers wounded in the war.

Source: Torok S. *Howard Florey—The Story: Maker of the Miracle Mould.* http://www.abc. net.au/science/slab/florey/story.htm [accessed June 2, 2002].

In the late 1970s, development of recombinant DNA products utilizing knowledge of cellular and molecular biology commenced. The biotechnology industry became a reality.

The pharmaceutical industry, together with the advances in gene therapy and understanding of mechanisms of causes of diseases, and the research results from the Human Genome Project have opened up a plethora of opportunities and made possible the development and use of drugs specifically targeting the sites where diseases are caused.

A1.6   FURTHER READING

Smith CG, O'Donnell JT, eds. *The Process of New Drug Discovery and Development*, 2nd ed., Informa Healthcare, 2006.

The Next Pharmaceutical Century: Ten Decades of Drug Discovery, 2007. http://pubs. acs.org/journals/pharmcent/Ch10.html [accessed September 29, 2007].

APPENDIX 2

CELLS, NUCLEIC ACIDS, GENES, AND PROTEINS

A2.1	Cells	398
A2.2	Nucleic Acids	400
A2.3	Genes and Proteins	404
A2.4	Further Reading	410

A2.1 CELLS

Cells are the basic units for all living organisms. All cells are bounded by a membrane, and bacterial and plant cells have a cell wall. The membrane protects the cell from the outside environment. It consists of a lipid bilayer (Fig. A2.1). The function of the membrane is to control materials that enter and exit the cell and enable biochemical reactions to take place within the cell.

A2.1.1 Prokaryote Cell

Simple single-cell organisms, such as bacteria and blue-green algae, are called prokaryotes (see Fig. A2.2). Prokaryotes do not have a well-defined nucleus.

The genetic material, deoxyribonucleic acid (DNA), is concentrated in the nuclear region. DNA controls the functions of the cell. Ribosomes, granular

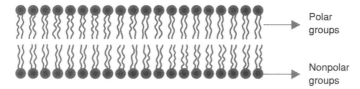

Figure A2.1   Lipid bilayer cell membrane.

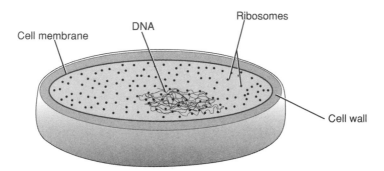

Figure A2.2   A prokaryotic cell.

structures that consist of ribonucleic acid (RNA) and proteins, are distributed in the cytosol (soluble part of the cell excluding the nuclear region).

Prokaryote cells divide and grow into two daughter cells. In the division process, the DNA replicates and each daughter cell receives one copy.

A2.1.2   Eukaryote Cell

Complex multicellular cells, such as those of plants and humans, are termed eukaryotes. The cell structure is considerably more complex than that of the prokaryote cells (see Fig. A2.3 for a human eukaryote cell; plant cells are not shown: they have a well-defined cell wall and different structure).

Within the cell membrane is the cytoplasm. This is where many biochemical reactions take place. The most important structure within the human cell is the nucleus. It is bounded by a nuclear membrane and is separated from other organelles (noncellular structures in a cell that serve specific functions) in the cytoplasm.

DNA is organized into strands within the chromosomes inside the nucleus. There are 46 chromosomes in the human cell, 23 from maternal (egg) and 23 from paternal (sperm) origin (see Exhibit A2.1). All human cells contain a full set of chromosomes and identical genes. However, different sets of genes are expressed, or turned on, in different cells, leading to the various types of cells, such as nerve cells and muscle cells.

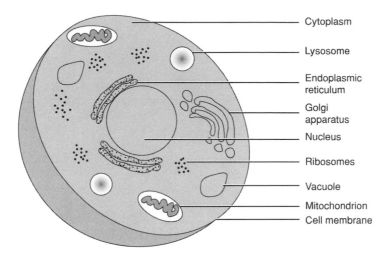

Figure A2.3   A human eukaryotic cell.

The following organelles are in the cytoplasm:

- *Mitochondrion:* The powerhouse of the cell, this is where oxidation processes take place to provide energy for the cell.
- *Endoplasmic Reticulum (ER):* A single membrane system of two distinct types, rough and smooth. The rough ER has ribosomes attached to the membrane, whereas the smooth ER does not.
- *Ribosomes:* These are sites where protein synthesis takes place.
- *Golgi Apparatus:* This is involved in gathering and dispatching of proteins and lipids.
- *Lysosomes:* These membrane-bound sacs are filled with enzymes for processing nutrients.
- *Vacuole:* This is a space within the cytoplasm that consists of wastes and materials taken in by the cell, for example, bacteria engulfed by a white blood cell.

A2.2   NUCLEIC ACIDS

A2.2.1   DNA

DNA is a polymer composed of monomeric nucleic acids called nucleotides. A nucleotide consists of a nitrogenous base, sugar, and phosphoric acid (whereas a nucleoside consists of only the base and sugar; see Fig. A2.4).

There are two types of bases: pyrimidines and purines (Fig. A2.5). The pyrimidine bases include cytosine, thymine, and uracil. Cytosine is found in

Exhibit A2.1   Cells and Chromosomes

Cells can be divided into germ cells and somatic cells. Germ cells are reproductive cells, for example, ova or sperm. Germ cells contain genetic characteristics that are passed on to the next generation. Somatic cells do not contribute their genes to future generations; they are the tissue cells such as nerve cells and muscle cells.

Within the cell is the nucleus with the chromosomes. DNA strands are housed within the chromosomes, together with some proteins. The 46 human chromosomes are grouped into 22 pairs and two sex chromosomes. Numbering of chromosomes is based on size, chromosome 1 being the largest and 22 the smallest.

In addition to the 22 pairs, a female cell contains two X chromosomes and a male cell contains an X and a Y chromosome. When a female egg (carrying an X chromosome) combines with male sperm having an X chromosome, a female offspring is born. When the egg combines with a sperm having a Y chromosome, a male offspring results.

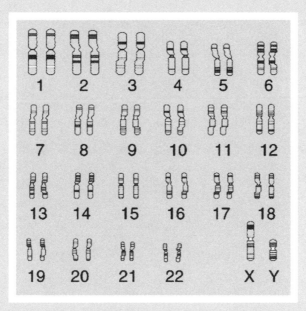

Source: Access Excellence Resource Center. *Understanding Gene Testing.* http://www.access-sexcellence.org/AE/AEPC/NIH/gene03.html [accessed January 30, 2003].

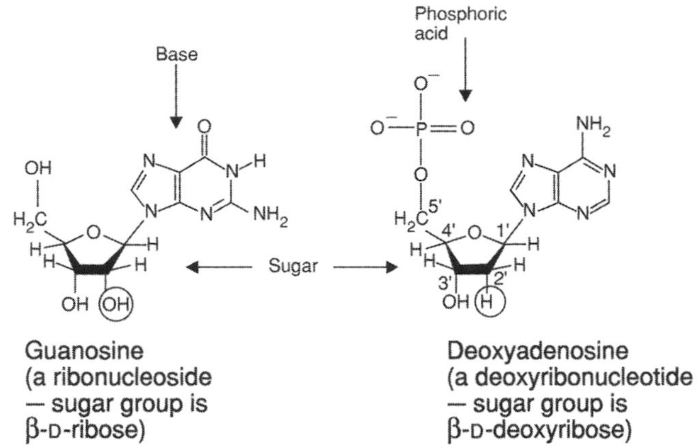

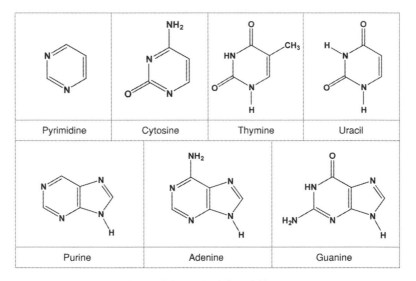

Figure A2.4   Nucleoside and nucleotide. Circled areas show the presence and absence of oxygen atom in the ribose and deoxyribose sugars.

Figure A2.5   Nucleic acid bases.

both DNA and RNA. Thymine only occurs in DNA, and uracil is substituted for thymine in RNA. The purine bases are adenine and guanine, both of which are found in DNA and RNA.

Nucleotides are joined into a chain formation, as illustrated in Fig. A2.6a. In DNA, two nucleotide chains intertwine around each other in a double helix formation (Fig. A2.6b). The backbone of the two strands is the phosphate–sugar linkage.

Alignment of the two strands is via the interactions of the bases: adenine (A) of one strand pairs up with thymine (T) of the complementary strand

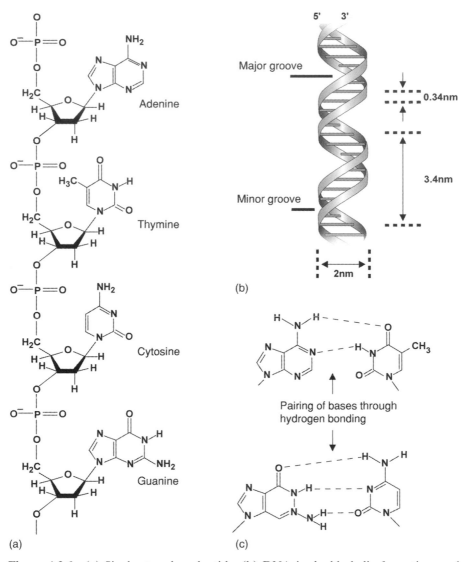

Figure A2.6 (a) Single strand nucleotide, (b) DNA in double helix formation, and (c) the pairing up of bases on two separate strands via hydrogen bonding to form the double helix.

(with two hydrogen bonds); similarly, guanine (G) pairs up with cytosine (C) (with three hydrogen bonds) as shown in Fig. A2.6c.

There are 10 base pairs in a complete turn of the helix, which spans a distance of 3.4 nm. The outside diameter of the helix is about 2 nm. By convention, the double stranded DNA sequence is written from left to right; the 5′ end (position 5 of the sugar group) is assigned to the top left-hand strand.

TABLE A2.1   The 20 Naturally Occurring Amino Acids

Name	Abbreviation	Name	Abbreviation
Alanine	Ala	Leucine	Leu
Arginine	Arg	Lysine	Lys
Asparagine	Asn	Methionine	Met
Aspartic acid	Asp	Phenylalanine	Phe
Cysteine	Cys	Proline	Pro
Glutamic acid	Glu	Serine	Ser
Glutamine	Gln	Threonine	Thr
Glycine	Gly	Tryptophan	Trp
Histidine	His	Tyrosine	Tyr
Isoleucine	Ile	Valine	Val

A2.2.2   RNA

RNA is made up of nucleotides similar to DNA, except that the sugar is β-D-ribose compared to DNA's β-D-deoxyribose (Fig. 2.4). There are two-stranded RNAs, but normally RNA exists in single strand form.

There are three kinds of RNA: messenger RNA (mRNA), transfer RNA (tRNA), and ribosomal RNA (rRNA). All three RNAs are involved in the synthesis of proteins using amino acids.

Information for making a protein is passed from the DNA to the mRNA. This is likened to the master copy (DNA) of a building plan residing in a document room (nucleus) being photocopied onto a duplicate (mRNA). The duplicated plan (mRNA) is then taken to a building site (ribosome) for protein construction. DNA determines the nucleotide sequence of the mRNA; the process of transferring the order of sequence is called transcription. Amino acids (there are 20 naturally occurring amino acids; see Table A2.1) for the construction of protein are brought to the ribosome by tRNAs. The role of rRNA is to combine with protein to form ribosomes, the site where protein synthesis takes place. The order of amino acids in the protein is controlled by mRNA via a process called translation. A schematic representation of the protein synthesis process is shown in Fig. A2.7.

A2.3   GENES AND PROTEINS

A2.3.1   Genes

Genes are our hereditary units. Each gene contains the instruction for the synthesis of protein. The summation of all the genes within a cell is called the genome. Instructions for synthesis of proteins are stored in the DNA through specific arrangements of the base sequence. The sequence is made up of the four bases: adenine, thymine, guanine, and cytosine (A, T, G, and C). These instruction codes are arranged in three-letter words called codons. Each word

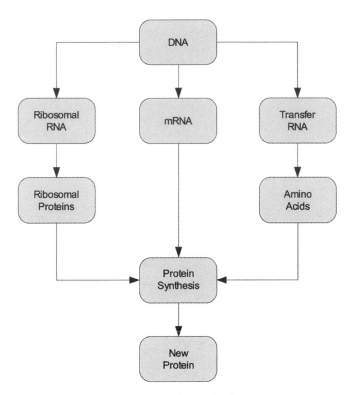

Figure A2.7   Protein synthesis process.

specifies which amino acid is to be used for constructing the proteins. Genes are not found in a continuous fashion in the DNA sequence. Sequences that are expressed (used to make proteins) are called exons. Intervening sequences, which do not code for proteins, are called introns. Promoters and repressors are present along the DNA sequence to control the expression or suppression of a gene. Information from the exons is transcribed from DNA to mRNA. There are many ways to process transcription, and so one gene can code for multiple versions of mRNA, leading to multiple proteins.

To transcribe information from DNA to mRNA, one strand of the DNA is used as a template. This is called the anticoding, or template, strand and the sequence of mRNA is complementary to that of the template DNA strand (Fig. A2.8) (i.e., C→G, G→C, T→A, and A→U; note that T is replaced by U in mRNA). The other DNA strand, which has the same base sequence as the mRNA, is called the coding, or sense, strand. There are 64 ($4 \times 4 \times 4$) possible triplet codes of the four bases; 61 are used for coding amino acids and three for termination signals. As there are 20 amino acids for the 61 codes, some triplets code for the same amino acid. A table of the genetic code is presented in Exhibit A2.2.

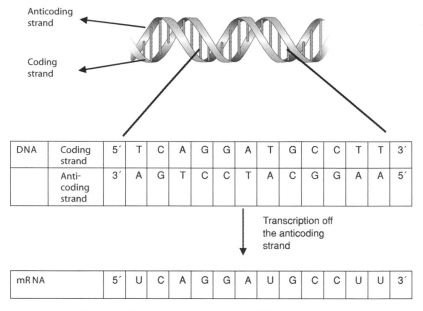

Figure A2.8 Transcription from DNA to mRNA.

Exhibit A2.2   The Genetic Code Dictionary

The triplet codon genetic codes of mRNA are translated into amino acids as shown below.

UUU→Phe	UCU→Ser	UAU→Tyr	UGU→Cys
UUC→Phe	UCC→Ser	UAC→Tyr	UGC→Cys
UUA→Leu	UCA→Ser	UAA→Stop	UGA→Stop
UUG→Leu	UCG→Ser	UAG→Stop	UGG→Trp
CUU→Leu	CCU→Pro	CAU→His	CGU→Arg
CUC→Leu	CCC→Pro	CAC→His	CGC→Arg
CUA→Leu	CCA→Pro	CAA→Gln	CGA→Arg
CUG→Leu	CCG→Pro	CAG→Gln	CGG→Arg
AUU→Ile	ACU→Thr	AAU→Asn	AGU→Ser
AUC→Ile	ACC→Thr	AAC→Asn	AGC→Ser
AUA→Ile	ACA→Thr	AAA→Lys	AGA→Arg
AUG→Met	ACG→Thr	AAG→Lys	AGG→Arg
GUU→Val	GCU→Ala	GAU→Asp	GGU→Gly
GUC→Val	GCC→Ala	GAC→Asp	GGC→Gly
GUA→Val	GCA→Ala	GAA→Glu	GGA→Gly
GUG→Val	GCG→Ala	GAG→Glu	GGG→Gly

Source: Klegerman ME, and Groves MJ. *Pharmaceutical Biotechnology, Fundamentals and Essentials*, Interpharm Press, Buffalo Grove, IL, 1992.

Exhibit A2.3   Examples of Selected Genetic Disorders

Down Syndrome: People with Down syndrome have 3 chromosomes, a trisomy, of chromosome 21. The result is lower cognitive ability and physical stature. The disorder also causes affected people to have a higher incidence of heart, intestinal, and thyroid problems.

Edwards' Syndrome: This is a genetic disorder with 3 chromosomes in chromosome 18 due to an abnormality in the egg or sperm before conception. Most fetuses do not survive and about 50% die *in utero*, and of those born, only 5–10% survive beyond one year. Babies have abnormally shaped head and other physical characteristics. In addition, sufferers of Edwards' syndrome have heart defects.

Klinefelter's Syndrome: This condition affects males where there are 3 chromosomes, XXY, instead of XY in the sex chromosome. These males are normally infertile and have some female characteristics. There may be a greater risk in developing germ cell tumors.

Turner's Syndrome: Instead of the XX female sex chromosome, female sufferers have only one X chromosome. Most fetuses do not survive. Those born are infertile and carry congenital defects in heart, kidney, and thyroid.

In situations where there are errors in the chromosomes or genes, genetic disorders may result as shown in Exhibit A2.3.

A2.3.2 Proteins

Proteins are the workhorses in our bodies and carry out all the essential processes and functions. They may come in the forms of enzymes for catalyzing reactions, hormones for transmitting information between cells, receptors for receiving signals, and antibodies for defending us from invading organisms. Proteins are made up of chains of amino acids. Amino acids have a carboxyl group (COO^-) at one end and an amino group (NH_3^+) at another. Peptide bonds are formed by joining one carboxyl group with an amino group (Fig. A2.9).

The translation of genetic information from the mRNA data in Fig. A2.8 gives rise to the amino sequence shown below, using the translation table from Exhibit A2.2 as a guide:

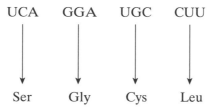

UCA	GGA	UGC	CUU
Ser	Gly	Cys	Leu

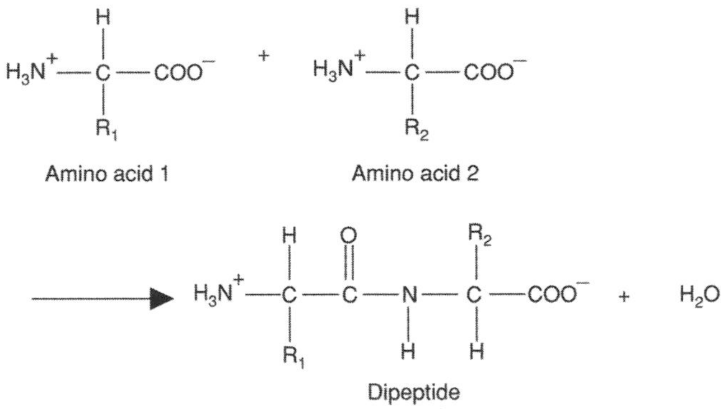

Figure A2.9   Formation of a peptide bond.

Through the formation of polypeptide bonds between amino acids, very long chains of sequences are obtained. Generally, proteins consist of hundreds and thousands of amino acids. For example, human hemoglobin has four polypeptide chains, of which two are α-chains and two are β-chains. There are 141 amino acids in each α-chain with a sequence of

Val Leu Ser Pro Ala···Thr Ser Lys Tyr Arg

The β-chain has 146 amino acids with the sequence

Val His Leu Thr Pro···Ala His Lys Tyr His

Proteins are not linear molecules. From the linear sequences (primary structures) of amino acids, the interactions of various components of the amino acids via H-bonding, disulfide bonding, and electrostatic influences can result in the proteins forming into helices or sheets (secondary structures). They can further become folded into large three-dimensional structures called tertiary structures. These tertiary structures may aggregate into even larger units through noncovalent bonding and give rise to quaternary structures (Figs. A2.10 and A2.11).

A2.3.3   Human Genome Project, Genomics, and Proteomics

The Human Genome Project was launched in 1990. It is a US$3 billion project involving 350 laboratories around the world. In 2001, a draft copy of the sequence of the three billion base pairs was published. By April 2003, 99% of the human genome had been sequenced. The project goals are as follows:

- *Identify* all the approximately 30,000 genes in human DNA
- *Determine* the sequences of the three billion chemical base pairs that make up human DNA
- *Store* this information in databases

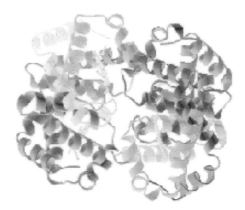

Figure A2.10   Protein structure: human deoxyhemoglobin. (*Source*: Protein Data Bank, PDB ID: 1A3N. Tame J, Vallone B. http://www.rcsb.org/pdb/cgi/explore.cgi?job=graphics;pdbId=1A3N;page=0&opt=show&size=250 [accessed April 16, 2003]. Used with permission.)

Figure A2.11   Protein structure: HIV-1 protease complexed with a tripeptide inhibitor. (*Source*: Protein Data Bank, PDB ID:1A30, Louis JM, Dyda F, Nashed NT, et al. Hydrophilic peptides derived from the transframe region of Gag-Pol inhibit the HIV-1 protease, *Biochemistry* 37:2105(1998). http://www.rcsb.org/pdb/cgi/explore.cgi?job=graphics;pdbId=1A30;page=0;pid=45021053070530&opt=show&size=250 [accessed April 16, 2003]. Used with permission.)

- *Improve* tools for data analysis
- *Transfer* related technologies to the private sector
- *Address* the ethical, legal, and social issues that may arise from the project

Scientists are continuing to determine the 30,000–40,000 genes that control our lives. Genomics helps scientist to understand biochemistry in our body and thereby develop better drugs to treat diseases.

The real value in the genome sequence is to find out the regions of the genome that encode proteins. Proteomics, the study of the structures and functions of proteins, further enhances our understanding of proteins and their functions, leading to insights on how they are affected in normal and disease conditions. Exhibit A2.3 shows some of the medical conditions due to genetic problems.

A2.4 FURTHER READING

Campbell JJ. *Understanding Pharma: A Primer on How Pharmaceutical Companies Really Works*, Pharmaceutical Institute, Raleigh, NC, 2005.

Campbell MK, Farrell SO. *Biochemistry*, 5th ed., Thomson Brooks/Cole, Belmont, CA, 2006.

SELECTED DRUGS AND THEIR MECHANISMS OF ACTION

Generic Name (Brand Name)	Class of Drug	Mechanism of Action
Acyclovir (Zovirax)	Antiviral agent	Inhibits the replication of viral DNA
Alendronate (Fosamax)	Treatment for osteoporosis	Binds to hydroxyapatite in bone and inhibits osteoclast-mediated bone resorption
Amoxicillin/clavulanate (Amoxil/Augmentin)	Antibiotic	Inhibits bacterial cell wall synthesis
Atenolol (Tenormin)	α-Adrenergic receptor blocker	Competitive blocker of α-adrenergic receptors in heart and blood vessels
Atorvastatin (Lipitor)	Antilipidemic agent	Inhibits the enzyme HMG-CoA reductase and reduces the biosynthesis of cholesterol
Candesartan (Atacand)	Angiotensin II receptor antagonist	Acts as an angiotensin II receptor antagonist
Celecoxib (Celebrex)	Anti-inflammatory, COX-2 inhibitor	Inhibits the synthesis of prostaglandins via the selective inhibition of the enzyme cyclooxygenase-2

Generic Name (Brand Name)	Class of Drug	Mechanism of Action
Chloroquin (Aralen)	Antimalarial	Inhibits protein synthesis by inhibiting DNA and RNA polymerase
Cimetidine (Tagamet)	H_2 receptor antagonist	Blocks H_2 receptor and reduces secretion of gastric acid and pepsin output
Cisplatin (Platinol)	Platinum-containing anticancer agent	Binds to DNA and prevents separation of the helical strands
Clomipramine (Anafranil)	Tricyclic antidepressant	Affects neuronal transmissions
Codeine (Codral Forte)	Narcotic analgesic	Binds to opiate receptors and blocks pain pathway
Diazepam (Valium)	Antianxiety agent, hypnotic	Acts as central nervous system depressant
Digoxin (Lanoxin, Lanoxicaps)	Cardiac glycoside	Inhibits Na/K/ATPase, increases intracellular calcium, and increases ventricular contractibility
Diphenhydramine (Benadryl)	H_1 receptor blocker	Blocks the actions of histamine on H_1 receptor
Doxazosin (Cardura)	α-Adrenergic blocker, antihypertensive agent	Blocks α_1-adrenergic receptor, resulting in decreased blood pressure
Fluoxethine (Prozac)	Selective serotonin reuptake inhibitor	Inhibits reuptake of 5-hydroxytryptamine (serotonin) into central nervous system neurons
Ibuprofen (Motrin)	Nonsteroidal anti-inflammatory	Inhibits cyclooxygenase, inhibition of inflammatory mediators
Interferon alpha (Roferon-A)	Antineoplastic	Inhibits replication of viruses or tumor cells
Lamivudine (Epivir, Epivir-HBV)	Antiviral	Inhibits HIV reverse transcriptase and DNA polymerase
Loratadine (Claritin)	H_1 receptor blocker	Antagonizes histamine effects
Methylphenidate (Ritalin)	Therapeutic agent for attention deficit hyperactivity disorder	Blocks reuptake of norepinephrine
Omeprazole (Prilosec)	Gastric proton pump inhibitor	Inhibits H^+K^+-ATPase
Paclitaxel (Taxol)	Antineoplastic	Inhibits tumor cell division

Generic Name (Brand Name)	Class of Drug	Mechanism of Action
Ramipril (Altace)	Angiotensin converting enzyme (ACE) inhibitor	Inhibits ACE, decreases peripheral arterial resistance
Tamoxifen (Nolvadex)	Antiestrogen agent	Inhibits DNA synthesis by binding to estrogen receptors on tumor cells
Zidovudine, azidothymidine, AZT (Retrovir, Combivir)	Antiretroviral agent	Inhibits HIV replication by blocking reverse transcriptase

Source: Ehrenpreis S, Ehrenpreis ED. *Clinician's Handbook of Prescription Drugs*, McGraw Hill, New York, 2001. (Used with permission from The McGraw Hill Companies.)

APPENDIX 4

A DHFR PLASMID VECTOR

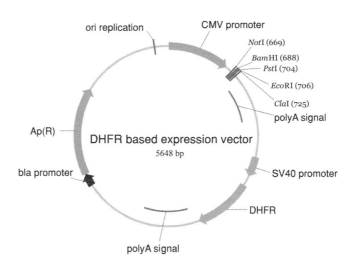

This DHFR-based expression vector has 5468 base pairs of double stranded DNA. The functions of the components are:

- *ori Replication:* This is the starting point for DNA replication.
- *CMV Promoter:* Promoter for the expression of gene of interest in mammalian cells; it allows for the binding of RNA polymerase to initiate transcription.

Drugs: From Discovery to Approval, Second Edition, By Rick Ng
Copyright © 2009 John Wiley & Sons, Inc.

TABLE A4.1 Restriction Endonucleases

Restriction Endonuclease	Bacteria	Sequence Recognized[a]	Ends[a]
NotI	*Nocardia otitis-caviarum*	GC^GGCCGC	Sticky
BamHI	*Bacillus amyloliquefaciens*	G^GATCC	Sticky
PstI	*Providencia stuartii 164*	CTGCA^G	Sticky
EcoRI	*Escherichia coli RY13*	G^AATTC	Sticky
ClaI	*Caryophanon latum L*	AT^CGAT	Sticky

[a]The symbol ∧ represents a cleavage site and "sticky end" means a stretch of unpaired nucleotides at the end of a DNA molecule.

Source: Dr. Matthias Brand, private communication (July 2007).

- *NotI—ClaI:* Multiple cloning sites for cloning of foreign genes of interest through the use of restriction endonucleases inserted at the locations shown; inserted foreign genes are joined to the plasmid through the use of ligases (see Table A4.1 for each type of endonuclease and the sequence recognized).

- *PolyA Signal:* This is the polyadenylation signal for attachment of the polyA tail to generate mature mRNA; it is important for transcription termination.

- *SV40 Promoter:* This is the promoter for expression of the *DHFR* gene.

- *DHFR:* The enzyme dihydrofolate reductase (DHFR) is required for nucleotide (thymine) synthesis and cell growth and serves as a selection marker in mammalian cells.

- *bla Promoter:* This is the promoter for expression of the beta-lactamase gene.

- *Ap(R):* This selectable marker gene is for the resistance of ampicillin (beta-lactamase), which degrades penicillins to enable selection of the desired DHFR vector.

Table A4.1 presents the origins and sequence recognition sites for the restriction endonucleases in the plasmid vector.

APPENDIX 5

VACCINE PRODUCTION METHODS

Diseases[a]	Vaccines	Methods of Production
Dipthera[1]	DTaP, DTaP/Hib, Tdap	Cell based
Hemophilus influenza type B (Hib)[1]	Hib, DTaP/Hib, Hib/Hepatitis	Cell based
Hepatitis A[2]	Hepatitis A (pediatric)	Cell based
Hepatitis B[2]	Hepatitis B preservative free (pediatric/adolescent)	Cell based
Hepatitis B[2]	Hepatitis B 2 dose (adolescent)	Cell based
Hepatitis B[2]	Hib/Hepatitis B	Cell based
Influenza (Flu)[2]	Influenza	Egg based
Measles[2]	MMR, measles	Egg based
Meningococcal[1]	Meningococcal conjugate (quadravalent)	Cell based
Meningococcal[1]	Meningococcal polysaccharide (quadravalent)	Egg based
Mumps[2]	MMR, Mumps	Egg based
Pertussis (whooping cough)[1]	DTaP, DTaP/Hib, Tdap	Cell based
Pneumococcal[1]	Pneumococcal conjugate (7-valent)	Cell based
Poliomyelitis (polio)[2]	IPV (Inactivated poliovirus vaccine) OPV (Oral polio vaccine)	Cell based

Drugs: From Discovery to Approval, Second Edition, By Rick Ng
Copyright © 2009 John Wiley & Sons, Inc.

Diseases[a]	Vaccines	Methods of Production
Rotavirus[2]	Rotavirus	Cell based
Rubella (German measles)[2]	MMR, rubella	Egg based
Tetanus[1]	DTaP, DTaP/Hib, Tdap	Cell based
Varicella (chickenpox)[2]	Varicella	Egg based

[a]Causative agent: 1 = bacteria, 2 = virus.

Source: Center for Disease Control and Prevention. VFC: Approved Vaccines and Biologicals. www.cdc.gov/vaccines/programs/vfc/parents/apprvd-vaccs.htm [accessed September 8, 2007].

Fertilized chicken eggs have been used for more than 50 years to produce vaccines. However, this method is slow; also, some virus may be deadly to the chicken embryos and hence kill off the production machine. Cell-based production is now used as it is more efficient and amenable to better control.

APPENDIX 6

PHARMACOLOGY/TOXICOLOGY REVIEW FORMAT

Information for the following pages is adapted from the Food and Drug Administration (2001), *Guidance for Reviewers, Pharmacology/Toxicology Review Format*. It provides in standardized formats the information that reviewers examine in reviewing Investigational New Drugs (INDs) and New Drug Applications (NDAs).

Pharmacology

Primary pharmacodynamics
    Mechanism of action
    Drug activity related to proposed indication
Secondary pharmacodynamics
Pharmacology summary
Pharmacology conclusions
Safety pharmacology
    Neurological effects
    Cardiovascular effects
    Pulmonary effects
    Renal effects

Drugs: From Discovery to Approval, Second Edition, By Rick Ng
Copyright © 2009 John Wiley & Sons, Inc.

    Gastrointestinal effects
    Abuse ability
    Other
Safety pharmacology summary
Safety pharmacology conclusions

Pharmocokinetics/Toxicokinetics

Pharmacokinetic parameters
    Absorption
    Distribution
    Metabolism
    Excretion
    Other studies
Pharmacokinetic/toxicokinetic summary
Pharmacokinetic/toxicokinetic conclusions

Toxicology

Study title
Key study finding
Methods
Dosing
    Species/strain
    Number/sex/group or time point (main study)
    Satellite groups used for toxicokinetics or recovery
    Age
    Weight
    Doses in administered units
    Route, form, volume, and infusion rate
Observations and times
    Clinical signs
    Body weights
    Food consumption
    Ophthalmoscopy
    Electrocardiography
    Hematology
    Clinical chemistry
    Urinalysis
    Gross pathology

    Organs weighed
    Histopathology
    Toxicokinetics
    Other
  Results
    Mortality
    Clinical signs
    Body weights
    Food consumption
    Ophthalmoscopy
    Electrocardiography
    Hematology
    Clinical chemistry
    Urinalysis
    Organ weights
    Gross pathology
    Histopathology
    Toxicokinetics
Summary of individual findings
Toxicology summary
Toxicology conclusions

Genetic Toxicology

Study title
Key findings
Methods
  Strains/species/cell line
  Dose selection criteria
    Basis of dose selection
    Range finding studies
  Test agent stability
  Metabolic activation system
  Controls
    Vehicle
    Negative controls
    Positive controls
    Comments
  Exposure conditions
    Incubation and sampling times

      Doses used in definitive study
      Study design
    Analysis
      Number of replicates
      Counting method
    Criteria for positive results
Summary of individual study findings
    Study validity
    Study outcome
Genetic toxicology summary
Genetic toxicology conclusions

Carcinogenicity

Study title
Key study findings
Study type
Species/strain
Number/sex/group; age at start of study
Animal housing
Formulation/vehicle
Drug stability/homogeneity
Methods
    Doses
    Basis of dose selection
    Restriction paradigm for dietary restriction studies
    Route of administration
    Frequency of drug administration
    Dual controls employed
    Interim sacrifices
    Satellite pharmacokinetic or special study groups
    Deviations from original study protocol
    Statistical methods
Observations and times
    Clinical signs
    Body weights
    Food consumption
    Hematology
    Clinical chemistry
    Organ weights

Gross pathology
Histopathology
Toxicokinetics
Results
    Mortality
    Clinical signs
    Body weights
    Food consumption
    Hematology
    Clinical chemistry
    Organ weights
    Gross pathology
    Histopathology
        Nonneoplastic
        Neoplastic
    Toxicokinetics
Summary of individual study findings
    Adequacy of the carcinogenicity study and appropriateness of the test
    model
    Evaluation of tumor findings
Carcinogenicity summary
Carcinogenicity conclusions
    Recommendations for further analysis
Labeling recommendations

Reproductive and Developmental Toxicology

Study title
Key study findings
Methods
    Species/strain
    Doses employed
    Route of administration
    Study design
    Number/sex/group
    Parameters and endpoints evaluated
Results
    Mortality
    Clinical signs
    Body weights

Food consumption
Toxicokinetics
For fertility studies
   In-life observations
   Terminal and necroscopic evaluations
For embryo–fetal development studies
   In-life observations
   Terminal and necroscopic evaluations: dams and offspring
For peripostnatal development studies
   In-life observations: dams and offspring
   Terminal and necroscopic evaluations: dams and offspring
Summary of individual study findings
Reproductive and developmental toxicology summary
Reproductive and developmental toxicology conclusions
Labeling recommendations

APPENDIX 7

EXAMPLES OF GENERAL BIOMARKERS

Laboratory Test	Normal Range
Laboratory Tests for Hematology	
WBC (white blood cell) count	$3.6-9.8 \times 10^3/\mu L$
RBC (red blood cell) count	Male: $4.2-6.2 \times 10^6/\mu L$
	Female: $3.7-5.5 \times 10^6/\mu L$
Hemoglobin	Male: 12.9–17.9 g/dL
	Female: 11.0–15.6 g/dL
Hematocrit	Male: 38–53%
	Female: 33–47%
WBC classification	
Band neutrophil	0–3% or 0–5%
Segmented neutrophil	45–70%
Lymphocyte	25–40% or $2.4 \pm 0.8 \times 10^3/\mu L$
Monocyte	2–8%
Eosinophil	1–3% or 70–400/μL
Basophil	0–0.5%
MCV (mean corpuscular volume)	82–98 fL (i.e., 10^{-15} L)
MCH (mean corpuscular hemoglobin)	27–32 pg (i.e., 10^{-12} g)
MCHC (mean corpuscular hemoglobin concentration)	31–36%
Platelet count	$120-400 \times 10^3/\mu L$
RDW (red cell distribution width)	11.5–14.5

Drugs: From Discovery to Approval, Second Edition, By Rick Ng
Copyright © 2009 John Wiley & Sons, Inc.

Laboratory Test	Normal Range
MPV (mean platelet volume)	9.8 ± 1.2 fL
Reticulocyte count	0.5–1.5%
ESR (erythrocyte sedimentation rate)	Male: <10 mm/h
	Female: <20 mm/h
Bleeding time	3–10 min
Clotting time	8–10 min
PT (prothrombin time)	10–13 s
APTT (activated partial thromboplastin time)	26–36 s
G-6-PD (glucose-6-phosphatase dehydrogenase)	4.10–7.90 IU/g Hb
Fibrinogen	200–400 mg/dL
FDP (fibrinogen degradation product)	<10 μg/mL
Total eosinophil count	70–400/mm^3

Laboratory Tests for Clinical Chemistry

Liver function tests	
ALP (alkaline phosphatase)	65–272 IU/L
AST/SGOT (serum glutamic oxaloacetin transaminase)	15–35 IU/L
ALT/SGPT (serum glutamic pyruvate transaminase)	3–30 IU/L or 8–45 IU/L
	5–40 IU/L
γ-GT (gamma glutamyl transferase)	0.3–1.0 mg/dL
Bilirubin	150–400 IU/dl
LDH (lactic acid dehydrogenase)	6.6–8.1 gm/dL
Total protein	3.9–5.1 gm/dL
Albumin	2.3–3.5 gm/dL
Globulin	
Renal function tests	
BUN (blood urea nitrogen)	5–20 mg/dL
Creatinine	0.7–1.5 mg/dL
Creatinine clearance	Male: 62–108 mL/min
	Female: 57–78 mL/min
Electrolytes	
Sodium (Na$^+$)	135–140 mmol/L
Potassium (K$^+$)	3.5–5.0 mmol/L
Chloride (Cl$^-$)	98–108 mmol/L
Calcium (Ca^{2+})	2.1–2.6 mmol/L
Phosphorus (P)	2.5–4.5 mg/dL
Magnesium (Mg^{2+})	1.9–2.5 mg/dL
Uric acid	Male: 3.5–7.9 mg/dL
	Female: 2.6–6.0 mg/dL
CPK (creatinine phosphokinase)	37–289 IU/L
Aldolase	1.7–4.9 units/L
Amylase	Serum: 30–200 IU/L
	Urine: 4–30 IU/2 h
Lipase	<200 units/L

Laboratory Test	Normal Range
Cholesterol	
Total cholesterol	130–200 mg/dL
HDL cholesterol	35–65 mg/dL
LDL cholesterol	<130 mg/dL
Apo A-1 (apolipoprotein)	Male: 66–151 mg/dL
	Female: 75–170 mg/dL
Apo B (apolipoprotein B)	Male: 49–124 mg/dL
	Female: 26–119 mg/dL
Triglycerides	<250 mg/dL
Glucose	
AC glucose	70–110 mg/dL
30 PC glucose	90–160 mg/dL
1-h PC glucose	90–160 mg/dL
2-h PC glucose	75–125 mg/dL
3-h PC glucose	70–110 mg/dL
HbA_{1C} (glycosylated hemoglobin)	4–7%
Serum iron	Male: 89–200 µg/dL
	Female: 70–180 µg/dL
Ferritin	Male: 27–300 ng/mL
	Female: 10–130 ng/mL
Acid P-tase (acid phosphatase)	Male: 4.7 IU/L
	Female: <3.7 IU/L
Protein electrophoresis	
Total protein	5.9–8.0 g/dL
Albumin	4.0–5.5 g/dL
Alpha-1 globulin	0.15–0.25 g/dL
Alpha-2 globulin	0.43–0.75 g/dL
Beta globulin	0.50–1.00 g/dL
Gamma globulin	0.60–1.30 g/dL
Lipoprotein electrophoresis	
Pre-beta	$20 \pm 6\%$
Beta	$50 \pm 5\%$
Alpha	$36 \pm 7\%$
Hemoglobin electrophoresis	
H_bA	97%
H_bA_2	1.5–3.5%
H_bF	<2.0%
H_bC	0
H_bS	0
Osmolality	280–295 mOsm/kg

Laboratory Test	Normal Range
Laboratory Tests for Urinalysis	
Dipstick test	
pH	4.6–8.0
Protein	<8 mg/dL
Glucose	0.1–1.0 EU/dL
Ketone	<10^5 colony/mL
Occult blood	<5/HPF
Urobilinogen	<5/HPF
Leukocyte esterase	0
Nitrite	0/LPF
Sediment	
RBC	
WBC	
Epithelial cells	
Casts	
Crystal	
Microorganisms	
Parasites	
Spermatozoa	
Specific gravity	1.016–1.022
Gram stain	
Bence–Jones protein	
Paraguat test	
Porphobilinogen	
Myoglobin	
Pregnancy tset	
Fractional urinalysis	

APPENDIX 8

TOXICITY GRADING

The FDA has set out guidance documents for grading toxicity in the conduct of clinical trials. The following is an adaptation from a recent document: *Guidance for Industry—Toxicity Grading Scale for Healthy Adult and Adolescent Volunteers Enrolled in Preventive Vaccine Clinical Trials* (2007).

TABLE A8.1 Clinical Abnormalities

Local Reaction to Injectable Product	Mild (Grade 1)	Moderate (Grade 2)	Severe (Grade 3)	Potentially Life Threatening (Grade 4)
Pain	Does not interfere with activity	Repeated use of nonnarcotic pain reliever >24h or interferes with activity	Any use of narcotic pain reliever or prevents daily activity	Emergency room (ER) visit or hospitalization
Tenderness	Mild discomfort to touch	Discomfort with movement	Significant discomfort at rest	ER visit or hospitalization

Drugs: From Discovery to Approval, Second Edition, By Rick Ng
Copyright © 2009 John Wiley & Sons, Inc.

TABLE A8.1 *Continued*

Local Reaction to Injectable Product	Mild (Grade 1)	Moderate (Grade 2)	Severe (Grade 3)	Potentially Life Threatening (Grade 4)
Erythema/ redness	2.5–5 cm	5.1–10 cm	>10 cm	Necrosis or exfoliative dermatitis
Induration/ swelling	2.5–5 cm and does not interfere with activity	5.1–10 cm or interferes with activity	>10 cm or prevents daily activity	Necrosis

Vital Signs	Mild (Grade 1)	Moderate (Grade 2)	Severe (Grade 3)	Potentially Life Threatening (Grade 4)
Fever (°C)	38.0–38.4	38.5–38.9	39.0–40	>40
(°F)	100.4–101.1	101.2–102.0	102.1–104	>104
Tachycardia— beats per minute	101–115	116–130	>130	ER visit or hospitalization for arrhythmia
Bradycardia— beats per minute	50–54	45–49	<45	ER visit or hospitalization for arrhythmia
Hypertension (systolic)— mmHg	141–150	151–155	>155	ER visit or hospitalization for malignant hypertension
Hypertension (diastolic)— mmHg	91–95	96–100	>100	ER visit or hospitalization for malignant hypertension
Hypotension (systolic)— mmHg	85–89	80–84	<80	ER visit or hospitalization for hypotensive shock
Respiratory rate—breaths per minute	17–20	21–25	>25	Intubation

TABLE A8.1   *Continued*

Systemic Illness	Mild (Grade 1)	Moderate (Grade 2)	Severe (Grade 3)	Potentially Life Threatening (Grade 4)
Illness or clinical adverse event (as defined according to applicable regulations)	No interference with activity	Some interference with activity not requiring medical intervention	Prevents daily activity and requires medical intervention	ER visit or hospitalization

Systemic (General)	Mild (Grade 1)	Moderate (Grade 2)	Severe (Grade 3)	Potentially Life Threatening (Grade 4)
Nausea/ vomiting	No interference with activity or 1–2 episodes/24 h	Some interference with activity or >2 episodes/24 h	Prevents daily activity, requires outpatient IV hydration	ER visit or hospitalization for hypotensive shock
Diarrhea	2–3 loose stools or <400 g/24 h	4–5 stools or 400–800 g/24 h	6 or more watery stools or >800 g/24 h or requires outpatient IV hydration	ER visit or hospitalization
Headache	No interference with activity	Repeated use of nonnarcotic pain reliever >24 h or some interference with activity	Significant; any use of narcotic pain reliever or prevents daily activity	ER visit or hospitalization
Fatigue	No interference with activity	Some interference with activity	Significant; prevents daily activity	ER visit or hospitalization
Myalgia	No interference with activity	Some interference with activity	Significant; prevents daily activity	ER visit or hospitalization

TABLE A8.2   Laboratory Abnormalities

Serum	Mild (Grade 1)	Moderate (Grade 2)	Severe (Grade 3)	Potentially Life Threatening (Grade 4)
Sodium— hyponatremia mEq/L	132–134	130–131	125–129	<125
Sodium— hypernatremia mEq/L	144–145	146–147	148–150	>150
Potassium— hyperkalemia mEq/L	5.1–5.2	5.3–5.4	5.5–5.6	>5.6
Potassium— hypokalemia mEq/L	3.5–3.6	3.3–3.4	3.1–3.2	<3.1
Glucose— hypoglycemia mg/dL	65–69	55–64	45–54	<45
Glucose— hyperglycemia Fasting—mg/dL Random—mg/dL	100–110 110–125	111–125 126–200	>125 >200	Insulin requirements or hyperosmolar coma
Blood urea nitrogen (BUN)—mg/dL	23–26	27–31	>31	Requires dialysis
Creatinine—mg/dL	1.5–1.7	1.8–2.0	2.1–2.5	>2.5 or requires dialysis
Calcium— hypocalcemia mg/dL	8.0–8.4	7.5–7.9	7.0–7.4	<7.0
Calcium— hypercalcemia mg/dL	10.5–11.0	11.1–11.5	11.6–12.0	>12.0
Magnesium— hypomagnesemia mg/dL	1.3–1.5	1.1–1.2	0.9–1.0	<0.9
Phosphorus— hypophosphatemia mg/dL	2.3–2.5	2.0–2.2	1.6–1.9	<1.6
CPK—mg/dL	1.25–1.5 × ULN	1.6–3.0 × ULN	3.1–10 × ULN	>10 × ULN
Albumin— hypoalbuminemia g/dL	2.8–3.1	2.5–2.7	<2.5	—
Total protein— hypoproteinemia g/dL	5.5–6.0	5.0–5.4	<5.0	—

TABLE A8.2 *Continued*

Serum	Mild (Grade 1)	Moderate (Grade 2)	Severe (Grade 3)	Potentially Life Threatening (Grade 4)
Alkaline phosphate—increase by factor	1.1–2.0 × ULN	2.1–3.0 × ULN	3.1–10 × ULN	>10 × ULN
Liver function tests—ALT, AST increase by factor	1.1–2.5 × ULN	2.6–5.0 × ULN	5.1–10 × ULN	>10 × ULN
Bilirubin—when accompanied by any increase in liver function test increase by factor	1.1–1.25 × ULN	1.26–1.5 × ULN	1.51–1.75 × ULN	>1.75 × ULN
Bilirubin—when liver function test is normal; increase by factor	1.1–1.5 × ULN	1.6–2.0 × ULN	2.0–3.0 × ULN	>3.0 × ULN
Cholesterol	201–210	211–225	>226	—
Pancreatic enzymes—amylase, lipase	1.1–1.5 × ULN	1.6–2.0 × ULN	2.1–5.0 × ULN	>5.0 × ULN

Hematology	Mild (Grade 1)	Moderate (Grade 2)	Severe (Grade 3)	Potentially Life Threatening (Grade 4)
Hemoglobin (female)—g/dL	11.0–12.0	9.5–10.9	8.0–9.4	<8.0
Hemoglobin (female) change from baseline value—g/dL	Any decrease —1.5	1.6–2.0	2.1–5.0	>5.0
Hemoglobin (male)—g/dL	12.5–13.5	10.5–12.4	8.5–10.4	<8.5
Hemoglobin (male) change from baseline value—g/dL	Any decrease —1.5	1.6–2.0	2.1–5.0	>5.0
WBC increase—cells/mm^3	10,800–15,000	15,001–20,000	20,001–25,000	>25,000
WBC decrease—cells/mm^3	2,500–3,500	1,500–2,499	1,000–1,499	<1,000
Lymphocytes decrease—cells/mm^3	750–1,000	500–749	250–499	<250
Neutrophils decrease—cells/mm^3	1,500–2,000	1,000–1,499	500–999	<500

TABLE A8.2 *Continued*

Hematology	Mild (Grade 1)	Moderate (Grade 2)	Severe (Grade 3)	Potentially Life Threatening (Grade 4)
Eosinophils— cells/mm³	650–1500	1501–5000	>5000	Hypereosi- nophilic
Platelets decreased —cells/mm³	125,000– 140,000	100,000– 124,000	25,000–99,000	<25,000
PT—increase by factor (prothrombin time)	1.0–1.10 × ULN	1.11–1.20 × ULN	1.21–1.25 × ULN	>1.25 ULN
PTT—increase by factor (partial thromboplastin time)	1.0–1.2 × ULN	1.21–1.4 × ULN	1.41–1.5 × ULN	>1.5 × ULN
Fibrinogen increase —mg/dL	400–500	501–600	>600	—
Fibrinogen decrease —mg/dL	150–200	125–149	100–124	<100 or associated with gross bleeding or disseminated intravascular coagulation (DIC)

Source: Food and Drug Administration. Guidance for Industry—Toxicity Grading Scale for Healthy Adult and Adolescent Volunteers Enrolled in Preventive Vaccine Clinical Trials. http://www.fda.gov/cber/gdlns/toxvac.htm [accessed September 28, 2007].

APPENDIX 9

HEALTH SYSTEMS IN SELECTED COUNTRIES

Factors	Australia	Canada	China	France	Germany	India	Japan	United Kingdom	United States
Population	19,731,000	31,510,000	1,311,709,000	60,144,000	82,478,000	1,065,462,000	127,654,000	59,251,000	294,043,000
GNI[a]/capita, US$	21,950	24,470	1,100	24,730	25,270	540	34,180	28,320	37,870
% GDP on healthcare	9.5	9.6	5.8	9.7	10.9	6.1	7.9	7.7	14.6
Per capita total health expenditure, US$	1,995	2,222	63	2,348	2,631	30	2,476	2,031	5,274
Hospital bed/1000 population	4.1	3.3	2.4	8.3	9.6	0.78	14.8 (incl. geriatric care beds)	2.9	3.0
Doctors/1000 population	2.5	1.9	1.6	3.3	3.6	0.48	2.0	1.6	2.8
Healthcare system	**Public:** Federal government funds universal medical services scheme (Medicare). State government finances health services, including hospitals. **Private:** Individuals purchase private insurance through government subsidies and regulation.	**Public:** Medicare funded through tax to provide comprehensive medical services. **Private:** Out-of-pocket payments for pharmaceuticals, home and community-based services, dental and optical services.	**Public:** Government Insurance Scheme, Labor Insurance Scheme, and Cooperative Medical are public systems. Most hospitals are government funded. **Private:** Private insurance introduced in 1982 and now growing.	**Public:** Assurance Maladies universally covers most population through salary-related social contributions. **Private:** Private insurance supplements charges and services not covered under Assurance Maladies.	**Public:** Statutory Health Insurance covers 90% of population. Government responsible for hospital services. Health Technology Assessment recommends reimbursement for medicines, devices, and surgeries. **Private:** Remaining population has private insurance.	**Public:** State Insurance Scheme and Government Health Scheme cover state and government employees and dependants. Hospitals and services vary in quality from state to state. **Private:** Voluntary private health insurance is limited. Majority is based on out-of-pocket expenses.	**Public:** Health schemes are segmented and comprehensive. **Private:** Hospitals are mostly privately owned, 70%.	**Public:** National Health Service (NHS) provides free and comprehensive care, finance is by taxation revenue. NHS owns >2000 hospitals. **Private:** 12% of population under private schemes, approximately 300 private hospitals.	**Public:** Medicare covers old and young disabled. Medicaid for low income people. About 25% of 5810 hospitals publicly owned. **Private:** Private insurance provided by employers or individually purchased. Approximately 75% of people <65 years old covered. About 16% of residents have no insurance.

[a]GNI = GDP + net revenue (interests and dividends) from overseas.

Source: World Health Organization report. http://www.who.int/tdr/publications/publications/pdf/tbdi/tbdi_profiles.pdf [accessed August 28, 2007].

ACRONYMS

ABS	acrylonitrile butadiene styrene
ACE	angiotensin converting enzyme
ADME	absorption, distribution, metabolism, and excretion
AIDS	acquired immune deficiency syndrome
ANDA	Abbreviated New Drug Application
APC	antigen-presenting cell
API	active pharmaceutical ingredient
ARTG	Australian Register of Therapeutic Goods
AUC	area under curve
BGTD	Biologics and Genetic Therapies Directorate (Canada)
BHK	baby hamster kidney
BLA	Biologics License Application
BPC	bulk pharmaceutical chemical
BSL	biosafety level
CAM	complementary or alternative medicine
CBER	Center for Biologics Evaluation and Research (FDA)
CD	cluster of differentiation
CDE	Center for Drug Evaluation (China)
CDER	Center for Drug Evaluation and Research (FDA)
CDSCO	Central Drugs Standard Control Organization (India)
cDNA	complementary DNA
CFR	Code of Federal Regulations (FDA)

Drugs: From Discovery to Approval, Second Edition, By Rick Ng
Copyright © 2009 John Wiley & Sons, Inc.

CFTR	cystic fibrosis transmembrane conductance regulator
CFU	colony forming unit
cGMP	current Good Manufacturing Practice
CIP	clean-in-place
CMC	chemistry, manufacturing, and control
COP	clean-out-of-place
CPAC	Central Pharmaceutical Affairs Council (Japan)
CHMP	Committee for Human Medicinal Products (EMEA)
CRF	case report form
CRO	clinical research organization
CTC	Clinical Trial Certificate
CTD	Common Technical Document
CTN	Clinical Trial Notification
CTX	Clinical Trial Exemption
CVMP	Committee for Veterinary Medicinal Products (EMEA)
DDR	Department of Drug Registration (China)
DED	Drug Evaluation Division (China)
DIN	Drug Identification Number (Canada)
DMF	Drug Master File
DNA	deoxyribonucleic acid
DQ	design qualification
ED	effective dose
EIR	Establishment Inspection Report
ELA	Establishment License Application
ELISA	enzyme linked immunosorbent assay
EMEA	European Medicines Agency
EPA	Environmental Protection Agency
EPAR	European Public Assessment Report
EPCB	end of production cell bank
EPO	erythropoietin
ERMS	European Risk Management Strategy
EST	expressed sequence tag
ESTRI	Electronic Standards for Transmission of Regulatory Information
EU	European Union
Fab	antigen-binding fragment
FACS	fluorescence-activated cell sorter
FBS	fetal bovine serum
Fc	constant fragment
FDA	Food and Drug Administration (United States)
FDCA	Food, Drug and Cosmetic Act (United States)
Fv	variable fragment
GAMP	Good Automated Manufacturing Practice
GAP	Good Agricultural Practice
GCP	Good Clinical Practice

GDEA	Generic Drug Enforcement Act (United States)
GLP	Good Laboratory Practice
GM-CSF	granulocyte–macrophage colony-stimulating factor
GMP	Good Manufacturing Practice
GPCR	G-protein coupled receptor
GQP	Good Quality Practice
GRAS	generally recognized as safe
GVP	Good Vigilance Practice
HAMA	human anti-mouse antibody
HEPA	high efficiency particulate air
hGH	human growth hormone
HGP	Human Genome Project
HIV	human immunodeficiency virus
HPLC	high performance liquid chromatography
HTS	high throughput screening
HVAC	heating, ventilation, and air-conditioning
ICDRA	International Conference of Drug Regulatory Authorities
ICH	International Conference on Harmonization
IDDM	insulin-dependent diabetes mellitus
IEC	independent ethics committee
IFN	interferon
IGF	insulin-like growth factor
IgG	immunoglobulin
IHR	International Health Regulations
IL	interleukin
IMP	investigational medicinal product
IMPACT	International Medicinal Products Anti-Counterfeiting Taskforce
IND	Investigational New Drug
INTERPOL	International Criminal Police Organization
IOC	International Olympic Committee
IPR	intellectual property right
IQ	installation qualification
IRB	Independent Review Board
ISO	International Organization for Standardization
LAL	limulus amebocyte lysate
LD	lethal dose
LIMS	laboratory information management system
MA	marketing authorization
MAb	monoclonal antibody
MAC	maximum allowable carryover
MCB	master cell bank
M-CSF	macrophage colony-stimulating factor
MedDRA	Medical Dictionary for Regulatory Activities Terminology
MHLW	Ministry of Health, Labor and Welfare (Japan)

MHPD	Marketed Health Products Directorate (Canada)
MHRA	Medicines and Healthcare Products Regulatory Agency (United Kingdom)
mRNA	messenger RNA
MRSA	methicillin-resistant *Staphylococcus aureus*
MSE	bovine spongiform encephalopathy
NCE	New Chemical Entity
NDA	New Drug Application
NDS	New Drug Submission
NHPD	Natural Health Products Directorate (Canada)
NICPBP	National Institute for the Control of Pharmaceutical & Biological Products (China)
NIDDM	non-insulin-dependent diabetes mellitus
NIH	National Institutes of Health (United States)
NME	new molecular entity
NMR	nuclear magnetic resonance
NOC	Notice of Compliance (Canada)
NOE	nuclear Overhauser effect
NSAID	nonsteroidal anti-inflammatory drug
OOS	out of specification
OPSR	Organization for Pharmaceutical Safety and Research (Japan)
OQ	operational qualification
OSHA	Occupational Safety and Health Administration
ORA	Office of Regulatory Affairs
OTC	over-the-counter
PAFSC	Pharmaceutical Affairs and Food Sanitation Council (Japan)
PAI	preapproval inspection
PAT	process analytical technology
PCR	polymerase chain reaction
PCT	Patent Cooperation Treaty
PD	pharmacodynamics
PDGF	platelet-derived growth factor
PDUFA	Prescription Drug User Fee Act (United States)
PFSB	Pharmaceutical and Food Safety Bureau (Japan)
PGHS	prostaglandin H_2 synthase
PHSA	Public Health Service Act (United States)
PI	Pharmaceutical Inspectorate
PIC/S	Pharmaceutical Inspection Cooperation Scheme
PK	pharmacokinetics
PLA	Product License Application
PMDA	Pharmaceutical and Medical Devices Agency (Japan)
POU	point of use
PQ	performance qualification

PTC	point to consider
QA	quality assurance
QC	quality control
QPCR	quantitative PCR
rDNA	recombinant DNA
RFID	radiofrequency identification
RLD	reference listed drug
RNA	ribonucleic acid
RO	reverse osmosis
SAR	structure–activity relationship
SARS	severe acute respiratory syndrome
SCID	severe combined immune deficiency
SFDA	State Food and Drug Administration (China)
SNP	single nucleotide polymorphism
SOD	superoxide dismutase
SOP	standard operating procedure
SPA	scintillation proximity assay
SPC	summary of product characteristics (EMEA)
SSM	standard safety margin
TCM	traditional Chinese medicine
TGA	Therapeutic Goods Administration (Australia)
TM	traditional medicine
TNF	tumor necrosis factor
TOC	total organic carbon
TPD	Therapeutic Products Directorate (Canada)
tRNA	transfer RNA
UHTS	ultrahigh throughput screening
UNESCO	United Nations Educational, Scientific and Cultural Organization
UNICEF	United Nations Children Fund
URS	user requirement specification
WADA	World Anti-Doping Association
WCB	working cell bank
WFI	water-for-injection
WHO	World Health Organization
WTO	World Trade Organization

GLOSSARY

adrenaline: A hormone that prepares the body for 'fright, flight or fight'; also called epinephrine.

adverse event: An unanticipated event that involves risk to the subject and that results in harm to the subject or others.

affinity: A measure of the binding of an antibody to an antigen.

amide: An organic compound containing an (O—C—N) group.

amine: An organic compound with the general formula of $R_{3-x}NH_x$ where R is a hydrocarbon group and $0 < x < 3$.

amino acid: An organic compound containing an amino group (—NH_2) and a carboxyl group (—COOH).

aminotransferase: An enzyme that catalyzes the transfer of an amino group to an acid.

amyloid: A glycoprotein that is deposited extracellularly in tissues.

angiotensin: A peptide. There are two forms of angiotensin: I and II. Angiotensin I is converted to angiotensin II by an enzyme, angiotensin-converting enzyme. Angiotensin II constricts blood vessels to increase blood pressure.

antibody: A protein secreted by B cells when they are stimulated by an antigen. Antibodies act specifically against particular antigens in an immune response.

Drugs: From Discovery to Approval, Second Edition, By Rick Ng
Copyright © 2009 John Wiley & Sons, Inc.

assay: A test or trial.

avidity: Strength of binding, especially the binding of an antibody to an antigen.

Bacillus: Rod-shaped bacteria.

bacteriophage: A type of virus that destroys bacteria; also called phage.

cell line: A collection of cells that will proliferate indefinitely when provided with appropriate space to grow and fresh medium to feed on.

cyclase: An enzyme that forms a cyclic compound.

cytochrome: A substance that contains iron and acts as a hydrogen carrier for the eventual release of energy in aerobic respiration.

cytometry: The method of counting cells using a cytometer.

Dalton: A unit of measurement equal to the mass of a hydrogen atom.

electrophoresis: The differential movement of molecules through a gel under the influence of an electric field.

endotoxin: A poison release by a bacterium when the cell wall is broken.

entropy: A measure of disorder.

epinephrine: See *adrenaline*.

esophagitis: Inflammation of the esophagus.

ethical drugs: Patented prescription drugs.

ex vivo: Outside a living body.

expression: Information from a gene is transcribed and translated, which results in the production of a protein.

generics: Copies of drugs for which the patents have expired.

genome: The entire DNA of a cell.

genomics: The study of genes and gene function.

glycoconjugate: A carbohydrate that is linked to a lipid or protein.

glycoprotein: See *glycosylation*.

glycosylation: During and after protein synthesis, the protein molecule can undergo modifications. Glycosylation is the attachment of a carbohydrate to the —OH group of serine and threonine (*O*-glycosylation) or the amide —NH_2 group of asparagine (*N*-glycosylation) of the protein, to form a glycoprotein.

hepatic: Relating to the liver.

hydrophilic: Soluble in water.

hydrophobic: Insoluble in water.

IgG: Immunoglobulin G, a class of antibody.

in vitro: Within a glass, in a test tube—an artificial environment.

in vivo: Within a living body.

intercellular: Between cells.

intracellular: Within cell.

ischemia: A low supply of oxygen due to low blood flow.

kinase: An enzyme that catalyzes the transfer of a phosphate group.

ligand: A molecule that binds to another molecule.

ligase: An enzyme involved in DNA replication.

Limulus amebocyte lysate: A reagent for determining the quantity of bacterial endotoxins. It is obtained from the aqueous extracts of circulating amebocytes of the horseshoe crab.

liposome: A spherical vesicle formed by a lipid enclosing an aqueous part.

log P: Logarithmic function of the partition coefficient.

lupus erythematosus: A chronic inflammatory disease of connective tissue, affecting the skin and internal organs.

lymphoma: A malignant tumor of the lymph nodes.

multiple sclerosis: A disease of the nervous system.

myclodysplasia: Abnormal or defective formation of the bone marrow.

Mycoplasma: Minute primitive bacteria without a rigid cell wall. *Mycoplasma pneumoniae* causes atypical pneumonia in humans.

myeloma cells: Malignant tumor cells.

nucleic acid: A molecule composed of nucleotides joined together.

nucleotide: A compound consisting of a nitrogen-containing base, a sugar and a phosphate group.

oligonucleotide: Molecule containing up to 20 nucleotides joined by phosphodiester bonds. Above this length, the term 'polynucleotide' is used.

otitis media: Inflammation of the middle ear.

pathogenesis: The mechanism and cellular events leading to the development of a disease.

peroxidase: An enzyme that catalyzes the oxidation of substances in the presence of hydrogen peroxide.

pH: The negative logarithm of H_3O^+ ion concentration. The scale ranges from 1 to 14; less than 7 is acidic and more than 7 is basic.

phagocytosis: The engulfment of a particle or a microorganism by leukocytes.

pharmacogenomics: The study of how an individual's genetic inheritance affects the body's response to drugs.

phosphatase: An enzyme that catalyzes the hydrolysis of phosphoric acid esters.

phosphorylation: The addition of a phosphate group to a compound.

plasmid: A cytoplasmic DNA that is capable of autonomous replication.

Pneumococcus: The bacterium *Streptococcus pneumoniae*, which is associated with pneumonia.

polypeptide: A molecule consisting of many joined amino acids, but not as complex as a protein.

prophylactic: An agent that is used to prevent the development of a disease or condition.

prostaglandin: A protein that has many functions, including mediation of the inflammatory process.

protease: An enzyme that acts on proteins.

proteomics: The study of protein expression of normal and diseased cells and tissues.

pyrogen: A substance or agent that causes fever.

re-stenosis: Recurrent stenosis, a condition where the blood vessel or heart valve is narrowed.

restriction enzyme: An enzyme that cuts DNA into short segments.

retrovirus: An RNA virus.

saccharide: A carbohydrate.

single nucleotide polymorphism: Difference at one nucleotide in a DNA sequence among individuals.

subcutaneous: Beneath the outer skin.

target: A specific protein or enzyme upon which a drug acts.

therapeutic index: A measure of the relative desirability of a drug, the ratio is given by LD_{50}/ED_{50}.

therapeutic: Treatment and healing of disease.

transcription: The process of transfer of genetic information from DNA to RNA.

translation: The process of transfer of information from RNA into manufacture of protein.

vector: A carrier.

xenotransplantation: The transplantation of cells, tissues or organs from non-human animal sources into humans.

INDEX

Abbreviated New Drug Application
(ANDA):
FDA license applications, 239–248
generic drugs, 249–250
Ab initio calculations, computational
chemistry, small molecule drug
development, 71
Absorption, pharmacokinetics, 145,
147–149
Absorption, distribution, metabolism
and excretion (ADME)
algorithms:
clinical trials, microdosing, 181
drug delivery systems, 165–168
generic drug approval, 249
investigational new drugs, FDA
review, 237
in silico studies, 158–161
Accelerated development/review:
EU marketing authorization,
262–263
FDA clinical trials guidelines, 238
Access and Benefit-Sharing Agreement
(ABA), 56

Accuracy, GMP analytical methods
validation, 306–307
Acquired immunodeficiency syndrome
(AIDS):
in developing countries, 364–365
vaccines, 104
Acronyms, table of, 436–440
Active pharmaceutical ingredient
(API):
defined, 8
Good Manufacturing Practice,
287–288
small molecule drugs, conventional
synthesis techniques, 332–337
Active transport, pharmacokinetics,
145
Adenocarcinoma *in situ* (AIS), Gardasil
therapy, clinical trial, 201–204
Adenosine diphosphate (ADP), enzyme
reactions, 35
Adenosine triphosphate (ATP), enzyme
reactions, 35
Adjuvant vaccines, characteristics and
applications, 102–103

Adverse events:
  clinical trial guidelines, 195
  in U.S. history, 280
Affinity chromatography, large
      molecule drug purification, 347
Aging, future drug research and, 369–371, 373
Agonists, defined, 31
Airborne environmental cleanliness
      requirements, Good
      Manufacturing Practice
      buildings and facilities
      guidelines, 290–293
Alimta, IND review, 238
Allogeneic transplant, defined, 129
Allosteric binding, drug interactions, 31–32
Alternative medicine, future research
      issues, 364–365
Alzheimer's disease:
  EGb 761® clinical trial, 183–184
  ethical issues in, 387–388
  future drug research on, 370
  vaccines, 103–104
Ames test, genotoxicity analysis, 157–158
Amino acids, DNA structure and
      properties, 403–404
Amyloid precursor protein (APP),
      Alzheimer's disease and, 387–388
Analytical methods:
  FDA guidelines, 244
  Good Manufacturing Practice
      validation, 305–307
  drug manufacturing guidelines, 324–325
Animal studies:
  preclinical trials, 158–161
  toxicity anlysis, 156–157
  transgenic technologies, 376–378
Antagonists, defined, 31–32
Anthrax, bioterrorism using, 377
Antibodies, large molecule drug
      discovery, 106–114
  chimeric, 111
  conjugate, 112–113
  human immune system, 107–108
  humanization, 110–112

monoclonal antibodies, 110
structure, 106–109
traditional antibodies, 109–110
Antidepressants, discovery and
      development, 22–23
Antigen-binding fragment, antibody
      molecules, 109
Antihyperlipidemic drugs, discovery
      and development, 22–23
Anti-inflammatory therapy, targets and
      receptors, 46–49
Antimicrobial drug resistance, future
      research, 379
Antiogensin converting enzyme (ACE),
      X-ray crystallography, 363
Antiretroviral drugs, Lopinavir/
      Ritonavir 2 clinical trial, 183,
      185–186
Antisense approach to drug discovery,
      13
  current clinical development, 82
  small molecule drug development,
      79–81
Antiulcerants, discovery and
      development, 22–23
Approvable Letter, NDA/BLA reviews,
      248
Approval Letter, NDA/BLA reviews,
      248
A priori target identification, drug
      discovery and, 20
Arabian medicine, history of, 394
Arachidonic acid, prostaglandin
      conversion, 46–49
Aranesp, preclinical trial, 169–171
Area under curve (AUC)
      measurements,
      pharmacokinetics, oral
      administration, 148
Aspirin (acetylsalicylic acid), target-
      drug interactions, 33
Assay development:
  drug receptors, 45–46
  FDA guidelines on, 245
  small molecule drugs, high
      throughput screening, 60
Atomic scattering factor, X-ray
      crystallography, small molecule
      drug development, 62–64

Atomic spectroscopy, drug
    manufacturing guidelines, 325
Atorvastatin, enzyme inhibition, 36
Attenuated vaccines, characteristics and
    applications, 97
Australia's Therapeutic Goods
    Administration (TGA):
    drug approval authority, 219–220
    marketing authorization by, 269–271
Autologous transplant, defined, 129
Avastin:
    large molecule drug formulation,
        166
    vascular endothelial growth factor,
        121
Avian influenza H5N1, vaccines, 99

Bcl-2, antisense development technique,
    81
Bioavailability, oral drugs, 167–168
Biodiversity prospecting, regulations
    for, 56
Bioethics, future issues in, 382–384
Biogenerics, manufacturing guidelines,
    353–355
Bioinformatics, small molecule drug
    development, 65, 68–69
Biological terrorism, research issues,
    376–377
Biologics:
    biogenerics manufacturing
        requirements, 353–355
    drug discovery, 13
    examples and applications, 96
    FDA license applications, 212,
        239–248
    FDA review, 246–248
    large molecule drug discovery,
        94–95
Bioluminescence, receptor assays,
    45–46
Biomarkers:
    clinical trial guidelines, 191
    examples, 424–427
Biopharmaceuticals:
    FDA guidelines on, 245
    Good Manufacturing Practice
        buildings and facilities
        guidelines, 292–293

large molecule drug discovery, 94–95,
    362–363
Biosafety levels, Good Manufacturing
    Practice buildings and facilities
    guidelines, 292–293
Bispecific antibodies, structure and
    function, 112–113
Blood-brain barrier (BBB), distribution
    pharmacokinetics, 150–151
Bone marrow transplant, overview,
    129
Botulinum toxin, future applications,
    372–373
Bovine spongiform encephalopathy,
    large molecule drug
    manufacturing, 344
Buccal and sublingual administration,
    pharmacokinetics, 148
Buckyballs, polymeric drug delivery
    systems, 168
Buildings and facilities:
    FDA guidelines on, 245
    Good Manufacturing Practice
        regulations, 289–293
Bupropion, future applications, 372

Caco-2 cell assays, preclinical trials,
    159–161
Canada's Health Canada:
    drug approval authority, 220
    marketing authorization by, 269
Cancer:
    p53 protein and, 24–25
    vaccines, 104
        FDA oversight, 381
Capillary array electrophoresis:
    drug manufacturing guidelines, 325
    small molecule drug development,
        76–77
CAPRIE clinical trial for Plavix,
    200–201
Capsules, manufacturing guidelines,
    349
Captopril, enzyme inhibition, 35
Carcinogenicity analysis, preclinical
    trials, 157
Case report form, clinical trial
    guidelines, 191–193
Celebrex, formulation, 163

Cell bank:
  FDA guidelines on, 245
  large molecule drug manufacturing, 343
Cell culture techniques, large molecule drug manufacturing, 340–348
Cell lines:
  FDA guidelines on, 245
  large molecule drug manufacturing, 343
Cells:
  chromosomes and, 401
  classification and structure, 398–400
Cell surface receptors, drug interactions, 30–33
Cell therapy:
  overview, 126–128
  Provenge development, 130
Center for Biologics Evaluation and Research (CBER) (FDA), 212–213
Center for Drug Evaluation and Research (CDER) (FDA), 210–212
Centralized authorization procedure, EU marketing authorization, 253–263
Cervical intraepithelial neoplasia (CIN), Gardasil therapy, clinical trial, 201–204
Change controls, Good Manufacturing Practice, 297
Chemical formulas:
  FDA Form 356h, 243–245
  selected drugs, 56–58
Chemical terrorism, research issues, 376
Chemistry, manufacturing, and controls (CMC) information, FDA guidelines, 235–237
Chickenpox, vaccines, 105
Chimeric antibodies, structure and function, 111
China's State Food and Drug Administration (SFDA):
  drug approval authority, 217–219
  marketing authorization by, 264, 266–268
Chinese medicine, early drug discovery and, 392–393

Chiral auxiliary, small molecule drug manufacturing, 338
Chiral building blocks, small molecule drug manufacturing, 338–339
Chiral drugs, development techniques, 83–84
Chiral synthesis, small molecule drug manufacturing, 338–339
Cholesterol:
  basic properties, 7–8
  formation reaction, 87–88
Cholesterol-lowering drugs, structure and function, 7–8
Chromosomes, cells and, 401
Circular dichroism, drug manufacturing guidelines, 325
Cleaning-in-place (CIP) approach, Good Manufacturing Practice, 300–301
Cleaning system, Good Manufacturing Practice, 299–301
Clean-out-of-place (COP), Good Manufacturing Practice, 300–301
Cleanroom pressure scheme, Good Manufacturing Practice buildings and facilities guidelines, 291–293
Clinical abnormalities, toxicity grading, 428–430
Clinical research organizations (CROs), clinical trial guidelines, 198
Clinical results, FDA guidelines on, 245–246
Clinical Trial Certificate (CTC) (EU), 252–253
Clinical Trial Exemption (CTX) (EU), 252–253
Clinical trials:
  defined, 177
  ethical issues, 177–180
  European Union applications, 252–253
  failures of, 159
  Gardasil, 201–204
  gene therapy, 126, 199–200
  government bodies and, 199
  investigational new drugs, FDA review, 237
  parameters and protocols, 4

phase I, 181–182
phase II, 182–184
phase III, 183, 185–186
phase IV, 183, 186
Plavix case study, 200–201
regulatory requirements, 186–198
    adverse events, 195
    biomarkers, 191
    case report form, 191–193
    clinical research organization, 198
    Good Clinical Practice
        requirements, 186, 189
    ICH clinical study efficacy
        guidelines, 187–188
    ICH protocol, 191–192
    inclusion and exclusion criteria,
        191
    informed consent, 190
    investigator, 190
    investigator's brochure, 190
    monitoring, 194
    protocol, 190–191
    randomization, placebo-controlled
        and double-blinded, 192–194
    serum tumor markers, 198
    sponsors, 196, 198
    statistics, 195–197
    surrogate markers, 198
Cloning, future drug development and,
    367–369
Code of Federal Regulations (CFR):
    FDA processes and controls, 213–214
    Good Manufacturing Practice,
        280–283
        electronic records maintenance,
            302–303
Cold chain, manufacturing, finished
    dosage, 351–352
Colony stimulating growth factors,
    structure and function,
    119–120
Combinatorial chemistry, small
    molecule drug development,
    71–74
    parallel synthesis, 73–74
    split and mix, 74
Committee for Medicinal Products for
    Human Use (CHMP), EU
    marketing authorization, 258,
    261–262

Comparability studies, current Good
    Manufacturing Practice
    initiative, 312
Complementary DNA (cDNA),
    microarray, 29
Computational chemistry, small molecule
    drug development, 69–71
Computer validation, Good
    Manufacturing Practice,
    301–304
Configurable software, Good
    Manufacturing Practice, 305
Conjugate antibodies, structure and
    function, 112–113
Consent decree, Good Manufacturing
    Practice, 330
Constant fragment (Fc), antibody
    molecules, 109
Control groups, clinical trials, phase II,
    181–182
Conventional synthesis techniques,
    small molecule API
    manufacture, 332–337
Convention on Biological Diversity
    (CBD), 56
Counterfeit drug case study,
    international regulatory
    authorities and, 269–273
Covalent bonds, target-drug
    interactions, 32–33
Creams, manufacturing guidelines,
    350–351
Crossover design, randomization
    guidelines, 194
Crystalline molecules, small molecule
    drug development, rational
    approach, 61–64
CURE clinical trial for Plavix, 201
Current Good Manufacturing Practice
    (cGMP), 4
    Food and Drug Administration
        initiative, 310–313
    regulatory authorities, 210
Custom (bespoke) software, Good
    Manufacturing Practice,
    305
Cyclooxygenase (COX) inhibitors:
    COX-2 inhibitor, enzyme inhibition,
        36
    targets and receptors, 46–49

Cys-loop superfamily, drug
development, 43–44
Cystic fibrosis, gene identification, 23,
26
Cytochrome P-450 (CYP),
pharmacokinetics, 151
Cytokines, large molecule drug
discovery, 113–121
growth factors, 118–121
interferons, 113, 115–116
interleukins, 115–118
lymphokines and monokines,
113–118
tumor necrosis factor, 118
Cytomegalovirus (CMV), Vitravene
therapy, 81

Data driven systems, *in silico* predictive
methods, 159–160
Data mining, small molecule drug
development, 68
Decoctions, defined, 54
Dendrimer structures, polymeric drug
delivery systems, 168
Design qualification (DQ), Good
Manufacturing Practice,
296–297
Development stage, of drug
development, 3–4
Dexetimide, chiral synthesis
manufacture, 339
Dextropropoxyphene, chiral synthesis
manufacture, 339
Diabetes mellitus, insulin and, 123
Diffusion, pharmacokinetics, 145–146
Digitalis, early history, 395
Dihydrofolate reductase (DHFR)
plasmid vector, structure and
properties, 414–415
Discovery stage of drug development,
3
Dissolution test, formulations, 164
Distribution mechanisms,
pharmacokinetics, 149–150
DNA, classification and structure,
400–404
DNA microarray, drug discovery, 26–29
DNA vaccines, characteristics and
applications, 100, 102

Docking simulation, computational
chemistry, small molecule drug
development, 70–71
Documentation procedures, Good
Manufacturing Practice, 293
Dose-effect curve, pharmacodynamics,
140–143
Double-blinded clinical trials,
guidelines for, 192–194
Double helix DNA, antisense therapy,
small molecule drug
development, 80–81
Double stranded RNAs (dsRNAs),
small molecule drug discovery,
81–83
Down syndrome, 407
Drug, FDA definition of, 1–2
Drug delivery systems:
polymeric compounds, 168
preclinical trials, 165–168
Drug discovery and development:
anti-inflammatory therapy, 46–50
assay development, 45–46
clinical trial stage, 4
development stage, 3–4
discovery stage, 3
early history, 391–394
economics of, 10–12
enzymes, 34–38
COX-2 inhibitor, 36
HIV drugs, 37–38
lipase inhibitor, 36
thermodynamics, 34–36
European Union regulation 726/2004,
254
evolution of, 396–397
FDA regulations concerning, 233
future challenges, 360
manufacturing stage, 4
marketing applications, 4–5
medical needs assessment, 21–23
medieval history of, 394
nuclear magnetic resonance and, 66–67
overview, 2–5
Pfizer Inc. case study, 14–15
pharmaceutical industry and, 5–10
process flow chart, 20–21
receptors and signal transduction,
38–45

G-protein coupled receptors, 39, 42, 44
  intracellular receptors, 44–45
  ion channel receptors, 42–44
  neurotransmitter binding, 44
  tyrosine kinases, 44
  target identification, 23–28
    DNA microarray, 26–29
    genes, 23–25
    radioligand binding, 25–27
  target-receptor interactions, 30–33
  target validation, 28–30
  trends in, 12–14
Drug interactions, targets or receptors, 30–33
Drug manufacturing, Good Manufacturing Practice:
  biogenerics, 353–355
  consent decree, 330
  finished dosage forms, 348–352
  generics, 352–353
  inspections, 325–332
  large molecule APIs, 340–348
    bovine spongiform enecephalopathy, 344
    cell culture techniques, 346
    cell lines/banks testing, 343
    chromatographic purification, 347
    etanercept case study, 348
    recombinant DNA techniques, 342
  overview, 320–322
  requirements, 322–325
  small molecule APIs, 332–339
    chiral synthesis, 338–339
    conventional synthesis techniques, 332–337
Drug master file (DMF), FDA guidelines, 246
Drug molecules, FDA guidelines, 243
Drug stability:
  FDA guidelines on, 245
  Good Manufacturing Practice, 309–310
Dry heat sterilization, Good Manufacturing Practice, 308–309
D values, sterilization and, 308–309

Economic issues, drug discovery and development, 10–12
Edema factor, anthrax, 377
Edible vaccines, 378
Edwards' syndrome, 407
Effectiveness, pharmacodynamics, 142
Efficacy testing, clinical trials:
  ICH guidelines, 183, 188
  phase III, 183
EGb 761® clinical trial, phase II study, 183–184
Egyptian medicine, early drug discovery and, 393
Electrostatic forces, target-drug interactions, 33
Enbrel (Etanercept):
  large molecule drug formulation, 166
  manufacturing techniques, 348
End of production cell bank (EPCB), large molecule drug manufacturing, 346
Enterococci bacteria, resistance research, 379
Entry inhibitors, HIV drug development, 38
Environmental factors, Good Manufacturing Practice, 321
  inspections, 330
Environmental Protection Agency (EPA), drug regulation and, 221
Enzyme catalysts, small molecule drug manufacturing, 338–339
Enzyme-linked immunosorbent assay (ELISA), subunit vaccines, 100–101
Enzymes:
  drug discovery and development, 34–38
  drug interactions, 31–33
  thermodynamic reactions, 34
Equipment, FDA guidelines, 244
Erythropoietin (EPO), structure and function, 119
Esomeprazole:
  development case study, 85
  enzyme inhibition, 35–38
  structure and function, 8

Etanercept (Enbrel):
large molecule drug formulation, 166
manufacturing techniques, 348
Ethical issues:
bioethics, 382–387
clinical trials, 177–180
defined, 2
Eukaryote cell, classification and
structure, 399–400
European Medicines Agency (EMEA):
application fees, 2007, 256–257
drug approval authority, 214–215
marketing authorization, 253–263
risk-based approach, 315
European Public Assessment Report
(EPAR), EU marketing
authorization, 258–263
European Union (EU):
Good Manufacturing Practice
regulations, 283
inspections guidelines, 332
regulation 726/2004, 254
regulatory authorities, 250–263
clinical trial applications, 252–253
marketing authorization, 253–263
Excipients, lyophilized formulations,
167
Exclusion criteria, clinical trial
guidelines, 191
Excretion, pharmacokinetics, 151–153
Expense/revenue curve, drug discovery
and development, 10–12
Expert reports, EU marketing
authorization, 258–263
Expert systems, *in silico* predictive
methods, 160
Expressed sequence tags, drug
discovery, 28
Extraction process, small molecule
potential drugs, 55–56
Exubera, large molecule drug
formulation, 166

Facilitated diffusion, pharmacokinetics,
145
Factory acceptance test (FAT), Good
Manufacturing Practice, 297
Fair subject selection, clinical trials,
178

False negatives and positives, clinical
trial guidelines, 197
Finished products:
Good Manufacturing Practice, 294
manufacturing guidelines, 348–352
inhalants, 350
liquids, 350
ointments and creams, 350–351
parenterals, 350
solids, 349–350
Firmware, Good Manufacturing
Practice, 305
Fluorescent activated cell sorter
(FACS), subunit vaccines,
100–101
Food and Drug Administration (FDA):
Center for Biologics Evaluation and
Research (CBER), 212–213
inspections guidelines, 327–332
Center for Drug Evaluation and
Research (CDER), 210–212
current Good Manufacturing
Practice initiative, 310–313
Investigational New Drugs,
234–235
chemistry, manufacturing, and
controls guidelines, 235–237,
243–245
clinical trials, 237
current Good Manufacturing Practice
initiative, 310–313
drug definition of, 1–2
drug development process, 233
Form 356h, license applications,
239–248
formulation guidelines, 164
generics, 249
ethical rules on, 383
gene therapy and cancer vaccine
oversight, 381
Good Manufacturing Practice
regulations, 279–283
computer validation, 301–304
risk-based approach, 314
history of, 211
inspections guidelines, 326–332
Investigational New Drug guidelines:
applications, 234–235
review, 237

New Drug Applications/Biologics
License Application, 239–248
organization and function, 210–214
orphan drugs, 248–249
over-the-counter drugs, 249–250
processes and controls, 213–214
review mechanisms, 238–239
Form 356h (FDA), 239–248
clinical results, 245–246
index of, 243
labeling requirements, 243
Formulations:
Aranesp case study, 170–171
preclinical trials, 161–165
Zeprexa case study, 170
Full human antibodies, 112
Fuzeon, manufacturing guidelines, 321

Gardasil, 100–101
clinical trial, 201–204
Gas sterilization, Good Manufacturing
Practice, 308–309
Gel filtration, large molecule drug
purification, 347
Generally recognized as safe (GRAS)
guidelines, formulations,
165–166
Generics:
defined, 2
ethical issues, 383–387
FDA approval, 249
manufacturing guidelines, 352–353
Genes, basic properties, 404–407
Gene targeting, drug discovery and
development, 23–25
Gene therapy:
clinical trials, 126, 199–200
drug discovery, 13
FDA oversight, 381
future research issues, 366–367
large molecule drug development,
124–126
vectors, 125
Genetic code dictionary, 406
Genetic disorders, 407
Genetic engineering, large molecule
drug manufacturing,
recombinant DNA techniques,
340–348

Genetic variations, future research and,
367
Genital warts, Gardasil therapy, clinical
trial, 201–204
Genomics:
Human Genome Project and,
408–410
small molecule drug development, 74,
76–77
Genotoxicity analysis, preclinical trials,
157
*GMP Guidance for Active
Pharmaceutical Ingredients*,
283–287
Good Agricultural Practice (GAP),
traditional medicine
development, 365
Good Clinical Practice (GCP),
regulatory authorities,
209–210
Good Laboratory Practices (GLPs):
animal studies, 158–161
clinical trial guidelines, 186–187, 189
regulatory authorities, 209–210
Good Manufacturing Practice (GMP):
analytical methods validation,
305–307
cleaning systems, 299–301
computer validation, 301–304
core elements, 287–297
buildings and facilities, 289–293
change controls, 297
documentation and records, 293
introduction and scope, 287–288
laboratory controls, 294–296
materials management, 293–294
packaging, identification and
labeling, 294
personnel, 289
process equipment, 293
production and in-process controls,
294
quality management, 288–289
validation, 296–297
drug manufacturing:
biogenerics, 353–355
consent decree, 330
finished dosage forms, 348–352
generics, 352–353

Good Manufacturing Practice (GMP):
(*cont'd*)
inspections, 325–332
large molecule APIs, 340–348
bovine spongiform
enecephalopathy, 344
cell culture techniques, 346
cell lines/banks testin, 343
chromatographic purification,
347
etanercept case study, 348
recombinant DNA techniques,
342
overview, 320–322
requirements, 322–325
small molecule APIs, 332–339
chiral synthesis, 338–339
conventional synthesis techniques,
332–337
FDA cGMP initiative, 310–313
generic drug approval, 249
maximum allowable carryover, 301
process validation, 304–305
regulatory requirements:
Europe, 283
International Conference on
Harmonization, 283–288
overview, 279
United States, 279–283
risk-based case study, 313–315
stability evaluation, 309–310
sterilization processes, 307–309
water systems, 297–299
G-protein coupled receptors (GPCRs):
drug development, 39, 42, 44
signal cascade, 40, 42
Granulocyte macrophage colony
stimulating factor (GM-CSF),
structure and function,
119–120
Greek medicine, early drug discovery
and, 393–394
Growth factors, 118–121
Guanosine diphosphate (GDP), drug
development, G-protein
coupled receptors, 39
Guanosine triphosphate (GTP), drug
development, G-protein
coupled receptors, 39

Half-life calculations, drug excretion
and, 152–153
Heating, ventilation, and air-
conditioning (HVAC) system,
Good Manufacturing Practice
buildings and facilities
guidelines, 290–293
Hepatitis C, interferons and, 115
Herceptin, development of, 128–131
Herpes zoster (shingles), vaccines,
106
High density lipoprotein (HDL),
structure and function, 7–8
High efficiency particulate air (HEPA)
filters, Good Manufacturing
Practice buildings and facilities
guidelines, 290–293
High performance liquid
chromatography (HPLC),
drug manufacturing guidelines,
324
High throughput screening (HTS):
small molecule drug development,
59–60
traditional medicine development,
364–365
"Hit" screening procedures, small
molecule drug development,
56–58
HMG-CoA reductase:
cholesterol formation, 87–88
cholesterol-lowering drugs, 8
Hormones, large molecule drug
development, 121–124
Horse antisera, antibodies from, 110
Human-caused particles, Good
Manufacturing Practice
buildings and facilities
guidelines, 289–293
Human donor antibodies, 110
Human equivalent dose (HED), animal
testing, 160–161
Human Fertilization and Embryology
Authority (HFEA), 383–384
Human Genome Project:
drug discovery and, 14
gene targeting, 23–25
genomics and proteomics research,
408–410

small molecule drug development, genomics and proteomics, 74, 76–77

Human growth hormone (hGH), structure and function, 122, 124

Human immune system, antibodies and, 107–108

Human immunodeficiency virus (HIV):
enzyme inhibition, 37–38
Fuzeon case study, 321
Lopinavir/Ritonavir 2 clinical trial, 183, 185–186
small molecule drug development, X-ray crystallography, 64
vaccines, 104

Human insulin, structure and function, 121–122

Humanization technologies, antibody molecules, 110–112

Human subjects, clinical trials, 179

Hybridization probing, DNA microarrays, 29

Hybridoma technique, monoclonal antibody production, 111

Hydrogels, polymeric drug delivery systems, 168

Hydrogen bonding, target-drug interactions, 33

Hydrophobic effects:
large molecule drug purification, 347
target-drug interactions, 33

Hyperbolic function, pharmacodynamics, 140–143

Hypertension, drug research on, 370

Identification of materials, Good Manufacturing Practice, 294

Imatinib mesylate (Gleevec):
enzyme inhibition, 35
FDA approval of, 214
rational approach to development of, 74–75

Immunoglobulins, antibodies and, 106–114

Inclusion criteria, clinical trial guidelines, 191

Income distribution, morbidity and mortality and, 385

Independent Review Board/Independent Ethics Committee (IRB/IEC), clinical trials, 179

India's Central Drugs Standard Control Organization (CDSCO):
drug approval authority, 219
marketing authorization by, 266–267

Indian (Ayurvedic) medicine, early drug discovery and, 393

Individualized medicine technology, future research issues, 366

Influenza viruses, vaccines and, 98–99

Informed consent:
clinical trial guidelines for, 190
clinical trials, 179

Inhalants:
manufacturing guidelines, 350
pharmacokinetics, 149

In-process controls, Good Manufacturing Practice, 294–295

In silico methods:
drug discovery, 28
preclinical trials, animal studies, 158–161
small molecule drug development, computational chemistry, 69–71

In situ methods, gene therapy, 124–126

Inspections, drug manufacturing, GMP guidelines, 325–332

Installation Qualification (IQ), Good Manufacturing Practice, 296–297
cleaning systems, 300–301
computer validation, 304

Insulin:
diabetes mellitus and, 123
inhalable insulin, 123
structure and function, 121–122

Intellectual property rights, future drug development and, 381–382

Interferons:
    hepatitis C and, 115
    structure and function, 113, 115
Interleukins, structure and function,
    115–118
Intermediate products, Good
    Manufacturing Practice, 294
International Conference on
    Harmonization (ICH)
    guidelines:
    clinical efficacy trial guidelines, 183,
        188
    clinical trial protocol, 192
    Good Manufacturing Practice
        regulations, 283–288
    regulatory activities, 222–223
    toxicology, 155–158
International Health Regulations (2005)
    (IHR), 225–227
International health systems, selected
    countries, 435
International Medicinal Products
    Anti-Counterfeiting
    Taskforce (IMPACT),
    272–273
International regulatory authorities,
    220–221, 225–227
Intracellular adhesion molecule 1
    (ICAM-1), antisense
    development technique, 81
Intracellular receptors:
    drug development, 44–45
    drug interactions, 31–33
Intramuscular administration:
    formulation guidelines, 165
    pharmacokinetics, 148–149
Intravenous administration:
    formulation guidelines, 165
    pharmacokinetics, 149
Investigational New Drugs (INDs):
    clinical trial failures, 159
    FDA guidelines, 211–212
        treament protocols, 238–239
    FDA regulations concerning, 234–
        235, 237
Investigators, clinical trial guidelines
    for, 190
Investigator's brochure, clinical trial
    guidelines for, 190

In vitro testing:
    DNA microarray technology, 30
    gene therapy, 124–126
    preclinical trials, animal studies,
        158–161
In vivo studies, DNA microarray
    technology, 30
Ion channel receptors, drug
    development, 42–44
Ion exchange chromatography, large
    molecule drug purification,
    347
Irrational approach to drug discovery,
    13
    small molecule drugs, 55–60
        high throughput screening, 59–60
        lead compound generation, 56–58
        lead compound purification and
            modifications, 58
        natural product collection, 55
        potential compound extraction,
            55–56
Isoelectric focusing, drug manufacturing
    guidelines, 324–325

Japan's Ministry of Health, Labor and
    Welfare:
    drug approval authority, 216–217
    marketing authorization, 263–266

Kidneys, drug excretion in, 151–153
Killed/inactivated vaccines,
    characteristics and
    applications, 97
Klinefelter's syndrome, 407
Knockout mice, DNA microarray
    technology, 30

Labeling requirements:
    FDA Form 356h, 243–244
    Good Manufacturing Practice, 294
    manufacturing guidelines, 351
Laboratory controls:
    Good Manufacturing Practice,
        294–296
    toxicity grading, 431–433
Lapatinib (Tykerb):
    development of, 131
    FDA approval, 248–249

Large molecule drug discovery:
  Alzheimer's disease, 388
  antibodies, 106–114
    chimeric, 111
    conjugate, 112–113
    human immune system, 107–108
    humanization, 110–112
    monoclonal antibodies, 110
    structure, 106–109
    traditional antibodies, 109–110
  biologics, 94–96
  biopharmaceuticals, 95, 362–363
  cytokines, 113–121
    growth factors, 118–121
    interferons, 113, 115
    interleukins, 115–118
    lymphokines and monokines, 113–118
    tumor necrosis factor, 118
  formulation guidelines, 164–166
  gene therapy, 124–126
  Good Manufacturing Practice, 321–322
  Herceptin and Tykerb case study, 128–131
  hormones, 121–124
    human growth hormone, 122, 124
    insulin, 121–122
  manufacturing techniques, 340–348
  overview, 93–95
  stem cells and cell therapy, 126–128
  vaccines, 95–106
    adjuvants, 102–103
    attenuated vaccines, 97
    avian influenza H5N1, 99
    DNA vaccines, 100, 102
    Gardasil, 101
    influenza viruses, 98–99
    killed/inactivated vaccines, 97
    peptide vaccines, 102
    recent research and clinical activities, 102
    subunit vaccines, 100–101
    table of, 103–106
    toxoids, 97, 100
    vector-based vaccines, 100
Latin square design, randomization guidelines, 194

Lead compounds, small molecule drug development:
  purification and modification, 58
  screening, 56–58
Lethal factor, anthrax, 377
Lifestyle drugs, future research on, 371–373
Ligand fitting, combinatorial chemistry, small molecule drug development, 71–74
Ligand-gated ion channels, drug development, 43–44
Limit of detection, GMP analytical methods validation, 306–307
Limit of quantitation, GMP analytical methods validation, 306–307
Limulus amebocyte lysate (LAL) test, drug manufacturing guidelines, 325
Linearity, GMP analytical methods validation, 306–307
Lipase inhibitor, enzyme inhibition, 36
"Lipinski Rule of 5," computational chemistry, small molecule drug development, 69–71
Lipitor (atorvastatin calcium):
  development case study, 84–89
  formulation, 162
Liquid dosage forms, manufacturing guidelines, 350
Local government authorities, drug regulation and, 221
Lopinavir/Ritonavir, clinical trial, phase III study, 183, 185–186
Low density lipoprotein (LDL), structure and function, 7–8
Lucentis, vascular endothelial growth factor, 121
Lymphokines:
  interferons, 113, 115
  interleukins, 115–118
  structure and function, 113–118
  tumor necrosis factor, 118
Lyophilized formulations, excipients for, 167

Macrophage colony stimulating factor (M-CSF), structure and function, 119–120

Malaria, vaccines, 105
Manufacturing:
    FDA guidelines, 245
    guidelines for, 4
Marine bioprospecting, small molecule
        drug discovery, 361
Marketing applications:
    drug development, 4–5
    European Union procedures, 253–263
Mass spectroscopy, drug manufacturing
        guidelines, 325
Master cell bank (MCB), large
        molecule drug manufacturing,
        341–348
Materials management, Good
        Manufacturing Practice,
        293–294
Maximum allowable carryover (MAC),
        Good Manufacturing Practice,
        301
Maximum recommended starting dose
        (MRSD), animal testing,
        160–161
Mechanisms of action, selected drugs,
        39, 41, 411–413
Medical needs identifcation, drug
        discovery, 21–23
Medicines and Healthcare Products
        Regulatory Agency (MHRA),
        clinical trial applications,
        252–253
Medieval drug discovery, history of, 394
Messenger RNA (mRNA), microarray
        targeting, 29
Metabolism mechanisms,
        pharmacokinetics, 150–151
    excretion, 151–153
Metabolomics, small molecule drug
        development, 77
Methicillin-resistant *Staphylococcus
        aureus* (MRSA), 379
Microbial contamination limits, Good
        Manufacturing Practice
        buildings and facilities
        guidelines, 290–293
Microdosing, clinical trials, 181
Molecular complexity:
    gene targeting and, 24
    target-drug interactions, 32–33

Monitoring, clinical trial guidelines,
        194
Monoclonal antibodies:
    cancer treatment, 114
    clinical trial case study, 195
    hybridoma technique, 111
    large molecule drug manufacturing,
        345–348
    structure and function, 110
    vascular endothelial growth factor,
        120–121
Monokines, large molecule drug
        discovery, 113–118
Monte Carlo simulations, computational
        chemistry, small molecule drug
        development, 71
Morbidity and mortality, global
        statistics on, 385
Multiple doses, pharmacokinetics
        analysis, 154–155
Mutual recognition procedure, EU
        marketing authorization,
        254–255

Nanotechnology, preclinical trials,
        168–169
National authorization procedure, EU
        marketing authorization,
        255–256
Natural product collection, small
        molecule drug discovery, 55
    future challenges, 360–361
Needleless injection, nanotechnology,
        168–169
Neurotransmitters, GPCR and ion
        channel binding, 44
New chemical entities (NCEs), large
        molecule drug discovery, 94
New Drug Applications (NDAs):
    FDA general guidelines, 211–212
    FDA license applications, 239–248
    FDA review, 246–248
New molecular entity (NME), Tykerb
        approval, 248–249
Nexium, basic properties, 8
Nonclinical pharmacoloty, FDA
        guidelines on, 245
Nonenzyme catalysts, small molecule
        drug manufacturing, 338

No observed adverse effect level (NOAEL), animal testing, 160–161

Not Approvable Letter (FDA), NDA/BLA reviews, 247

Nuclear magnetic resonance (NMR) spectroscopy:
basic principles, 66
small molecule drug development, 65–67

Nucleic acids, classification and structure, 400–404

Null hypothesis, clinical trial guidelines, 196–197

Obesity management, drug development and, 372

Observational techniques, drug discovery and, 86

Occupational Safety and Health Administration (OSHA), drug regulation and, 221

Ointments:
defined, 54
manufacturing guidelines, 350–351

Old age diseases, future drug research on, 369–371

Oligonucleotides, antisense therapy, small molecule drug development, 80–81

Omeprazole, development case study, 85

Ontak, proleukin and, 117

Open reading frames (ORFs), subunit vaccines, 100–101

Operating systems (computer), Good Manufacturing Practice, 305

Operational qualification (OQ), Good Manufacturing Practice, 296–297
cleaning systems, 300–301
computer validation, 304

Oral administration:
bioavailability, 167–168
pharmacokinetics, 145, 147–148

Organic chemistry synthesis, small molecule drug discovery, 321

Orlistat, obesity management, 372

Orphan drugs, FDA guidelines for, 248–249

Oseltamivir (Tamiflu), X-ray crystallography and dvelopment of, 64

Osteoporosis, future drug research on, 371

Out of specification (OOS) results, Good Manufacturing Practice, 295
inspections, 330

Over-the-counter (OTC) drugs:
defined, 2
FDA approval, 249–251

Packaging control:
Good Manufacturing Practice, 294
manufacturing guidelines, 351

Paclitaxel (Taxol):
conventional synthesis manufacturing, 337
purification and modification, 58

Parallel group techniques, randomization guidelines, 194

Parallel synthesis, combinatorial chemistry, 73–74

Parallel track, in clinical trials, 239

Parenteral drugs:
manufacturing guidelines, 350
sterilization, Good Manufacturing Practice, 307–309

Parkinson's disease, future drug research on, 371

Paroxysmal nocturnal hemoglobinuria (PNH), Soliris therapy, EU marketing authorization, 263

Passive diffusion, pharmacokinetics, 145

Patent issues, drug discovery and development, 11–12

PEG-intron, large molecule drug formulation, 166

Penicillin, history of, 396–397

Peptide vaccines, characteristics and applications, 102

Performance-enhancing drugs:
future development of, 373–375
growth factors and, 120

Performance qualification (PQ), Good
    Manufacturing Practice, 297
  cleaning systems, 300–301
  computer validation, 304
Personnel, Good Manufacturing
    Practice regulations, 289
Perspiration, drug excretion and,
    152–153
Pfizer Inc., drug discovery and
    development case study,
    14–15
*Pharmaceutical cGMPs for the 21st
    Century: A Risk-Based
    Approach*, 310–313
Pharmaceutical industry:
  biopharmaceuticals, best-selling
    products, 7
  drug development and discovery,
    5–10
  early history of, 395–396
  ethical issues, 383–387
  global sales by region, 6
  research and development investment
    statistics, 9–10
  restrictions in use and availability,
    380–381
  top best-selling products, 6–7
  top ten companies, 8–9
Pharmaceutical Inspection Cooperation
    Scheme (PIC/S), regulatory
    activities, 223–224
Pharmacodynamics:
  Aranesp case study, 170–171
  preclinical studies, 139–143
    effectiveness, 142
    potency, 141–142
    safety margin, 143
    therapeutic index, 142
  Zeprexa case study, 170
Pharmacogenomics, bioinformatics and,
    68–69
Pharmacokinetics:
  Aranesp case study, 171
  preclinical trials, 143–155
    absorption, 145, 147–149
    applications, 154–155
    clinical applications, 154–155
    distribution, 149–150
    excretion, 151–153

  metabolism, 150–151
  toxicology, 155–158
  transport mechanism, 144–146
  Zeprexa case study, 170
Pharmacology review format, 418–423
pH levels, pharmacokinetics, 147
Pinocytosis, pharmacokinetics, 145
Placebo-controlled clinical trials,
    guidelines for, 192–194
Placental barrier, distribution
    pharmacokinetics, 150
Plant studies, transgenic technologies,
    376–378
Plasma concentration, pharmacokinetics
    analysis, 154–155
Plavix, clinical trial case study, 200–201
Pluripotent cells, overview, 127
Pneumococcal disease, vaccines, 104
Polio vaccine, clinical trial, 199
Polyclonal antibodies, structure and
    function, 109–110
Polymeric compounds, drug delivery
    systems, 168
Postmarketing approval, clinical trials,
    phase IV, 183–184
Potency, pharmacodynamics, 141
p53 protein, cancer and, 24–25
Precision, GMP analytical methods
    validation, 306–307
Preclinical trials:
  drug delivery systems, 165–168
  FDA Good Laboratory Practices,
    138
  formulations and delivery systems,
    161–168
  nanotechnology, 168–169
  overview, 137–139
  pharmacodynamics, 139–143
    effectiveness, 142
    potency, 141–142
    safety margin, 143
    therapeutic index, 142
  pharmacokinetics, 143–155
    absorption, 145, 147–149
    applications, 154–155
    clinical applications, 154–155
    distribution, 149–150
    excretion, 151–153
    metabolism, 150–151

toxicology, 155–158
transport mechanism, 144–146
Zeprexa/Aranesp case study,
 169–171
Prescription Drug User Fee Act
 (PDUFA), 248–249
Prilosec:
 formulation, 162
 legal issues, 382
 structure and function, 8
Process analytical technology (PAT),
 current Good Manufacturing
 Practice initiative, 312–313
Process equipment, Good
 Manufacturing Practice, 293
Process validation, Good
 Manufacturing Practice,
 304–305
Production controls:
 FDA guidelines, 245
 Good Manufacturing Practice,
 294–295
 vaccine production, 416–417
Product release criteria, FDA
 guidelines on, 245
Product sterility problems, Good
 Manufacturing Practice
 inspections, 330
Prokaryote cell, classification and
 structure, 398–399
Proleukin, Ontak and, 117
Propranolol, chiral synthesis
 manufacture, 339
Prostaglandins, arachidonic acid
 conversion, 46–49
Protective agent, anthrax, 377
Proteins:
 basic properties, 407–408
 crystallization, small molecule drug
 development, 62–64
 extraction, small molecule drug
 development, 76–78
 synthesis optimization:
 gene structure and, 404–407
 large molecule drug manufacturing,
 321–322
Proteomics:
 Human Genome Project and,
 408–410

small molecule drug development, 74,
 76–77
Protocol, clinical trial guidelines for,
 190–191
Proton pump inhibitors:
 nexium, 8
 omeprazole and esomeprazole
 development, 85
Provenge, development of, 130
Prozac, formulation, 162
Purification techniques:
 large molecule drug manufacturing,
 chromatographic techniques,
 347
 small molecule API manufacturing,
 337
Purified water (PW) systems, Good
 Manufacturing Practice,
 297–299

Quality management, Good
 Manufacturing Practice
 regulations, 288–289
Quantitative polymerase chain reaction
 (QPCR), drug manufacturing
 guidelines, 325

Rabies management, early history,
 395
Radiation sterilization, Good
 Manufacturing Practice,
 308–309
Radioligand binding, drug discovery
 and, 25–27
Randomization, clinical trial guidelines,
 192–194
Range intervals, GMP analytical
 methods validation, 306
Rational approach to drug discovery:
 defined, 13
 small molecule drugs, 60–79
 bioinformatics, 65–69
 combinatorial chemistry, 71–74
 computational chemistry, 69–71
 future research issues, 361–362
 genomics and proteomics, 74–77
 metabolomics, 77
 nuclear magnetic resonance
 spectroscopy, 65

Rational approach to drug discovery:
(*cont'd*)
systems biology, 77–79
X-ray crystallography, 61–64
Raw materials:
FDA guidelines, 244
Good Manufacturing Practice
manufacturing guidelines,
320
Reaction vessel design, small molecule
API manufacturing, 332–337
Reagent characteristics, small molecule
API manufacturing, 335–337
Receptors:
basic properties, 26
drug development, 38–45
G-protein coupled receptors, 39, 42,
44
intracellular receptors, 44–45
ion channel receptors, 42–44
neurotransmitter binding, 44
tyrosine kinases, 44
drug interactions, 30–33
pharmacodynamics, 140–143
Recombinant DNA techniques, large
molecule drug manufacturing,
340–348
Records management, Good
Manufacturing Practice, 293
Rectal administration,
pharmacokinetics, 148
Regulatory authorities:
applications overview, 232–233
Australia's Therapeutic Goods
Administration, 219–220,
268
Canada's Health Canada, 220, 268
China's State Food and Drug
Administration, 217–219, 264,
266–268
clinical trial guidelines, 199
counterfeit drug case study, 269–273
European Medicines Agency, 214–
215
European Union, 250–263
clinical trial applications, 252–253
marketing authorization, 253–263
Food and Drug Administration, 210–
214, 233–250

chemistry, manufacturing, and
controls, 235–237
clinical trials, 237
drug development process, 233
generics, 249
Investigational New Drug
applications, 234–235
Investigational New Drug review,
237
New Drug Applications/Biologics
License Application, 239–248
orphan drugs, 248–249
over-the-counter drugs, 249–250
review mechanisms, 238–239
ICH guidelines, 223
India's Central Drugs Standard
Control Organization, 219,
266–267
international agencies, 220–221
International Conference on
Harmonization, 222
International Health Regulations
(2005), 225–227
Japan's Ministry of Health, Labor,
and Welfare, 216–217,
263–2656
non-drug regulatory agencies,
221–222
Pharmaceutical Inspection
Cooperation Scheme,
223–224
role of, 209–210
World Health Organization, 222–223,
227
Regulatory requirements:
clinical trials, 186–198
adverse events, 195
biomarkers, 191
case report form, 191–193
clinical research organization, 198
Good Clinical Practice
requirements, 186, 189
ICH clinical study efficacy
guidelines, 187–188
ICH protocol, 191–192
inclusion and exclusion criteria,
191
informed consent, 190
investigator, 190

investigator's brochure, 190
monitoring, 194
protocol, 190–191
randomization, placebo-controlled
and double-blinded, 192–194
serum tumor markers, 198
sponsors, 196, 198
statistics, 195–197
surrogate markers, 198
future issues, 380–381
Good Manufacturing Practice:
Europe, 283
International Conference on
Harmonization, 283–288
overview, 279
United States, 279–283
Renaissance period, drug discovery
history and, 394–395
Repeatability, GMP analytical methods
validation, 306–307
Repeated dose toxicity studies,
preclinical trials, 156–157
Reproducibility processes, Good
Manufacturing Practice, 321
Reproductive toxicology, preclinical
trials, 157–158
Research and development,
pharmaceutical industry
statistics on, 9–10, 15
Restriction endonucleases,
dihydrofolate reductase
(DHFR) plasmid vector,
414–415
Retrovir vaccine, development, 87
Reversed phase chromatography,
large molecule drug
purification, 347
Reverse osmosis, water purification
systems, 298–299
Reverse transcriptase, HIV drug
development, 37–38
Rheumatoid arthritis, tumor necrosis
factor, 118
Risk-based development approach,
Good Manufacturing Practice,
313–315
Risk-benefit ratios, clinical trials, 179
RNA, classification and structure,
404

RNA-induced silencing complex
(RISC), small molecule drug
development, 82–83
RNA interference (RNAi) technique:
basic principles, 13
small molecule drug development,
81–83
Robustness, GMP analytical methods
validation, 307
Roman medicine, early drug discovery
and, 393
Ruggedness intervals, GMP analytical
methods validation, 307

Safety studies:
Good Manufacturing Practice
manufacturing guidelines,
320
pharmacodynamics, 143
toxicology, 155–158
San Qi, legend of, 392
Scientific validity, clinical trial, 178
Screening of compounds, small
molecule drug development,
56–58
high throughput screening, 59–60
Scurvy drugs, early history, 395
Serpentine receptors, drug
development, 39
Sertraline, enzyme inhibition, 36
Serum tumor markers, clinical trial
guidelines, 198
Severe combined immunodeficiency
(SCID), gene therapy research,
367–368
Short interfering RNA (siRNA):
to drug discovery, 13
small molecule drug development,
81–83
Sickle cell anemia, gene targeting, 25
Signal transduction:
drug development, 38–45
G-protein coupled receptors, 39, 42,
44
intracellular receptors, 44–45
ion channel receptors, 42–44
neurotransmitter binding, 44
tyrosine kinases, 44
pharmacodynamics, 140

Single nucleotide polymorphisms (SNPs):
  bioinformatics and, 68–69
  future research, 367–368
Single strand nucleotide, structure and properties, 402–404
Sleeping sickness, ethical issues concerning, 383–387
Small molecule drug discovery:
  Alzheimer's disease, 388
  antisense approach, 79–81
  chiral drugs, 83–84
  formulation guidelines, 164–165
  future challenges, 360–362
  Good Manufacturing Practice, 321
  irrational approach, 55–60
    high throughput screening, 59–60
    lead compound generation, 56–58
    lead compound purification and modifications, 58
    natural product collection, 55
    potential compound extraction, 55–56
  Lipitor case study, 84–89
  manufacturing techniques, 332–339
    chiral synthesis techniques, 338–339
    conventional synthesis techniques, 332–337
  natural products, 360–361
  overview, 53–54
  rational approach, 60–79
    bioinformatics, 65–69
    combinatorial chemistry, 71–74
    computational chemistry, 69–71
    future challenges, 361–362
    genomics and proteomics, 74–77
    metabolomics, 77
    nuclear magnetic resonance spectroscopy, 65
    systems biology, 77–79
    X-ray crystallography, 61–64
  RNA interference approach, 81–83
Smallpox, vaccines, 105–106
  early history, 395
Social value issues, clinical trial, 178
Sodium dodecyl sulfate-polyacrylamide gel electrophoresis (SDS-PAGE), drug manufacturing guidelines, 324

Software categories, Good Manufacturing Practice, 305
Solid dosage forms, manufacturing guidelines, 349–350
Soliris case study, EU marketing authorization, 263
Specificity, GMP analytical methods validation, 306–307
Specific pathogen-free (SPF) conditions, animal studies, 158–161
Spectroscopy, drug manufacturing guidelines, 325
Split and mix techniques, combinatorial chemistry, small molecule drug development, 74–75
Sports medicine, performance-enhancing drugs, 373–375
Stability test:
  formulations, 164
  Good Manufacturing Practice, 309–310
Standard safety margin (SSM), pharmacodynamics, 143
State health authorities, drug regulation and, 221
Statistics, clinical trial guidelines, 195–197
Steam under pressure sterilization, Good Manufacturing Practice, 308–309
Stem cell therapy:
  drug discovery, 13–14
  future research, 367–369
  overview, 126–128
Sterilization, Good Manufacturing Practice, 307–309
Stroke medicine, drug research on, 370–371
Structure-activity relationships (SARs), small molecule drug development, 69–71
Structure factor, X-ray crystallography, small molecule drug development, 62–64
Subcutaneous administration:
  formulation guidelines, 165
  pharmacokinetics, 148–149
Subunit vaccines, 100–101

Superoxide dismutase (SOD), aging drug development, 373
Surrogate markers, clinical trial guidelines, 198
Syngeneic transplant, defined, 129
Syrups, defined, 54
Systems biology, small molecule drug development, 77–79

Tablets, manufacturing guidelines, 349
Target-drug interactions, 30–33
Target identification, drug discovery and development, 23–28
  DNA microarray, 26–29
  expressed sequence tags and *in silico* methods, 28
  genes, 23–25
  radioligand binding, 25–26
Target receptors, selected drugs, 39–40
Target validation, drug discovery, 28, 30
Tetramethylsilane (TMS) standard, nuclear magnetic resonance, 66
TGN1412 monoclonal antibody, clinical trial case study, 195
Thalidomide:
  chiral synthesis manufacture, 339
  regulation of, 209–210
Therapeutic index, pharmacodynamics, 141
Therapy classes, drug discovery and development, 22–23
Thermodynamics, enzyme reactions, 34–35
Three-dimensional (3D) structure, small drug development, 60–79
Tinctures, defined, 54
Topical administration, pharmacokinetics, 149
Toxicity analysis:
  grading protocols, 428–433
  preclinical trials, 155–157
Toxicology:
  Aranesp case study, 171
  FDA guidelines on, 245
  pharmacokinetics, 155–158

review format, 418–423
Zeprexa case study, 170
Toxoids, characteristics and applications, 97, 100
Trade/proprietary drug name, defined, 8
Traditional medicine:
  early Chinese medicine, 392
  forms of, 54
  future research issues, 364–365
Transdermal administration, pharmacokinetics, 149
Transport mechanism, pharmacokinetics, 144–146
Trasgenic animals and plants, future applications, 376–378
Trastuzumab, development of, 129–131
Tumor necrosis factor, structure and function, 118
Tumor suppressor gene, p53 protein, 24–25
Turner's syndrome, 407
Tykerb:
  development of, 128–131
  FDA approval, 248–249
Type I/type II errors, clinical trial guidelines, 197
Tyrosine kinase receptors, drug development, 44–45

Ultra-high throughput screening (UHTS), small molecule drug development, 59–60
United States, Good Manufacturing Practice regulations, 279–283
Unrealated allogeneic transplant, defined, 129
User requirement specifications (URS), computer validation, 304

Vaccines:
  cancer, 104, 381
  early history of, 395
  edible vaccines, 378
  large molecule drug discovery, 95–106
    adjuvants, 102–103
    attenuated vaccines, 97

Vaccines: (*cont'd*)
  avian influenza H5N1, 99
  DNA vaccines, 100, 102
  Gardasil, 101
  influenza viruses, 98–99
  killed/inactivated vaccines, 97
  peptide vaccines, 102
  recent research and clinical
    activities, 102
  subunit vaccines, 100–101
  table of, 103–106
  toxoids, 97, 100
  vector-based vaccines, 100
  polio vaccine, clinical trial, 199
  production methods, 416–417
Validation Master Plan (VMP),
    components of, 296
Validation protocols, Good
    Manufacturing Practice,
    296–297
Vancomycin resistant *Enterococci*
    (VSE), 379
Van der Waals forces, target-drug
    interactions, 33
Variable fragment (Fv), antibody
    molecules, 109
Varicell-zoster virus (VZV), vaccines,
    106
Vascular endothelial growth factor
    (VEGF), structure and
    function, 120–121
Vector-based vaccine, characteristics
    and applications, 100
Viable cell concentration and viability,
    large molecule drug
    manufacturing, 345–348
Viagra, discovery of, 86
Viral proteins, HIV drug development,
    37–38
Virtual screening process,
    computational chemistry, small
    molecule drug development,
    72–74

Vitravene, antisense development
    technique, 81
Voltage-gated ion channels, drug
    development, 43–44
Vulvar intraepithelial neoplasia (VIN),
    Gardasil therapy, clinical trial,
    201–204

Water-for-injection (WFI) systems,
    Good Manufacturing Practice,
    297–299
Water systems, Good Manufacturing
    Practice, 297–298
Working cell bank, large molecule drug
    manufacturing, 341–348
World Health Organization (WHO):
  International Medicinal Products
    Anti-Counterfeiting Taskforce
    (IMPACT), 272–273
  regulatory activities, 222–223, 227
  Traditional Medicine Strategy, 2002–
    2005, 365
World Medical Association Declaration
    of Helsinki, clinical trials,
    180

X-ray crystallography:
  angiotensin converting enzyme, 363
  rational drug development, small
    molecule drug development,
    61–64

Zanamivir (Relenza), X-ray
    crystallography and
    development of, 64
Zaprexa, preclinical trial, 169–170
Z distribution, clinical trial guidelines,
    196–197
Zelnorm, IND review, 238
Zidovudine:
  development history, 87
  HIV therapy, 37–38
Z values, sterilization and, 308–309